Advanced Radiation Sources and Applications

NATO Science Series

A Series presenting the results of scientific meetings supported under the NATO Science Programme.

The Series is published by IOS Press, Amsterdam, and Springer (formerly Kluwer Academic Publishers) in conjunction with the NATO Public Diplomacy Division.

Sub-Series

I. **Life and Behavioural Sciences**	IOS Press
II. **Mathematics, Physics and Chemistry**	Springer (formerly Kluwer Academic Publishers)
III. **Computer and Systems Science**	IOS Press
IV. **Earth and Environmental Sciences**	Springer (formerly Kluwer Academic Publishers)

The NATO Science Series continues the series of books published formerly as the NATO ASI Series.

The NATO Science Programme offers support for collaboration in civil science between scientists of countries of the Euro-Atlantic Partnership Council. The types of scientific meeting generally supported are "Advanced Study Institutes" and "Advanced Research Workshops", and the NATO Science Series collects together the results of these meetings. The meetings are co-organized by scientists from NATO countries and scientists from NATO's Partner countries — countries of the CIS and Central and Eastern Europe.

Advanced Study Institutes are high-level tutorial courses offering in-depth study of latest advances in a field.
Advanced Research Workshops are expert meetings aimed at critical assessment of a field, and identification of directions for future action.

As a consequence of the restructuring of the NATO Science Programme in 1999, the NATO Science Series was re-organized to the four sub-series noted above. Please consult the following web sites for information on previous volumes published in the Series.

http://www.nato.int/science
http://www.springeronline.com
http://www.iospress.nl

Series II: Mathematics, Physics and Chemistry – Vol. 199

Advanced Radiation Sources and Applications

edited by

Helmut Wiedemann
Stanford University, Applied Physics Department and SSRL/SLAC, Stanford, CA, U.S.A.

 Springer

Published in cooperation with NATO Public Diplomacy Division

Proceedings of the NATO Advanced Research Workshop on
Advanced Radiation Sources and Applications
Nor-Hamberd, Yerevan, Armenia
August 29–September 2, 2004

A C.I.P. Catalogue record for this book is available from the Library of Congress.

ISBN-10 1-4020-3449-0 (PB)
ISBN-13 978-1-4020-3449-3 (PB)
ISBN-10 1-4020-3448-2 (HB)
ISBN-10 1-4020-3450-4 (e-book)
ISBN-13 978-1-4020-3448-0 (HB)
ISBN-13 978-1-4020-3450-7 (e-book)

Published by Springer,
P.O. Box 17, 3300 AA Dordrecht, The Netherlands.

www.springeronline.com

Printed on acid-free paper

Figure 1.

ACADEMICIAN M.L. TER-MIKAELIAN
(1923 – 2004)

These proceedings are dedicated by his students and colleagues
to the memory of M.L. Ter-Mikaelian, who passed away on January 30, 2004.

Mikael Levonovich Ter-Mikaelian was born in Tbilisi, Georgia, November 10, 1923. Finishing the Physical-Mathematical faculty of Yerevan State University (YSU) in 1948 he went to Moscow where he completed his candidate dissertation at the Lebedev Physical Institute in 1953. In 1954-1963 he was affiliated with the Yerevan Physics Institute (YerPhI), first as Head of the Theoretical Department, then as Deputy Director. In 1963 Ter-Mikaelian left YerPhI and began new activities in quantum electronics at YSU working as the Dean of the Physics Department, establishing four new Chairs, the Joint Radiation Laboratory of the YSU and the Academy of Sciences of Armenia (ASA). In 1968 he founded the Institute of Physical Research (IPhI) of ASA and was its director up to 1994. In 1994 he became the honorary director of IPhI continuing to work as head of IPhI's theoretical department until his death in 31 January 2004. Ter-Mikaelian's devotion to and research results in lasers and quantum electronics are compiled in his monograph with A.L. Mikaelian and Yu. Turkov "Solid State Optical Generators" (Soviet Radio, 1967) as well as in many review articles. His main scientific achievements in high-energy physics are the content of his classic monograph "High Energy Electromagnetic Processes in Condensed Media" published in 1969 and 1972, in Russian and in English (Wiley Interscience) and address: 1) the discovery of the coherence length in high-energy particle interactions (1952), 2) the development of the theory of coherent bremsstrahlung (CB) and pair production (CPP) in single crystals (1952-1954), 3) the prediction of a longitudinal density (Ter-Mikaelian) effect suppressing soft bremsstrahlung due to medium polarization (1952-1953), 4) the formulation of the theory of X-ray transition radiation (XTR) produced in multilayers taking into account the interference phenomena (1960-1961), and 5) the prediction of parametric X-ray radiation (PXR) produced in crystals due to Bragg diffraction of pseudophotons accompanying high energy charged particles (1965). His theoretical results have been experimentally studied and confirmed after many years of their prediction and were the main topics of many international conferences including the latest two Nor-Amberd Workshops devoted to radiation of high energy particles. In his last years, Ter-Mikaelian returned to high energy physics problems, published overviews and review articles and planned to prepare the second edition of his famous monograph on radiation processes.

R.O. Avakian, R.B. Fiorito, K.A. Ispirian and H. Wiedemann

Contents

ix

Preface

A NATO Advanced Research Workshop on "Advanced Radiation Sources and Applications" was held from August 29 to September 2, 2004. Hosted by the Yerevan Physics Institute, Yerevan, Armenia, 30 invited researchers from former Soviet Union and NATO countries gathered at Nor-Hamberd, Yerevan, on the slopes of Mount Aragats to discuss recent theoretical as well as experimental developments on means of producing photons from mostly low energy electrons.

This meeting became possible through the generous funding provided by the NATO Science Committee and the programme director Dr. Fausto Pedrazzini in the NATO Scientific and Environmental Affairs Division. The workshop directors were Robert Avakian, Yerevan Physics Institute, Armenia and Helmut Wiedemann, Stanford (USA). Robert Avakian provided staff, logistics and infrastructure from the Yerevan Physics institute to assure a smooth execution of the workshop. Special thanks goes to Mrs. Ivetta Keropyan for administrative and logistics support to foreign visitors. The workshop was held at the institute's resort in Nor-Hamberd on the slopes of Mount Aragats not far from the Yerevan cosmic ray station. The isolation and peaceful setting of the resort provided the background for a fruitful week of presentations and discussions.

Following our invitations, 38 researchers in this field came to the workshop from Armenia, Belarus, Romania, Russia, Ukraine, Denmark, France, Germany and the USA. Commuting from Yerevan local scientists joined the daily presentations. Over a five day period 40 presentations were given.

The main themes of this conference focused on new theoretical developments and recent experimental results in the area of channeling and transition radiation as well as the production of parametric X-rays and Smith-Purcell radiation. New technologies are available which become relevant to a variety of aspects in electron-photon interactions in and with dense media. Construction of periodically strained superlattices result in high field crystalline microundulators. Research and development in this area is vibrant and active although more experimental efforts and results would greatly benefit progress towards practical radiation sources.

Finally, the development and establishment of a synchrotron radiation source in Armenia has been discussed. Research and development of electron beam sources, radiators and instrumentation offer great opportunities for creative and productive activity.

December 27, 2004

HELMUT WIEDEMANN

OVERVIEW OF ADVANCED PHOTON SOURCES AND THEIR APPLICATIONS

R. Avakian, K. Ispirian
Yerevan Physics Institute, Alikhanyan Br. Str. 2, 375036, Yerevan, Armenia

Abstract: The existing synchrotron radiation sources and fourth generation X-ray sources, which are projected at SLAC (USA) and DESY (Germany) are very expensive. Foe this reason the search of the novel and cheaper sources using various types of radiation produced by 5-20 MeV electrons available at many hospitals, universities and firms in various countries is of great interest.

Keywords: Bremsstrahlung, Transition Radiation, Channeling Radiation, Parametric X-ray Radiation, X-ray Cherenkov Radiation, Laser Compton Scattering.

1. INTRODUCTION

The requirements to the beam parameters and irradiation dose depend on various factors. It is known that for one of the most difficult applications, namely, for the K-edge subtraction angiography, in order to provide good contrast it is necessary 50-100 ms irradiation using photon beams with two values of just above and below the iodine K-edge $\omega_K = 33.17$ keV, with / $\Delta\omega/\omega \approx 1\%$ and with flux density 3.10^{11} ph/(s.mm^2). For successful mammography over an area of 8-15 cm^2 of 8 cm thick breast it was required irradiation with dose density 2.10^9 ph/cm^2 using photon beams with $\omega = 15$-25 keV, $\Delta\omega/\omega \approx 5$-10 %.

Only the modern SR facilities, the SR sources of third generation as ELETTRA in Trieste, Italy, and ESRF in Grenoble, France, SPEAR3 SLAC, CANDLE in Yerevan can provide monochromatic beams with the above

1

H. Wiedemann (ed.), Advanced Radiation Sources and Applications, 1–16.
© 2006 *Springer. Printed in the Netherlands.*

described severe parameters. However small accelerator 5-20 MeV could provide X-ray beam with modest parameters using processes of Bremsstrahlung (BS), Transition Radiation (TR), Channeling Radiation (ChR), Parametric X-ray Radiation (PXR) and X-ray Cherenkov Radiation (XCR) as well as Laser Compton Scattering (LCS).

2.   BREMSSTRAHLUNG EXPERIMENT

Recent bremsstrahlung experiments have been reported by W. Mondelaers et al. [1].

Linear electron accelerator providing electron energies from 1.75 MeV up to 15 MeV, beam intensity from a few electron up to 2 mA average current with 2% duty factor and about 4 mm thick graphite or tantalum target and an optimized diffractometer consisting of a 4 mm thick bent crystal (Si or Ge or Cu) in Cauchois geometry which is hit from its convex side by bremsstrahlung beam, a system of collimators on a Rowland circle with 10 m diameter. The authors claim that they obtain ~ 10^5 ph/(s.cm^2), which gives an electron-photon conversion coefficient 8.10^{-12}. That is sufficient for performing many experiments.

3.   TRANSITION RADIATION

Ordinary TR - is the consequence of the readjustment of the field associated with the charged particle, when it moves in a material showing a sudden change of polarization. The emission takes place in a cone centered on the trajectory of the particle with an opening angle practically equal to 2/ , where   is the Lorenz factor of the particle.

If the arrangement of foils is periodic, coherent X-ray emission, called resonant transition radiation (RTR), resulting from constructive interference of the waves emitted by each foil can be observed. Standard RTR was observed by a Japanese team [2], which has irradiated a Ni/C multi layer structure of period d = 397 nm with 15 MeV electrons.

When the electrons cross the periodic multi layer structure under Bragg condition this was called the Bragg Resonant Transition Radiation (BRTR). There is also Diffracted Transition Radiation (DTR). This radiation is similar to PXR. In the experiment a relatively intense hard X-ray (around 15 keV) emission from 500 MeV electrons crossing a W/B$_4$C multi layer with a period d = 1.24 nm has been observed by N. Nasonov et al. [3].

PXR and DTR yields have different dependencies on multi layer thickness. In contrast to PXR the DTR yield is not proportional to the sample thickness. The relative contribution of PXR and DTR to the total yield depends on the electron energy. If it is smaller than the critical energy $m^* = m_B/_p$, then PXR dominates. The PXR efficiency of a multi layer source can exceed that from a crystal for these lower electron energies. This occurs due to a larger number of multi layer's giving a coherent contribution to the formation of X-rays. The number of emitted X-ray photons is about 10^{-4}-10^{-3} per electron. Crystals are limited to 10^{-6}-10^{-5} photons per electron.

Efficiency of DTR source was approximately the same as that of PXR radiator: 10^{-4} ph/electron (Fig. 1).

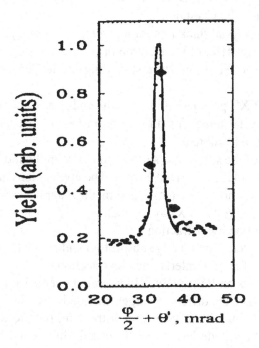

Figure 1. The comparison of the measured and calculated (only DTR) orientation dependence of the collimated X-ray yield.

4. PARAMETRIC X-RAYS PRODUCTION

The PXR have several interesting and useful features:

- The emitted X-ray energy depends on the angle between the crystal and the electron beam, which makes the source energy tunable.

- The natural line width (energy distribution) is very small (less than 2 eV) and thus the practical line width depends on the experimental geometry.
- The radiation can be emitted at large angles relative to the electron beam, which helps to reduce background from bremsstrahlung radiation.
- The X-rays are emitted with a narrow angular distribution and are simultaneously reflected to several angles around the crystal. This effect can be used in order to run several experiments simultaneously.
- The X-rays are linearly polarized.
- The efficiency of converting electrons to photons is high in comparison to other mechanisms that produce monochromatic X-rays.
  For PXR application we need:
- To find the most suitable crystal material (graphite, Silicone, Germanium, Metallic Crystals).
- To optimize the crystal thickness for a given X-ray energy.
- To investigate the effect of the electron beam divergence.
- To calculate the effect of electron straggling in the Crystal using Monte Carlo Method
- To investigate PXR production in the Laue and Bragg geometries.
- To design a crystal target for high electron beam currents address heat and charge transfer problems.

The properties of the source are suitable to many applications that require an intense monochromatic polarized X-rays. The applications are:

- Phase imaging for none destructive and medical applications.
- Real-time imaging (pulsed X-ray source).
- Explosive detection by diffraction
- Fissile material detection by edge absorption and fluorescent.
- Measurements of photon interaction cross-sections.

Theoretical results on PXR yield [4] integrated over a 1 mm^2 surface at 1 m from the crystal for different target materials is shown in Fig. 2. Calculation predicts that intensity of bremsstrahlung produced from Cu and W is an order of magnitude larger than from LiF. In Cu and W the ratio of signal to noise of PXR is 0.003. This ratio for the low Z materials like graphite is 0.4. It was shown that for photon energy less than 20 keV LiF and graphite are the best. Between LiF and graphite LiF is better due to the mosaicity of graphite.

An attempt to increase of intensity of PXR was done in the experiment [5] The PXR yield from quartz (SiO2) crystal under influence of ultrasonic waves with frequency 5.7 MHz was investigated. The increasing of intensity by factor of ~3 was observed (Fig. 3). The effect has been clearly seen and was repeated many times. Similar effect for increasing an intensity of scattered X-ray photons under influence of ultrasonic waves in distorted crystal was observed earlier. All these observed phenomena have the same nature and are connected to the increasing of the reflectivity in crystal.

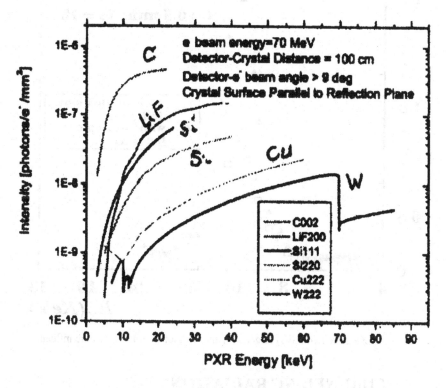

Figure 2. Integrated PXR yield for LiF, graphite, Si, Cu and W[4].

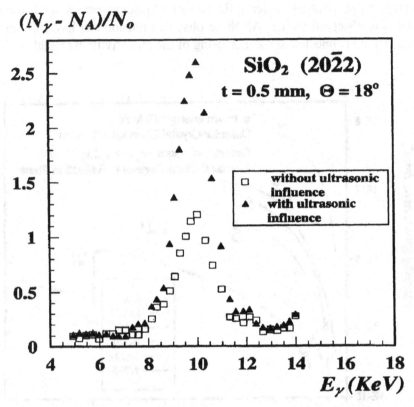

Figure 3. PXR spectra on SiO_2 target with and without external ultrasonic influence.

5.   CHANNELING RADIATION

Channeling radiation is a strong candidate for the application of small electron accelerators in many fields of science and applied physics. In medical physics monochromatic and tunable radiation offers the possibility of monochromatic X-ray imaging with the advantages of reducing dose and improving the contrast of image.

Channeling radiation is energetic, bright and tunable, and has narrow line width in the spectral peaks. All these qualities make channeling radiation a unique photon source in the X-ray region, especially for small energy accelerator. For possible applications, an intense source delivering of the order of 10^{12} ph/s with a narrow bandwidth of about 10% (FWHM), which is collimated and tunable about 10 and 40 keV is achievable. Since the crystal will also have to stand high electron beam current of several hundred μA, it has to have a high thermal conductivity.

At the Darmstadt [6] supercontacting accelerator electron beam with 10 MeV and angular divergence of 0.05 mrad (which is two orders of magnitude less than critical channeling angle) following parameters for beam was achieved.

Axial channeling: Electron energy 5.4 MeV, beam current I=3 nA, crystal -diamond, thickness 50 m with 4 mm diameter, photon energy E=10 keV, $\Delta E/E$=0.1, N_γ =10^6 ph/s, and for I=30μ A, N_γ=2x10^{10}ph/s, Ich/Ibr=5:2.

To achieve the desirable intensity 10^{12}ph/s we would have to increase the current by two orders of magnitude, which is not reasonable. So we have to go to higher electrons energies.

Planar channeling: Electron energy 9 MeV, photon energy E=8 keV, diamond (110) plane, crystal thickness 13 μm (Fig. 4). For a current of I=30 μA, the intensity is N_γ =10^{11} ph/s per 10% bandwidth. It has been found that the crystal withstands a beam current of 30 μA for several hours. Photon peak energy dependence on electron energy was confirmed as $\gamma^{3/2}$. Therefore at E_γ^{peak} =30 keV an electron energy of 22.5 MeV is needed. The intensity of channeling radiation shows a $\gamma^{5/2}$ dependence

To make channeling radiation from crystal useful for new applications we needs X-ray intensities of $10^{11} \div 10^{12}$ ph/sec at photon energies of 10-40 keV.

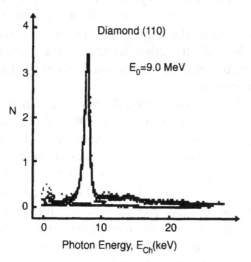

Figure 4. Spectral distribution of ChR.

6.   ENHANCEMENT

New ideas to use piezoelectric crystal as radiator and to apply ultrasonic waves have been published since 1980. They show that all the above-considered processes of ChR, CB and pair production in single crystals can be varied if the crystallographic planes are deformed properly. The first publications [7-9] reported on studies of the effect of external electromagnetic waves, transversal ultrasonic waves and superlattice structure on the increase of intensity of ChR. Later it has been shown that in the field of transversal acoustic standing waves one can significantly enhance the intensity of ChR in piezoelectric crystals. In view of this it was reasonable to carry out some investigations on the spectral and angular characteristics of electron ChR in piezoelectric crystals without external waves. Our experiment was performed with a 4.5 GeV electron beam extracted from the Yerevan Electron Synchrotron [10]. The ChR spectra and integral yield of photons in piezoelectric crystals SiO_2, $LiNbO_3$ and CdS have been experimentally measured under axial and planar channeling orientations. The experimental investigation of ChR in piezoelectric crystals has shown that one can use these crystals in order to transform electron beam energy into the photon beam. When the high-energy electron or positron enters the crystal in channeling orientation its trajectory is determined within the framework of classical theory. It was shown [11] that by applying transversal ultrasonic wave the ChR spectra of high-energy positrons have resonance behavior. For the transversal ultrasonic waves the radiation intensity of the first and the other harmonics has a maximum when the ultrasonic wavelength approaches the wavelength of mechanical oscillations of positrons in the channel.

The wavelength of the particle oscillations at channeling is proportional to the square root of energy. For instance, 100 MeV electrons channeled in the (110) plane of a Si crystal have an oscillation period of about 1 mkm.

The frequency of ultrasonic waves for a wavelength of 1 mkm oscillation period of channeled particles should be of 3.5 GHz. For the electrons of 10 GeV in Si crystal for plane (110) the period of mechanical oscillations will be about 10 mkm and the resonance frequency of ultrasonic wave will be of 0.35 GHz. In case of low acoustic power when the amplitude of displacement of the atoms under influence of acoustic wave is about 0.2 of the distance between the crystallographic planes, the integrated radiation intensity of the positrons with energy of 16 GeV in quartz crystal increases by factor of about 2.32.

When the amplitudes of the displacement will be increased this factor will be increased until the dechanneling will play the dominant role.

Unfortunately there is no calculation for high-energy electrons. However for high energies when the classic approach is good enough the mentioned effect should exist also for electrons. There is a difference connected with atomic potential and as consequence different type mechanical motion of channeled particle. Positrons entering the crystal under angles relative to the crystal axis or planes from zero up to channeling critical angle have the same wavelength of mechanical oscillation in the channel. This is also proven by our measurement of channeling radiation spectra of 2-6 GeV positrons in diamond. There is only one very good pronounced peak in the measured spectra. In Fig. 5 the electron [12] and positron [13] spectra are presented for comparison.

Therefore it is easy for positrons to choose the resonance wavelength of the ultrasonic wave. In the electron case, when the period of mechanical oscillations depends on the electron entrance angle with respect to the crystals axes or planes we have a "bunch" of wavelengths of mechanical oscillations of channeled electrons. In this case the applied ultrasonic wave should have the same "bunch" of wavelengths. On the other hand, in case of electrons it is possible to stimulate one single desired photon energy in the radiation spectra by applying a particular ultrasonic wavelength.

The analysis of the experimental data for positron production has shown that only a few part of the radiation of high-energy electron in crystals is due to channeling radiation. The main part of the radiation is due to the CB. Unfortunately we have not calculation on influence of ultrasonic waves on CB in crystal yet.

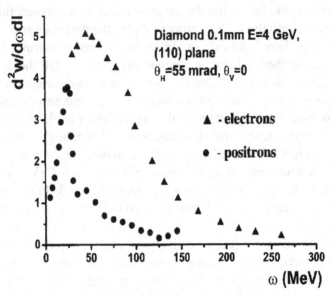

Figure 5. Planar channeling radiation spectra of electrons and positrons.

7.   X-RAY CHERENKOV RADIATION

Among interactions of relativistic electrons with a medium that cause emission of radiation, Cherenkov radiation is a promising candidate for a compact soft X-ray source.

The soft X-ray region the Cherenkov radiation is characterized by a single-line spectrum and by forwardly directed emission and only requires low- relativistic electrons from a laboratory-sized accelerator. For a long time, Cherenkov radiation was excluded in the soft X-ray region, because at these wavelengths materials are highly absorbing and the refractive index is generally smaller than unity, however, realized that at some inner-shell absorption edges the refractive index exceeds unity and Cherenkov radiation will be generated in a narrowband region, which was demonstrated for the carbon K-edge by using 1.2 GeV electrons.

Recently, it was observed [14] for the first time Cherenkov radiation emitted in the water-window spectral region (450 eV) from a titanium and vanadium foil generated by 5 MeV electrons.

The water window is the spectral region between the carbon K-absorption edge (284 eV) and the oxygen K-absorption edge (543 eV). This spectral region is ideal for X-ray microscopy of biological samples due to the relative transparency of water and the high natural contrast. The special characteristics of soft X-ray Cherenkov radiation make this source interesting for such applications. For the first experiments titanium and vanadium 10 μm thick was selected. This thickness is for both materials much larger than the absorption length (0.5 μm) at the Cherenkov photon energy.

In the experiment [14] using CCD detector it was determined the Cherenkov radiation energy for titanium 459± 2eV and for vanadium 519 ± 3eV.

Integrating the measured angular distribution obtained in the experiment over all emission angles a total yield of 3.5×10^{-4} phs/el was found for titanium and 3.3×10^{-4} ph/el for vanadium. For titanium this is slightly higher than the theoretical value of 2.4×10^{-4} ph/el.

On the basis of the measured yields [14] for titanium and vanadium, one can now evaluate the potential of a Cherenkov radiation-based compact source in the water window. Using a high-power, but laboratory-sized, 10 MeV accelerator of 1 mA average current the total output is 2.2×10^{12} ph/s (0.16 mW) for titanium and 2.1×10^{12} ph/s (0.17 mW) for vanadium. The corresponding brightness is 2.7×10^{9} ph/(s.mm^2.sr.0.1% BW) for titanium and 3.8×10^{9} ph/(s.mm^2.sr.0.1% BW) for vanadium, assuming a 100-μm electron-beam spot size. These brightness values are comparable to the values obtained from laser-produced plasma (1×10^{10} ph/(s.mm^2.sr.0.1% BW)) and high harmonic generation (5×10^{7} ph/(s.mm^2.sr.0.1% BW)). In contrast to the laser-based sources, the Cherenkov spectrum consists of only a single, isolated peak and that no debris formation occurs.

8. X-RAY DIFFRACTION RADIATION IN SUPERLATTICES (XDR)

To expand XDR spectrum into nanometer wave- length region one needs crystalline structures with appropriate periods, comparable or larger then the radiation wavelength. For this purpose artificial multi-layered structures are needed.

Recently the new kinds of superlattices were discovered in fullerenes [15]. It was demonstrated that fullerites and nanotube ropes may serve as good enough Bragg mirrors for soft X-ray radiation with wavelength up to 20 Å and higher. In this connection it is of interest to consider XDR from fullerene superlattices. It was develop the theory of XDR in superlattices and discuss the role of the dynamical effects in diffraction of virtual photons. The comparison of fullerene superlattices with ordinary crystals and MLS shows some important advantages of the former in the yield of radiation.

Spectral attention is paid to the case of small grazing incidence of charged particles with respect to the main directions of superlattices where effects of X-ray channeling may be of importance. Nanocrystals and artificial multi layered structures may have substantial advantage over ordinary crystals in the total of XDR photons. Moreover, they may produce relatively intense, monochromatic and linearly polarized radiation in nanometer wavelength region where ordinary crystals are unable to produce XDR at all. Compared to the multi layered structures, nanocrystals produce more monochromatic radiation with comparable or higher total yield.

9.   LASER COMPTON SCATTERING

During the head-on collisions of laser photons of energies of a few eV with relativistic electrons of energies higher than a few MeV the process of inverse laser Compton scattering (LCS) of photons results in an increase of photon energies at the expense of the electron energy and emission of highly directed (in the direction of electron beam), quasi monochromatic and polarized X- and γ-ray beams. LCS as a method for production of intense hard photon beams was first proposed in the works [16,17]. After the first experimental investigations [18] LCS has found wide application for the study of photo production processes at high energies. In the first experiments the conversion efficiency of the electrons into photons, $k = N_{ph} / N_e$, was very small, $\sim 10^{-7}$, however, it has been shown [19] that at future linear colliders one can achieve $k \approx 1$.

The recent progress in development of table-top-terawatt (T^3) lasers and of new moderate energy linear accelerators with photocathode RF guns allowed to expect that picosecond and even femtosecond X-ray pulses with intensities up to $\sim 10^{10}$ ph/bunch can be produced [20]. In [21] it has been considered the methods of obtaining high pulse and average X-ray yields. Moreover, it has been shown [22] that having a small, tabletop 8 MeV electron storage ring and moderate 100 W Nd;YAG laser one can obtain $9 \cdot 10^{14}$ ph/s CW source of keV photons. Such predictions are very attractive since they satisfy our requirement of X-ray sources, making LCS the most promising X-ray photon production mechanism.

Recent experiment was done at Idaho Accelerator center [23]. The 20 MeV electron beam of the Idaho Accelerator center is focused in the center of the interaction chamber with the help of quadrupoles and other monitoring devices. The electron macro bunch length was varied from 2 up to 20 ns with electron charge from 1.4 to 8 nC. No spot size is given, but the authors say that using slits they can vary the size of the spot, A_{int}, and the energy spread of the electron beam. After interaction the electron beam is deflected 45^0 to a fast toroid and Faraday cup.

The 7 ns long, 10 Hz and 100 Mw peak power pulses of the first and second harmonics of a Nd:YAG laser with $\lambda_L = 1024$ and 532 nm were first expanded and then focused in the center of interaction region with spot diameters 0.24 and 0.12 mm. The LCS photon spectral distributions were measured by a high resolution Si(Li) detector without (Fig. 6) and with a 25.4 μ m thick stainless steel foil absorbing almost completely photons with energy lower than 8 keV.

Compared with the above discussed other processes of X-ray photon beam production LCS has the advantage that it does not require solid radiators which can be destroyed with the increase of the electron beam intensities. The achieved yields and efficiencies are not limited and can be increases. Therefore, in the future LCS can become the most productive way of tabletop production of intense X-ray beams.

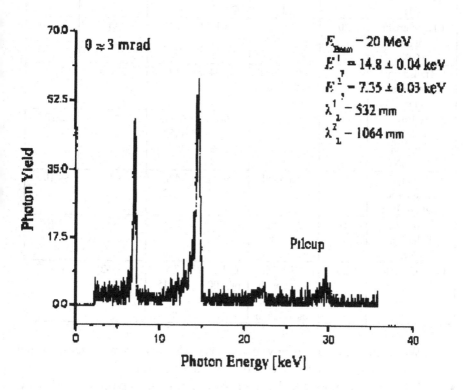

Figure 6. LCS spectra measured without the steel absorber.

10.   CONCLUSION

Without considering polarization and a series of other important properties one can conclude that the above considered types of radiation produced by electrons with energies of a few tens of MeV are sufficiently effective to provide beams of X-ray photons for many applications. The expected and observed photon fluxes per electron through a 1 cm^2 area at ~100 cm distance from the radiator, or a photon flux $N^i = (dN/d\Omega)\Delta\Omega$ (ph/e/mm^2) within a solid angle of $\Delta\Omega = S/R^2 = 10^{-4}$ sr. For comparison we summarize the characteristics of photon beams from a variety of different processes in Table 1.

Table 1

Type of radiation	E, (MeV)	Radiator	L, (cm)	w, (keV)	$\Delta w/w$, (%)	N^i, (ph/e)
Bremsstrahlung (B)	15	C+Rowland	0.4 C+ Si	20-100	<2	>8·10^{-12}
Characteristic radiation (CR)	20	Si	0.01	1.8	<1	2.5·10^{-11}
Transition radiation (TR)	15	10 pairs of ~200 nm W/C	0.0004	2-6	20-60	2·10^{-6}
X-Ray Cherenkov radiation (XCR)	10	V, Ti	0.001	0.45, 0.52	<0.1	2·10^{-6}
X-Ray parametric radiation (PXR)	8	Diamond (110)	L>L$_{abs}$	4.6 11.2	<0.1	~4·10^{-10}
Channeling radiation (ChR)	10	Diamond (110)	0.02	2-15	~20	4·10^{-6}
Laser Compton Scattering (LCS)	20	Nd:YAG Laser with λ =1064, and 532 nm	L=L$_e$= 60-600	7.5 and 15	~(1-2)	6·10^{-6}

- We need high current, low energy (10-25 MeV), low emitance super conducting CW accelerators for many applications;
- We need radiators which could withstand high electron currents (few hundred μA – 2 mA);

- We need new generation of detectors for detecting high-current photon fluxes.

References

1. W. Mondelaers, P.Cauwels, B.Masschaele, M.Dierick, P.Lahorte, J.Jolie, S.Baechlet, T.Materna, A 15 MeV Linear Accelerator Based Source of Tunable Monochromatic X-Rays (unpublished report THC16, 2001).
2. K. Yamada, T.Hosakawa, H.Takenaka, Observation of soft X-rays of single-mode resonsnt transition radiation from a multilayer target with a submicrometer period., *Phys. Rev. A* **59**, 3673-3679 (1999).
3. N.Nasonov, V.Kaplin, S.Uglov, M.Piestrup, C.Gary, X-ray relativistic electrons in a multilayer structure, *Phys. Rev. E* **68**, 036504-1 – 036504-7 (2003).
4. Y. Danon, B.A.Sones, R.C.Block, Novel X-ray Source at the RPI LINACc, in: Proceedings, American Nuclear Society Annual Meeting, 2002 (Hollywood, Florida).
5. R.O.Avakian, A.E.Avetisyan, H.S.Kizoghyan, S.P.Taroyan, Investigation of quasi-Cherenkov emission of 4.5 electrons in diamond and qurtz crystals, *Radiation Effects and Defects in Solids* **117**, 17-22 (1991).
6. H.Genz, L.Groening, P.Hofmann-Staschek, A.Richter et al., Channeling radiation of electrons in natural diamond crystals and their coherence and occupation lengths, *Phys. Rev. B* **53**, 8922-8936 (1996).
7. V.V. Kaplin. S.V.Plotnikov, S.A.Vorob'yov, The radiation of charged particles on the channeling in deform crystals, *Zh. Tekh. Fiz.* **50**, 1079-1081 (1980).
8. V G Baryshevsky et al, *Phys. Lett. A* **77**, 61 (1980).
9. H.Ikezi, Y.R.Lin-Liu, T.Ohakawa, Channeling radiation ina periodically distorted crystal, *Phys. Rev. B* **30**, 1567-1569 (1984).
10. R. Avakian et al., Radiation from 4.5 GeV electrons in piezoelectric crystals, *Nucl. Instr. and Meth. B* **48**, 266 (1990).
11. A.R. Mkrtchyan et al.,R.H.Gasparyan, R.G.Gabrielyan, A.G.Mkrtchyan, Channeled positron radiation in the longitudinal and transverse hypersonic field, *Phys. Lett. A* **126**, 528-530 (1988).
12. R.Avakian et al., Radiation of high-energy electrons near crystallographic axes and planes of diamond crystal, *Radiation Effects and Defects in Solids* **91**, 257 (1985).
13. R.O.Avakian, I.I.Miroshnichenko, J.J.Mirray and T.Vigut, Radiation of ultrarelativistic positrons moving in a crystal near crystallographic axes and planes, *Sov. Phys. JETP* **56(6)**, June 1982.
14. W.Knulst, O.J.Luiten, M.J.van der Wiel, and J.Verhoeven, Observation of narrow-band Si L-edge Cherenkov radiation generated by 5 MeV electrons, *App. Phys. Lett.* **79**, 2999-3001 (2001).
15. R.Saito, Physical Properties of Carbon Nanotubes, Imperial College Press, London.
16. F.R.Arutyunian and V.A.Tumanian, The Compton effect on relativistic electrons and the possibility of obtaining high-energy beams; *Phys. Lett.* **4**, 176-179 (1963).
17. R.H. Milburn, Electron scattering by an intense polarized photon field, *Phys. Rev. Lett* **10**, 75-77 (1963).
18. O.F. Kulikov, Y.Y.Telnov, E.I.Filippov. M.N.Yakimenko, Comptom effect on moving electrons, *Phys. Lett.* **13**, 344-346 (1964).

19. Y. Telnov, TESLA, Technical Design Report v. 6, DESY, 011, 2001/ ECFA, 209, 2001.
20. P. Sprangle, A.Ting, E.Esarey and A.Fisher, *J. Appl. Phys.* **72**, 5032 (1992).
21. E.Esarey, P. Sprangle, A.Ting, S.K.Ride, Laser synchrotron radiation as a compact source of tunable, short pulse hard X-rays, *Nucl. Instr. and Meth. A* **331**, 545-549 (1993).
22. Z.Huang and R.D.Ruth, SLAC-PUB-7556, 7677 (1997).
23. K.Chouffani, D. Wells, F. Harmon, J. Jones and G. Lancaster, Laser-Compton scattering from 20 MeV electron beam, *Nucl. Instr. and Meth. A* **495**, 95-106 (2002).

DYNAMICAL EFFECTS FOR HIGH RESOLUTION PARAMETRIC X-RADIATION

I.D. Feranchuk and O.M. Lugovskaya
Belarusian State University, Fr.Skaryny av., 4, 220050 Minsk, Belarus

Abstract: Coherent bremsstrahlung (CBS) and parametric X-radiation (PXR) are proposed to consider as two branches of pseudo-photon scattering. The role of the dynamical diffraction effects and the coherent bremsstrahlung is estimated in dependence on the crystal parameters and the particle energy. The conception of the High Resolution Parametric X-radiation (HRPXR) is introduced and the universal form for HRPXR spectral-angular distribution is considered.

Keywords: Parametric X-radiation; Coherent Bremsstrahlung; Dynamical Diffraction; Pseudo-photon method; High Resolution Diffraction.

1.   INTRODUCTION

Many problems connected with the X-ray production from the relativistic charge particles (for definiteness, electrons) in a crystal were discussed in the very last decade. They can be divided by the following groups: 1) the role of the dynamical diffraction effects in these processes; 2) what are the contributions of the various radiation mechanisms to the total intensity and 3) which applications of the X-ray radiation from the relativistic electrons in a crystal could be most essential?

Meanwhile, it seems important to remind that the analogous questions have been already considered quite long ago in the series of publications of Baryshevsky, Feranchuk and co-authors. In particular, the general theory of the X-ray radiation from relativistic electrons in a crystal taking into account the various mechanisms of the photon production and their dynamical

17

H. Wiedemann (ed.), Advanced Radiation Sources and Applications, 17–26.
© 2006 *Springer. Printed in the Netherlands.*

diffraction have been formulated in [1]. The influence of the dynamical effects on the formation of the fine structure of the PXR peaks was considered for the forward direction [2, 3], for the highly asymmetrical case [4], for the degenerate diffraction in the case of backward Bragg geometry [5]. The possible advantages and shortages of the PXR as the source of X-rays in comparison with other radiation mechanisms in the same wavelength range were discussed in [6]. It has been also shown that PXR could be mainly important for the spectral sensitive applications, that is, for the cases when the only high spectral intensity in the narrow wavelength range is important. Some of non-trivial physical problems of such type were considered in [7-9] and a series of them have been described in detail in [10, 11].

It is essential that the above-mentioned results concerning the PXR features were theoretically predicted before the experimental confirmation. Some of them initiated the first observations of the PXR peaks [12] and the angular distribution in them [13]. The simplified model for the description of the PXR characteristics was suggested in [14]. At present there are a lot of experimental works in this field (see [15-17] for recent ones and references therein) and the most of their results are in a qualitative agreement with the predictions of [14] with rather small deviation in the quantitative details. Essential contributions to the detailed analysis of this phenomenon were made in many theoretical works (for example, [18-24] and references therein).

The series of qualitatively new experimental works has been recently appeared (for example, [25]) which could open the essentially new side in the PXR investigations and applications. The two-crystal scheme of the detector was used for very high frequency resolution of the emitted photon spectrum. In some sense this is analogous to the essential extension of the applications of the conventional X-ray diffractometry, which was done due to high resolution X-ray diffraction (HRXRD) and now is the very important part of the modern semiconductor and nanostructure technology [26]. Therefore it becomes actually to discuss in more detail the general approach for the description of the dynamical effects in the formation of X-ray spectra from the charged particles in crystals. The fine structure of such spectra can be considered as the high resolution parametric X-radiation (HRPXR) and it should be represented in some universal form as it was made for the kinematical PXR [14]. It will help to analyze the experimental data and consider the possible applications of HRPXR.

The paper is organized as follows. In Sec.2 we'll discuss shortly physical mechanisms of PXR and CBS radiation in terms of pseudo-photon method. Interference between PXR and CBS is discussed in Sec.3. Analysis of the dynamical diffraction effects in the case of HRPXR and possible applications of HRPXR are discussed in Sec.4.

2. COHERENT BREMSSTRAHLUNG AND PARAMETRIC X-RADIATION IN A CRYSTAL AS THE TWO BRANCHES OF PSEUDOPHOTON SCATTERING

The intrinsic electromagnetic field of relativistic particle with the energy E_p and the velocity $\vec{v}$ ($v \approx c$) can be represented as superposition of pseudo-photons, which characteristics are close to characteristics of real photons with the wave vector

$$\vec{k} = \frac{\omega \vec{v}}{v^2} . \tag{1}$$

The pseudo-photon beam is concentrated in the cone of $\Delta\theta \sim \dfrac{m}{E_p} = \gamma^{-1}$

(γ - Lorentz-factor) with specific minimum along of $\vec{v}$ direction and with the spectrum

$$n(\omega) = \frac{e^2}{\pi\omega} \ln\left(\eta \frac{E_p}{\omega} \right), \tag{2}$$

where ω is the pseudo-photon frequency, $\eta \cong 1$, $\hbar = c = 1$.

These pseudo-photons are the potential source of real photons with "white" spectrum and any radiation mechanism could be presented as process of pseudophotons conversion into real photons within some spectral interval, which depends on nature of interaction between the external field and the charged particle.

One of the most known orientation effects in radiation processes in medium is Coherent Bremsstrahlung (CBS). It is important to determine clearly the essential physical difference between CBS and PXR and it is possible due to pseudo-photon concept.

The amplitudes of pseudo-photon scattering on periodically positioned atoms of crystal are coherent in certain directions, for which the Bragg condition is fulfilled:

$$\vec{k}' = \vec{k} + \vec{\tau}; \quad \frac{\omega}{c}\vec{n} = \frac{\omega\vec{v}}{v^2} + \vec{\tau}, \tag{3}$$

where $\vec{\tau}$ is the crystal reciprocal vector. Vectors $\vec{\tau}$ are specified by translation symmetry of the crystal and are the most important characteristics of diffraction processes. In Eq.(3) the initial wave vector corresponds to pseudo-photon and the wave vector $\vec{k}'$ describes real photon with frequency ω.

The expression for the Bragg condition should be considered as the equation for determination of emitted photon spectrum. At the given particle velocity and for specified $\vec{\tau} \neq 0$ desired frequencies could be determined as the roots of quadratic equation, following from (3)

$$\frac{\omega^2}{v^2\gamma^2} + 2\frac{\omega}{v^2}(\vec{v}\vec{\tau}) + \tau^2 = 0. \tag{4}$$

As a result we have two branches of coherent radiation spectrum from the charged particle in crystals

$$\omega_{1,2} = \gamma^2\{-(\vec{v}\vec{\tau}) \pm \sqrt{(\vec{v}\vec{\tau})^2 - \frac{v^2\tau^2}{\gamma^2}}\}. \tag{5}$$

In case of high particle energy ($\gamma \gg 1$) the frequencies are

$$\omega_1 \approx 2(\vec{v}\vec{\tau})\gamma^2 = \frac{4\pi}{d}\gamma^2,$$

$$\omega_2 \approx \frac{v^2\tau^2}{2|(\vec{v}\vec{\tau})|} = \frac{\pi\tau v}{\sin\theta_B}, \sin\theta_B = \frac{|(\vec{v}\vec{\tau})|}{v\tau}, \tag{6}$$

where d is the distance between crystallographic planes, corresponding to the vector $\vec{\tau}$. Thus, for the first case emitted photons wavelength ($\lambda_1 = 2\pi c / \omega$) is much smaller ($\sim \gamma^{-2}$) then lattice period, while the wavelength range for the second branch of radiation is the same order of the value d.

For each spectral branch the photons are emitted in rather narrow angular ranges, but they are concentrated in essentially different directions

$$\cos\theta_1 = \frac{(\vec{v}\vec{n})}{v} = \frac{c}{v} + \frac{(\vec{v}\vec{\tau})}{v\omega_1} \approx 1,$$

$$\cos\theta_2 = \frac{c}{v} + \frac{(\vec{v}\vec{\tau})}{v\omega_2} \approx 1 - 2\sin^2\theta_B, \tag{7}$$

$$\theta_2 \approx 2\theta_B.$$

The first spectral branch exactly corresponds to CBS. For this kind of radiation photons are emitted at small angles along vector $\vec{v}$ and wavelength is essentially smaller then the lattice period and quickly decreases with the particle energy growth. At the same time the second spectral branch determines spectrum of PXR, which is emitted at large angles in respect to vectors $\vec{v}$ with the frequencies determined mainly by the parameters of crystal elementary cell.

3.   INTENSITY OF THE PXR+CSB RADIATION

The expression for spectral-angular RXR+CBS intensity can be written (the intensities of the photons with different polarization $s = 1,2$ can be considered separately) in following form:

$$\frac{d^2 N_{\tau s}}{d\omega d\vec{O}} = \frac{e^2 \omega}{4\pi^2} \left| M_{PXR} + M_{CBS} \right|^2,$$

$$M_{PXR} = (\vec{e}_{\tau s} \vec{v}) \sum_{\mu=1,2} \gamma_{\mu s}^\tau [L_{0\tau} - L_{\mu\tau s}][e^{iL/L_{\mu\tau s}} - 1],$$

$$M_{CBS} = -\frac{V_{-\tau}}{(\vec{\tau}\vec{v})} \{ (\vec{e}_{\tau s}\vec{v}) L_{0\tau} [e^{iL/L_{\mu\tau s}} - 1] - $$

$$ - \frac{(\vec{e}_{0s}\vec{\tau})}{E_p} \left[\sum_{\mu=1,2} 2\gamma_{\mu s}^0 e^{i\omega n_{\mu s} L} L_{\mu\tau s} (e^{iL/L_{0\tau}} - 1) + L_{0\tau}(e^{iL/L_{0\tau}} + 1) \right] \} \qquad (8)$$

The matrix elements M_{CBS}, M_{PXR} for CBS and PXR were determined in detail before [27, 28].

The values $L_{0\tau}$ and $L_{\mu\tau s}$ are known as the radiation coherent lengths for the vacuum and crystal correspondingly. The PXR matrix element actually includes the coherent superposition of both coherent lengths. It was suggested in some papers to consider the contribution, which is proportional to $L_{0\tau}$ and conditioned by the crystal boundary, as a special radiation mechanism - diffraction transition radiation (DR).

One can see from the expression for PXR+CBS spectral-angular distribution that in general case the total radiation intensity is defined by the interference of the PXR and CBS amplitudes. As it was shown in our earlier paper [28] the PXR and CBS interference is the most essential for the non-relativistic electrons when the angular distributions for both radiation mechanisms are undistinguishable.

If one takes into account the expressions for the components of the potential V_τ and susceptibility χ_τ of the crystal the ratio of the CBS and PXR integral intensities in the same reflex can be estimated as

$$\xi = \frac{N_{CBS}}{N_{PXR}} \approx \left[\frac{Z - F(\vec{\tau})}{F(\vec{\tau})} \right]^2 \left(\frac{m^2}{E_p \tau} \right)^2 \frac{1}{16 \sin^4 \theta_B}, \qquad (9)$$

where Z, $F(\bar{\tau})$ are the atomic charge and scattering factor correspondingly.

In other words, the CBS contribution should be taken into account in the following electron energy range

$$E_p \le \left| \frac{Z - F(\bar{\tau})}{F(\bar{\tau})} \right| \left(\frac{m^2}{4\tau \sin^2 \theta_B} \right). \qquad (10)$$

This estimation shows that in the energy range $E_p \ge 50\text{MeV}$ this radiation mechanism is not very important except the case of the high-order harmonics when the atomic scattering factor can be exponentially small.

This estimation is in the qualitative agreement with the experimental results of the paper [29]. The figure 1 demonstrates the contribution of CBS to the integral intensity of the reflex calculated by numerical integration of spectral–angular distribution over frequencies together and without M_{CBS} for the different electron energies. Calculations were carried out for Bragg backward geometry. One can see that the effect of CBS for the electron energy E_p=80 MeV is negligible, but becomes important for E_p=10 MeV.

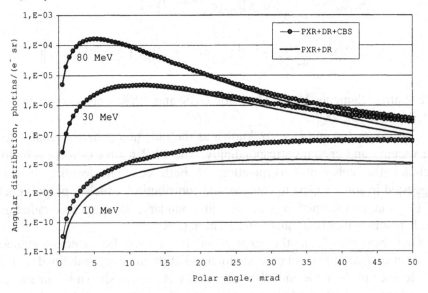

Figure 1. Angular distributions of PXR and CBS for the different electron energies

4. ANALYSIS OF DYNAMICAL DIFFRACTION EFFECTS IN THE CASE OF HRPXR

In this section we will analyze the general formula for spectral-angular PXR+CBS intensity (8) in the narrow angular-spectral range where the dynamical effects could be appeared.

The expression for PXR+CBS intensity can be used for any crystal thickness L. Particularly, in the case $L < L_{ext} \cong \dfrac{1}{\omega_B \chi_\tau}$ (L_{ext} is the extinction length, here and below χ_0, χ_τ, $\chi_{-\tau}$ are the Fourier-components of complex crystal susceptibilities), these formula takes the same form as in the kinematical diffraction theory which was analyzed earlier in our paper [14]. Therefore in this paper we consider the opposite case $L > L_{ext}$ when the dynamical effects are supposed to be essential for HRPXR analogously to the conventional high resolution X-ray diffraction (HRXRD) [26]. Moreover, in order to make more clear the specific features of the HRPXR peaks let us consider the case of the thick crystal with $L > L_{abs} \cong \dfrac{1}{2\omega_B \, \mathrm{Im} \chi_0}$ (L_{abs} is the absorption length) when the oscillation structure of the diffraction peaks is not appeared [26].

Thus, there are two essentially different angular and spectral scales for characterization the structure of the PXR reflex. One of them (low resolution scale - LRS) is defined by the angular dependence of the radiation coherent length [10]:

$$(\delta\theta)^2 \prec |\chi_\tau|, \quad (\delta\theta)_{LRS} \prec \sqrt{|\chi_\tau|} \sim 10^{-2} - 10^{-3}. \tag{11}$$

In the scope of LRS one can integrate spectral-angular intensity (8) over one of the variable (θ or ω) and find the universal forms of distributions have been considered earlier in [14] and have been investigated in many experiments. Actually these distributions are not dependent on the dynamical effects and their amplitudes are defined by the value L_{abs}. Analogous picture takes place in the conventional low-resolution diffraction when the intensity of the reflex is also proportional to L_{abs} [26].

But there is also high-resolution scale (HRS), which is defined by the Bragg condition for the emitted photons, that is

$$|\alpha_B| \prec |\chi_\tau|, \quad (\delta\theta)_{HRS} \sim \frac{\delta\omega}{\omega_B} \prec |\chi_\tau| \sim 10^{-5} - 10^{-6}, \tag{12}$$

where $\alpha_B = \dfrac{2\vec{k}\vec{\tau} + \tau^2}{k^2}$ is the deviation parameter from the Bragg condition.

Certainly, the experimental investigation of such thin structure of the PXR reflex should be connected with the two-crystal detector technique, which is widely used in modern HRXRD but was firstly applied recently [25] for HRPXR analysis.

The expression for spectral-angular PXR intensity in the case of high-resolution scale can be represented in some universal form that is analogous to kinematical PXR theory and the conventional dynamical diffraction theory. Let us introduce new variables having the scale of the order of unity:

$$\eta_s = \frac{-\beta_1\alpha_B + \chi_0(\beta_1 - 1)}{\kappa_s}, \quad x_s = \frac{\theta^2 + \theta_{ph}^2}{\kappa_s}, \quad \kappa_s = 2C_s\sqrt{\beta_1\chi_\tau\chi_{-\tau}}. \tag{13}$$

In this variables the distribution of the PXR photons can be represented as

$$\frac{\partial^3 N_{\tau s}}{\partial \eta_s \partial x_s \partial \varphi} = \frac{e^2 v_s}{4\pi^2 \beta_1 \sin \theta_B} I(\eta_s, x_s), \tag{14}$$

with the same universal function $I(\eta, x)$ for both polarizations:

$$I = \frac{x - u_{ph}}{x^2\left|\eta + \text{sign}(\eta)\sqrt{\eta^2 - 1}\right|^2}\left|1 - \frac{x}{x - \eta - \text{sign}(\eta)\sqrt{\eta^2 - 1} - u_0}\right|^2, \tag{15}$$

$$u_{ph} = \frac{\theta_{ph}^2}{\kappa_s}, \quad u_0 = \frac{\chi_0}{\kappa_s}, \quad x > u_{ph}, \quad -\infty < \eta < \infty.$$

In (16) $v_{1,2} = \sin^2\varphi(\cos^2\varphi)$, the values β_1, C_s, θ_{ph}, φ one can find in [27]. Here the function $\text{sign}(\eta)$ is (+1) for $\eta > 0$ and (-1) for $\eta < 0$. Analogously to HRXRD the scans of the distribution on the dimensionless variables for HRPXR can be realized by various experimental scans both on angles and frequency.

Figure 2 shows the characteristic form of the universal function $I(\eta, x)$. One can see that it includes the well-known Darvin step conditioned by the first factor in (15) and more narrow Cherenkov peak due to the last term in (15). The ratio of amplitudes of these peaks depend on the values x, u_{ph}, u_0.

It is easy to estimate with (15) the maximal dimensionless spectral intensity for PXR:

$$\left(\frac{\partial^3 N_{\tau s}}{\partial \eta_s \partial x_s \partial \varphi}\right)_{max} = \frac{e^2 v_s^2}{4\pi^2 \beta_1 \sin \theta_B} Q^2,$$

$$Q = \frac{\text{Re}\chi_0}{\text{Im}(\chi_0 - \kappa_s)} \sim 10 - 100, \tag{16}$$

which is achieved in the very narrow spectral range

$$\frac{\delta\omega}{\omega} \cong \mathrm{Im}\,\chi_0 \sim 10^{-6} - 10^{-7}.$$

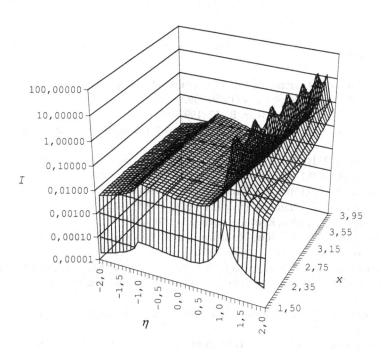

Figure 2. Universal function for the HRPXR spectral-angular distribution (u_{ph}=0.5; u_0=1.1+ i 0.05)

5. CONCLUSION

High spectral intensity of HRPXR could open new directions for application of such radiation for the crystal and nanostructure analysis. In particular, the estimation for the spectral intensity of the synchrotron radiation (SR) from one electron leads to the value $e^2/2\pi$ [6], which can be essentially less than for HRPXR. Certainly, in the most cases the integral number of SR photons from the storage ring with a great electron current is much larger than the quantity of PXR quanta from linear accelerator with a quite small average current. But the applications of HRPXR as the X-ray source for structure analysis could be of interest for the spectral-sensitive experiments where the spectral density of the radiation in the narrow spectral range is necessary. Some of possible applications HRPXR: a) the source of very monochromatic X-rays with high spectral intensity; b) spectral-sensitive experiments; c) the same fields as for HRXRD.

References

1. V.G.Baryshevsky, I.D.Feranchuk, J. de Phys. (Paris) 44 (1983) 913.
2. I.D.Feranchuk, Kristallographiya 24 (1979) 289.
3. V.G.Baryshevsky, Nucl. Instr. Meth. B 122 (1997) 13.
4. V.G.Baryshevsky, I.D.Feranchuk, A.O.Grubich, A.V.Ivashin, Nucl. Instr. Meth. A 249 (1986) 308.
5. I.D.Feranchuk, A.V.Ivashin, I.V.Polikarpov, J. of Phys D: Appl. Phys. 21 (1988) 831.
6. V.G.Baryshevsky, I.D.Feranchuk, Nucl. Instr. Meth. 228 (1985) 490.
7. V.G.Baryshevsky, I.D.Feranchuk, Phys. Lett. A 76 (1980) 452.
8. V.G.Baryshevsky, I.D.Feranchuk, Phys. Lett. A 102 (1984) 141.
9. I.D.Feranchuk, A.V.Ivashin, Krystallographiya 34 (1989) 39.
10. V.G.Baryshevsky, Channeling, Radiation and Reactions in Crystals under the High Energies, Belarusian University, Minsk, 1982.
11. I.D.Feranchuk, Coherent Phenomena for the Processes of X-ray and γ - Radiation from the Relativistic Charged Particles, Doctor of Science Thesis, Belarusian University, Minsk, 1985.
12. Yu.N.Adishchev, V.G.Baryshevsky, S.A.Vorob'ev et al., Pis'ma Zh. Eksp. Teor. Fiz. 41 (1985) 295.
13. V.G.Baryshevsky, V.A.Danilov, I.D.Feranchuk et al., Phys. Lett. A 110 (1985) 477.
14. I.D.Feranchuk, A.V.Ivashin, J. de Phys. (Paris) 46 (1985) 1981.
15. K.-H. Brenzinger, C.Herberg, B.Limburg, et al., Zh. Phys. A 358 (1997) 107.
16. J.Freudenberger, H.Genz, V.V.Morokhovskyi et al., Phys. Rev. Lett. 84 (2000) 270.
17. 5th Int. Symp. on Radiation from Relativistic Electrons in Periodic Structures, Topical Issue of Nucl. Instr. Meth. B 201 (2003) 1.
18. H.Nitta, Phys. Rev. B 45 (1992) 7621.
19. H.Nitta, Nucl. Instr. Meth. B 115 (1996) 401.
20. H.Nitta, J. of Phys. Soc. of Japan 69 (2000) 3462.
21. A.Caticha, Phys. Rev. B 45 (1992) 9641.
22. I.Ya.Dubovskaya, S.A.Stepanov, A.Ya.Silenko, A.P.Ulyanenkov, J. of Phys.: Condensed Matter 5 (1993) 7771.
23. N.Nasonov, V.Sergienko, N.Noskov, Nucl. Instr. Meth. B 201 (2003) 67.
24. A.V.Scchagin, Radiation Physics and Chemistry 61 (2001) 283.
25. H.Backe, C.Ay, N.Clawiter et al., Proc. Int. Symp. on Channeling - Bent Crystals – Radiation Processes, Frankfurt am Main (2003) 41.
26. A.Authier, S.Lagomarsino, B.Tanner (Ed), X-ray and Neutron Dynamical Diffraction, NATO ASI Seies, Plenum Press, New York, 1996.
27. Baryshevsky V.G., Lugovskaya O.M.. Physics of Atomic Nuclei. V.66 (2003) 411.
28. I.D.Feranchuk, A.P.Ulyanenkov, J.Harada, J.C.H.Spence, Phys. Rev. E 62 (2000) 4225.
29. V.V.Morokhovsky, J.Freudenberger, H.Genz et al., Phys. Rev. B 61 (2000) 3347.

DIFFRACTION, EXTRACTION AND FOCUSING OF PARAMETRIC X-RAY RADIATION, CHANNELING RADIATION AND CRYSTAL UNDULATOR RADIATION FROM A BENT CRYSTAL

A.V. Shchagin

Kharkov Institute of Physics and Technology, Kharkov 61108, Ukraine

Abstract: The paper describes the possibility of producing a focused parametric X-ray radiation (PXR) without applying any special X-ray optics. The PXR is emitted by relativistic charged particles that are channeling along a bent crystal. The PXR emitted from the whole length of the bent crystal is brought into focus. Some properties of focused PXR are estimated for typical experimental conditions, its possible applications are discussed. Besides, the feasibility of diffraction of parametric X-ray radiation emitted by relativistic particles in a bent crystal is described for the first time. The PXR generated at one set of crystallographic planes may be diffracted by another set of crystallographic planes in the same bent crystal. In the example being, the PXR is emitted in the backward direction from the crystallographic planes perpendicular to the trajectory of particles channeling along a thin long bent crystal. Then, it reaches the region of the bent crystal, where the Bragg condition is fulfilled for another crystallographic plane, and undergoes diffraction at a right angle. After diffraction, the PXR escapes from the thin crystal and becomes focused. Some properties of the diffracted focused PXR are estimated for typical experimental conditions. The experiment for observation of focused non-diffracted and diffracted PXR is proposed. Besides, possibilities for diffraction, extraction and focusing of crystal undulator radiation, as well as focusing of channeling radiation from a long crystal are proposed.

Keywords: parametric X-ray radiation, channeling radiation, crystal undulator radiation, diffraction of X-rays, bent crystal, channeling particles, focusing of X-rays

*E-mail of A.V. Shchagin: shchagin@kipt.kharkov.ua

27

H. Wiedemann (ed.), Advanced Radiation Sources and Applications, 27–45.
© 2006 *Springer. Printed in the Netherlands.*

1.   INTRODUCTION

The possibility of steering charged particle motion by a bent crystal has first been predicted by Tsyganov [1]. The steering can be realized for the particles which are traveling through a long bent crystal in the channeling regime [2]. The channeling particles may be deflected, if the crystal is bent smoothly enough. Experiments on proton beam deflection, focusing and extraction by bent crystals have been performed, e.g., in Gatchina [3] at proton beam energy 1 GeV, Serpukhov at 70 GeV [4,5], CERN at 120 GeV [6] and at 450 GeV [7,8], Fermilab at 900 GeV [9]. The steering of positive particles at planar channeling is most effective (see, e.g. [7]). Bent crystals of about several centimeters in length are generally used in the experiments. The crystals seem to be convenient for production, diffraction, extraction and focusing of parametric X-ray radiation (PXR), channeling radiation (ChR) and crystal undulator radiation (CUR) generated by channeling relativistic particles.

Experimental studies on the PXR from relativistic electrons moving through a thin crystal have been performed since 1985 to gain an insight into the PXR properties and to develop a new source of a quasimonochromatic polarized X-ray beam. The PXR has its maximum intensity in the vicinity of the Bragg direction relative to the crystallographic plane (PXR reflection). The PXR reflection from crystals was observed and investigated at electron beam energies ranging from a few MeV to several GeV, and at PXR energies from a few keV to hundreds of keV. The validity of kinematic theory [10] for the description of PXR properties has been demonstrated in the most of related publications. General information about the nature, properties and investigations of the PXR, as well as references to original papers can be found, e.g., in reviews [11-14]. The experiment on observation of PXR from protons has been described in Ref. [15]. At the same time it should be noted, that the PXR from particles channeling in a bent crystal has not been studied as yet, so far as we know.

In the present paper we will show the possibility of focusing the PXR generated in a bent crystal by channeling particles [16] and consider some of its properties. Besides, the consideration will be given to the diffraction and focusing of PXR emitted in the backward direction in bent crystal. The experiment will be proposed for observation of focused (both non-diffracted and diffracted) PXR. In the end, we will briefly discuss the possibilities for extraction and focusing of CUR and ChR generated by channeling positrons in a crystal.

2.   HOW TO FOCUS THE PXR

Let us suppose that relativistic charged particles are moving along a bent crystalline plate in the channeling regime, as it is shown in Fig. 1. The bent crystal is cylindrical in shape, R being the radius, and f is the axis of the cylinder. The particles are channeling along the crystallographic planes denoted by the reciprocal lattice vector $\vec{g}$. We will consider the PXR reflections from crystallographic planes aligned at 45° relative to the particle trajectory and perpendicular to the plane of Fig. 1a. These crystallographic planes are denoted by the reciprocal lattice vectors $\vec{g}_1$ and $\vec{g}_2$. The PXR reflections from the planes are going perpendicularly to the particle trajectory and the crystalline plate surface. The PXR reflections from $\vec{g}_1$ are going from the whole plate to the axis f, or, in other words, the PXR reflections are focused. The PXR reflections from $\vec{g}_2$ are going in opposite directions, i.e., from the axis f. Thus, the PXR generated on the crystallographic planes $\vec{g}_1$ in the whole bent crystal, several centimeters in length will be collected (focused) on the axis f.

The Huygens construction for formation of the PXR from a perfect crystal may be found in Refs. [12,14]. Here, Fig. 2 shows the Huygens construction for formation and focusing of the PXR wavetrain from a bent crystal. The spherical wavefronts are going from the points, where the particle crosses the crystallographic planes denoted by the reciprocal lattice vector $\vec{g}_1$. They are going to the focuses with a phase difference divisible by 2π and form the PXR wavetrain. For brevity, here we call such formation as the PXR focusing. Evidently, that the PXR energy (frequency) in the radial direction is the same as the energy (frequency) for the PXR at a right angle to the particle beam, $E = \hbar\omega = \dfrac{V\hbar g_1}{\sqrt{2}}$.

From the classical point of view, the train of PXR permanent-amplitude waves arrives at the focus. The direction of train propagation changes within the small angle $\dfrac{L}{R}$. The relative natural PXR spectral peak width may be estimated as $\left(\dfrac{\Delta E}{E}\right)_{nat} \sim \dfrac{1}{n}$, where n is a number of crystallographic planes crossed by the particle, that is equal to the number of wavelengths in the train.

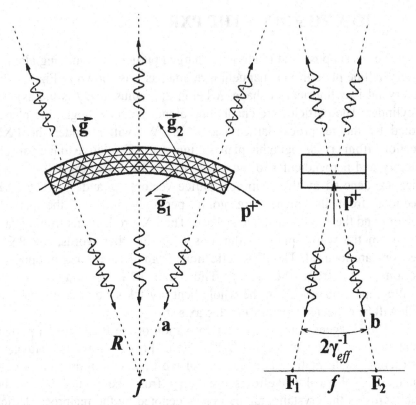

Figure 1. Production of focused PXR by particles channeling in a bent crystalline plate. The side and front views are shown in Figs. 1a and 1b, respectively. The particle beam is indicated by arrows p^+. The particles are channeling along the crystallographic planes denoted by the reciprocal lattice vector $\vec{g}$. The single-crystal plate is cylindrical in shape with a radius of curvature R around the axis f. The wavy lines with arrows show the direction of radiationpropagation at the maxima of the PXR reflections. The PXR reflections from the crystallographic planes denoted by the reciprocal lattice vector $\vec{g}_1$ are focused at points F_1 and F_2 on the axis f. The PXR reflections from the crystallographic planes denoted by the reciprocal lattice vector $\vec{g}_2$ are going in opposite directions and form virtual images of points F_1 and F_2. The PXR focused at points F_1 and F_2 is linearly polarized practically in the plane of Fig. 1b.

From the quantum point of view, the PXR quantum can arrive at the focus on the axis f for the time $\Delta t = \dfrac{L}{V}$ of the particle motion through the crystal. In this case, the PXR spectral peak width may be estimated from the Heisenberg principle $\Delta E \Delta t \sim \hbar$. With the use of the expression for the PXR frequency, this principle gives $\dfrac{\Delta E}{E} \sim \dfrac{1}{2\pi n}$. Thus, the classical and quantum approaches give practically the same, very narrow relative natural spectral peak width of about $1/n$. However, in a real experiment, the broadening of the PXR spectral peak width should be due to a limited angular resolution of the both detector and the whole experiment, to a non-ideal cylindrical shape of the crystal, radiation of non-channeling particles and other experimental factors. The properties of the PXR reflection generated at a right angle to the particle beam in thin crystals have been studied in several papers. The image of the whole PXR reflection at a right angle to the particle beam was first observed in Ref. [17], its polarization was considered in [18], and a detailed structure of the angular distribution of yield was studied in [19]. The shape of the angular distribution of the yield may be found, e.g., in figures of Refs. [17,19,14]. The shape of angular distribution of linear polarization directions in the PXR reflection at the right angle to the particle trajectory may be found, e.g., in figures of Refs. [18,14].

The angular distribution of the PXR yield in the reflection at a right angle to the particle trajectory has two maxima in the plane of Fig. 1b. Therefore the PXR will be focused at two points F_1 and F_2 in the axis f at an angular distance of about $2\gamma_{eff}^{-1}$ one from another. Here, γ_{eff} is the effective relativistic factor in the medium[14], $\gamma_{eff}^{-1} = \sqrt{\gamma^{-2} + |\chi_0|}$, where γ is the relativistic factor of incident particles, χ_0 is the dielectric susceptibility of the medium. The PXR is linearly polarized in the vicinity of points F_1 and F_2. The linear polarization direction is practically in the plane of Fig. 1b. Thus, the focused PXR is linearly polarized in the plane of Fig. 1b. The size of the PXR spot focused around the points F_1 and F_2, is about $\sim \gamma_{eff}^{-1} R$. The PXR energy (frequency) in the vicinity of points F_1 and F_2 varies as

$$\Delta E = E \frac{b}{R}$$

with a shift of the observation point at a distance b from the axis f in the horizontal direction, and remains unchanged at shifts of the observation point along the f axis [10,20,21].

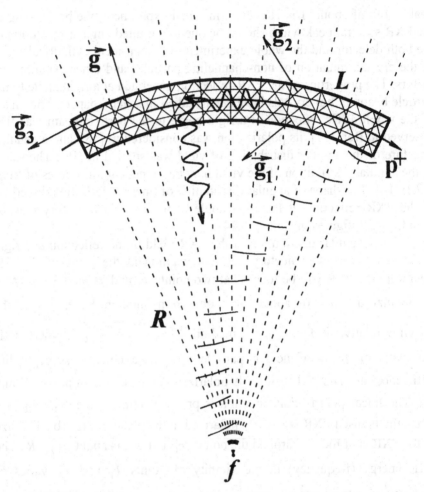

Figure 2. The Huygens construction for focusing the PXR wave train. Spherical surfaces of equal phase (spherical wave fronts) are going from the points, where the channeling particle crosses the crystallographic planes $\vec{g}_1$. Parts of spherical wavefronts are shown on the radial lines R at the moment, when the particle is at the end of the right arrow p^+. The wave train of length $l = \dfrac{cL}{V}$ with the wave length $\lambda = \dfrac{c2\pi\sqrt{2}}{Vg_1}$ and the number of periods $n = \dfrac{Lg_1}{2\pi\sqrt{2}}$ is focused in the vicinity of axis f. Besides, the backward-going PXR is generated on the crystallographic planes denoted by the reciprocal lattice vector $\vec{g}_3$. It is shown by a horizontal wavy line with the arrow. Then it is diffracted on the crystallographic planes denoted by the reciprocal lattice vector $\vec{g}_2$ and is focused at the axis f.

Apart from the PXR emission, the particles moving through the crystalline plate ionize the crystal atoms, and thus make them emit the characteristic X-ray radiation (CXR). The CXR is isotropic and not focused, but it exists at points F_1 and F_2, too.

Further, the backward-going PXR should be focused at the same points F_1 and F_2. This radiation is generated by channeling particles at the crystallographic planes denoted by the reciprocal lattice vector $\vec{g}_3$ in Fig. 2. It propagates in the vicinity of the angle 180° to the particle trajectory and is shown by a horizontal wavy line with the arrow in Fig. 2. At a certain path length, the PXR frequency will satisfy the Bragg condition for the crystallographic planes $\vec{g}_2$. Therefore, a part of the backward-going PXR polarized in the plane of Fig. 1b will undergo the Bragg diffraction. The direction of propagation of diffracted PXR is shown by a wavy line with the arrow along the radius R of the crystal curvature in Fig. 2. Thus, a part of backward PXR will be focused in the vicinity of the same points F_1 and F_2. The energy, the linear polarization and the spot size of diffracted backward PXR focused around the points F_1 and F_2 are practically the same as in the case of non-diffracted PXR.

3. ESTIMATION OF SOME PROPERTIES OF FOCUSED PXR

To understand the main features of focused PXR, let us estimate them for typical experimental conditions. As an example, we have performed calculations for 70 and 450 GeV protons channeling along the crystallographic plane (110) with the reciprocal lattice vector $\vec{g} =< \overline{1}\,\overline{1}\,0 >$ (see Fig. 1) in the Si single-crystal plate, 0.5 mm in thickness and $L = 5$ cm in length, with the radius of curvature $R = 5$ m. The PXR reflections under consideration are generated at the crystallographic planes parallel to the ones denoted by the reciprocal lattice vectors $\vec{g}_1 =< 100 >$ and $\vec{g}_2 =< 0\,\overline{1}\,0 >$. The calculations were performed using formulae from Refs. [12,14,22], derived within the framework of the Ter-Mikaelian theory [10]. The focused PXR reflections arrive at the points F_1 and F_2 from the crystallographic planes having a nonzero structure factor parallel to (100), these are the (400), (800), (12 00) planes. The PXR reflections from higher-index planes have substantially lower yields. The PXR reflections from similar crystallographic planes, that are parallel to $(0\,\overline{1}\,0)$, are going in opposite directions from the axis f. They form virtual images of focal points F_1 and F_2.

The $< \overline{1}\,10 >$ axis is aligned along the proton beam in the above-considered example. To avoid axial channeling in the experiment, one has to provide misorientation of the crystal lattice in the (110) plane for a small angle, which is somewhat larger than the critical channeling angle. The critical channeling angle for the Si single crystal at proton energies of 70 and 450 GeV is about $2.4 \cdot 10^{-5}$ and $1 \cdot 10^{-5}$ radian, respectively [2]. The main features of PXR emission, the positions of points F_1, F_2, and the properties of focused PXR will change only slightly at such a small misorientation.

The PXR properties are functions of the relativistic factor of the incident particle and the absolute value of particle charge [10,12,22]. Therefore, the results of present calculations might be valid for 38 and 245 MeV positrons and electrons, which have the same relativistic factors. However, the motion of these particles in a long crystal can differ from the motion of protons. For example, a significant part of high-energy protons can move through a long bent crystal in the channeling regime [2,4-9]. But electrons strongly scatter at channeling. So far as we know, there have been no publications about experimental research on channeling of positrons in bent crystals and on generation of PXR by positrons in any crystals.

4.   ESTIMATION OF SOME PROPERTIES OF CXR

The yield of CXR at 1.74 keV due to K-shell ionization of Si atoms by 70 and 450 GeV protons was estimated in accordance with the recommendations from Ref. [24]. In the calculations, the K-shell ionization cross section for Si atoms was taken to be 1654 and 1946 barn. These cross section values were calculated for 38 and 245 MeV electrons having the relativistic factor identical to that of 70 and 450 GeV protons. The effect of channeling was not taken into account in the present estimation of the CXR yield, as it is not expected to be very significant at high particle energies (see Refs. in [24]).

All calculations were performed for a uniform distribution of protons in the crystalline plate thickness. The radiation excited by channeling protons is going to the points F_1 and F_2 across the plate, and is attenuated by the Si crystal. The attenuation of PXR and CXR in the plate was taken into account in the calculations.

5. ESTIMATION OF SOME PROPERTIES OF BACKWARD-GOING DIFFRACTED FOCUSED PXR

Particles generate the backward-going PXR at the crystallographic planes with non-zero structure factors $(4\bar{4}0),(8\bar{8}0),(12\,\overline{12}\,0)$ that are perpendicular to the particle trajectory and parallel to the ones denoted by the reciprocal lattice vector $\vec{g}_3$ in Fig. 2. The maximum yield in the PXR reflection is at the angle $\pi - \gamma_{eff}^{-1}$ relative to the tangent to the particle trajectory. The radiation energy in the maximum of the PXR reflection is $E_{PXR}^{\pi-\gamma_{eff}^{-1}} = \dfrac{c\hbar g_3}{2\sqrt{\varepsilon_0}}$.

As the radiation propagates through the bent crystal it will reach the region, where it will be diffracted by the crystallographic planes $(0\bar{4}0),(0\bar{8}0),(0\,\overline{12}\,0)$ (they are parallel to the ones denoted by the reciprocal lattice vector $\vec{g}_2$) toward of cylinder axis, as the Bragg condition will be fulfilled. The angular distance between the regions, where the PXR is emitted and then diffracted, is about $\dfrac{\gamma_{eff}^{-2}}{2}$. This means that the backward-going PXR from the crystallographic planes $(4\bar{4}0),(8\bar{8}0),(12\,\overline{12}\,0)$ will be diffracted by the crystallographic planes $(0\bar{4}0),(0\bar{8}0),(0\,\overline{12}\,0)$ toward the axis f after covering backward path of about 507, 463, 455 μm at a proton energy of 70 GeV and 69, 25, 17 μm at a proton energy of 450 GeV respectively.

Only the PXR with its polarization direction parallel to the plane of Fig. 1b will be diffracted. Thus, backward PXR polarized in the plane of Fig. 1b will be extracted and focused in the vicinity of points F_1 and F_2. The attenuation of PXR over the path in a backward direction reduces the PXR intensity. The energies are practically equal, and the shapes of angular distributions are similar to the ones of the PXR generated at the crystallographic planes (400), (800), (12 00). The maximum possible yield of backward diffracted focused PXR at the focuses was estimated with taking into account the X-ray attenuation over tangent and radial paths in the crystal and on the assumption that all backward-going PXR is diffracted.

The forward going (dynamic) PXR, as well as the synchrotron radiation and bremsstrahlung linearly polarized in the plane of Fig. 2b, if they are sufficiently high at present conditions, can be diffracted by the crystallographic planes (400),(800),(12 00) to the same focuses.

6.   RESULTS

The results of calculations for non-diffracted PXR and CXR are presented in Table 1. The X-ray radiation spectrum at points F_1 and F_2 has four spectral peaks in the 1.7 – 19.4 keV energy range, this being convenient for registration by standard X-ray spectrometric detectors. The peak at 1.74 keV is due to a non-focused non-polarized CXR, which is produced mainly by protons moving in a 13.3 μm layer at a concave side of the plate. This layer thickness is determined by X-ray attenuation in the Si crystal (see Table 1). Thus, only a little part of the total number of protons passing through the plate produces the CXR that reaches the points F_1 and F_2. The other spectral peaks are due to a focused polarized PXR. The spectral peaks of focused PXR at $E = 6.46$, 12.91, 19.37 keV are produced mainly by protons moving in the 37 μm layer at a concave side of the plate, in the 270 μm layer at the same side of the plate, and in almost the whole crystal plate thickness, respectively.

The calculated data on the backward-going diffracted focused PXR are presented in Table 2. The energy, polarization and positions of focuses are practically identical to the ones for the non-diffracted focused PXR. It be seen from Table 2 that diffracted PXR intensity is much lower than the intensity of non-diffracted PXR at a proton energy of 70 GeV. However, the intensities become comparable at a proton energy of 450 GeV, especially for high-order PXR reflections. Therefore, the yield of focused PXR in high-order reflections at high proton energies may be increased due to the contribution from the diffracted backward PXR.

The radiation sources for different spectral peaks are specifically distributed across the crystal thickness. Therefore, the measurements and analysis of relative intensities of spectral peaks may give the estimates of beam proton distribution within the thickness of the crystal. The measurements can be done for the whole crystal plate or for its separate part, provided that other parts of the plate are screened, and the X-ray detector can see only this separate part of the crystal. The CXR from the 13.3 μm layer at a convex side of the bent crystal may be registered by an X-ray detector installed opposite to the convex side of the bent crystal. The CXR intensities at both sides of the crystal should be close to each other, as the CXR is isotropic and non-focused.

The angular divergence (convergence) of focused PXR may be controlled by varying the radius of curvature R of the bent crystal. Therefore, we have an X-ray source of several centimeters in size, with a provided smooth varying of the divergence (convergence). Such a monochromatic source of polarized X-rays with a smooth variation of divergence may be useful for calibration of large-aperture X-ray equipment, for example, X-ray space telescopes [25]. The divergence (convergence) of PXR reflections from channeling particles is L/R in the plane of Fig. 1a.

Table 1. Properties of CXR and focused PXR induced by 70 and 450 GeV protons channeling along crystallographic planes (110) in a bent Si single-crystal plate, 0.5 mm in thickness and $L = 5$ cm in length. The radius of crystal curvature is $R = 5$ m. The origin of radiation (CXR or focused PXR from the labled crystallographic planes) and polarization properties of the radiation are given in the first and second lines of the table, where E is the energy of spectral peaks of radiation, $(\Delta E/E)_{nat} = 1/n$ is the natural PXR spectral peak width on the axis f for a point-like detector, $(\Delta E/E)_D = b/R$ is the PXR spectral peak width for the horizontal size of the detector $b = 1$ cm, T_e is the e-fold attenuation length of radiation having the energy E in a Si single crystal, γ_{eff}^{-1} is the inverse effective relativistic factor (for comparison, the inverse relativistic factor γ^{-1} for incident 70 and 450 GeV protons is $1.34 \cdot 10^{-2}$ and $2.08 \cdot 10^{-3}$, respectively), ΔF is the distance between the points F_1 and F_2, I is the number of quanta per proton per cm^2 at the points F_1 and F_2 with X-ray attenuation in the crystal taken into account for channeling protons randomly distributed within the thickness of the plate

Radiation		CXR Si	PXR(400)	PXR(800)	PXR(12 00)
Polarization		No	Linear	Linear	Linear
E, keV		1.74	6.46	12.91	19.37
$(\Delta E/E)_{nat}$			$3.84 \cdot 10^{-9}$	$1.92 \cdot 10^{-9}$	$1.28 \cdot 10^{-9}$
$(\Delta E/E)_D$			$2 \cdot 10^{-3}$	$2 \cdot 10^{-3}$	$2 \cdot 10^{-3}$
T_e in Si, μm		13.3	37	270	865
Protons 70 GeV	γ_{eff}^{-1}	-	$1.42 \cdot 10^{-2}$	$1.36 \cdot 10^{-2}$	$1.35 \cdot 10^{-2}$
	ΔF, mm	-	142	136	135
	$I, \dfrac{quanta}{cm^2 \cdot p^+}$	$1.65 \cdot 10^{-7}$	$9.56 \cdot 10^{-8}$	$1.86 \cdot 10^{-8}$	$1.55 \cdot 10^{-9}$
Protons 450 GeV	γ_{eff}^{-1}	-	$5.24 \cdot 10^{-3}$	$3.18 \cdot 10^{-3}$	$2.63 \cdot 10^{-3}$
	ΔF, mm	-	52.4	31.8	26.3
	$I, \dfrac{quanta}{cm^2 \cdot p^+}$	$1.94 \cdot 10^{-7}$	$6.90 \cdot 10^{-7}$	$3.32 \cdot 10^{-7}$	$3.98 \cdot 10^{-8}$

The experiment on observation of focused PXR may be performed at a facility with a crystal bent for steering the proton beam having the channeling particles intensity of about or above $10^7 p/s$. It can be seen from Table 1 that all spectral peaks have comparable intensities at a proton energy of 450 GeV, and thus, can be measured simultaneously by a spectrometric X-ray detector. The intensity of high-order spectral peaks decreases with decrease in the proton energy down to 70 GeV, but the low-index peaks still may be observed in the experiment [26].

The X-ray detector, being about 1 cm^2 square, should be installed at the point F_1 or F_2 on the axis f. The PXR from non-channeled particles can be rejected by registering X-rays in coincidence with the particles that have passed through the bent crystal in the channeling regime. There should be a vacuum connection between the bent crystal and the X-ray detector in order to observe soft X-rays, because of their attenuation in air. The X-rays of energies 1.74 and 6.46 will be absorbed practically completely over a distance of 5 m in air. However, observation of higher-energy focused PXR is possible in air. The X-rays of energies 12.91 and 19.37 keV will be attenuated by factors of about 0.23 and 0.60, respectively, within 5 m in dry air. The energies of all spectral peaks are practically independent of incident relativistic particle energy. Yet, their intensities can increase (decrease) with an increasing (decreasing) incident proton energy [12,22].

Table 2. Some properties of backward-going diffracted and focused PXR induced by 70 and 450 GeV protons channeling along the crystallographic planes (110) in a bent Si single-crystal plate, 0.5 mm in thickness and 5 cm in length. The radius of crystal curvature is 5 m. The origin of backward radiation (PXR from the crystallographic plane) is given in the first column of the table, E is the energy of spectral peaks of radiation, the backward-going PXR is diffracted in the crystallographic planes shown in the third column. I is the estimated maximum number of quanta per proton per cm^2 at the points F_1 and F_2 of backward-going diffracted and focused PXR. Attenuation of X-rays over backward and radial paths in the crystal is taken into account for channeling protons randomly distributed within the thickness of the plate.

Radiation from $\bar{g}_3$	E, keV	Diffraction from $\bar{g}_2$	Protons 70 GeV $I, \dfrac{quanta}{cm^2 \cdot p^+}$	Protons 450 GeV $I, \dfrac{quanta}{cm^2 \cdot p^+}$
PXR ($4\bar{4}0$)	6.46	($0\bar{4}0$)	$3.30 \cdot 10^{-15}$	$3.36 \cdot 10^{-8}$
PXR ($8\bar{8}0$)	12.91	($0\bar{8}0$)	$3.86 \cdot 10^{-10}$	$3.56 \cdot 10^{-8}$
PXR ($12\,\overline{12}\,0$)	19.37	($0\,\overline{12}\,0$)	$6.38 \cdot 10^{-11}$	$2.78 \cdot 10^{-8}$

7.   DISCUSSION

Let us discuss briefly some possibilities for extraction and focusing of other kinds of radiation emitted by relativistic particles (positrons) channeling in a long crystal.

7.1   Extraction of channeling radiation

The Bragg diffraction of channeling radiation (ChR) in one of the crystallographic planes of the same crystal, where the ChR is generated, has been considered in Ref. [23]. The diffraction may be used for extraction of the ChR from a long crystal. The scheme of the extraction is shown in Fig. 3. Here, the ChR of frequency ω_{ChR}, equal to the Bragg frequency ω_B for the crystallographic planes denoted by the reciprocal lattice vector $\vec{g}_1$, may be extracted from a long crystal before the ChR is attenuated. The diffracted and extracted channeling radiation (DChR) is accompanied by the ordinary PXR reflection that is generated by the same relativistic particles at the crystallographic planes denoted by the reciprocal lattice vector $\vec{g}_1$.

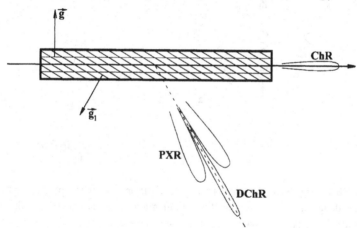

Figure 3. Extraction of channeling radiation from a perfect crystal. Positrons or other positive particles are channeling in a perfect crystal along the crystallographic planes denoted by the reciprocal lattice vector $\vec{g}$. They emit channeling radiation (ChR) going along the particle trajectory. A part of the ChR of frequency ω_{ChR} is diffracted to the Bragg direction relative to the crystallographic planes denoted by the reciprocal lattice vector $\vec{g}_1$. The Bragg direction is shown by a dotted line. The diffraction is possible on the condition that $\omega_{ChR} = \omega_B$, where ω_B is the Bragg frequency for $\vec{g}_1$. The angular distribution of ordinary PXR reflection of frequency close to ω_B is shown schematically around the Bragg direction. The angular distribution of diffracted and extracted DChR of frequency ω_B is shown along the Bragg direction.

7.2   Focusing of channeling radiation

In some cases, the extracted channeling radiation, shown in Fig. 3, may be focused if the crystal is smoothly bent. As an example, Fig. 4 shows the focusing of ChR extracted at a right angle from a cylindrically bent crystal. Positrons are channeling along the crystallographic planes denoted by the reciprocal lattice vector $\vec{g}$. A part of channeling radiation with the frequency ω_{ChR} equal to the Bragg frequency ω_B and with the polarization in the plane of Fig. 4b is diffracted in the crystallographic planes denoted by the reciprocal lattice vector $\vec{g}_1$ and is focused at point F_3 on the axis of the crystal curvature f. Besides, the tail of ChR having the frequency somewhat below ω_B can appear on the left of the axis f in Fig. 3a because of ChR diffraction at some distance from the particle trajectory. The ordinary PXR is focused at points F_1 and F_2, in much the same way as in Fig. 1.

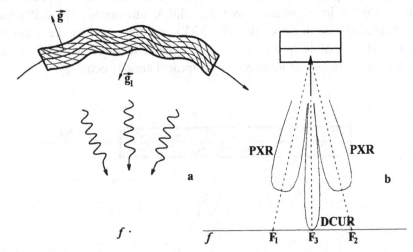

Figure 4. Focusing of channeling radiation. Positrons or other positive particles are channeling along the bent crystal in the crystallographic planes denoted by the reciprocal lattice vector $\vec{g}$. They emit channeling radiation of frequency $\omega_{ChR} = \omega_B$ along of particles trajectory (ω_B is the Bragg frequency relative to the crystallographic planes at $45°$ to the particle trajectory, that are denoted by the reciprocal lattice vector $\vec{g}_1$). The diffracted and extracted part of the channeling radiation (DChR) is focused at point F_3. The ordinary PXR reflection is focused at points F_1 and F_2 just as in Fig. 1.

7.3   The PXR from a crystal undulator

In recent years, experimental efforts have been in progress to observe the crystal undulator radiation (see, e.g., Ref. [24]). In experiments, relativistic particles have to move in the channeling regime along a periodically bent crystal and emit undulator radiation in the forward direction, as is shown in Fig. 5. We would like to draw attention to the fact that besides of the crystal undulator radiation (CUR), the particles should emit a number of PXR reflections in the vicinity of the Bragg directions of different crystallographic planes of the same crystal. The PXR should be rather intense, as it is generated in a long crystal. The observation of the PXR and characteristic X-rays at a significant observation angle may be useful for exact alignment of the crystal undulator and for diagnostics in the experiment. The PXR train should be frequency and phase modulated because of the periodical modulation of a crystal lattice orientation in the undulator.

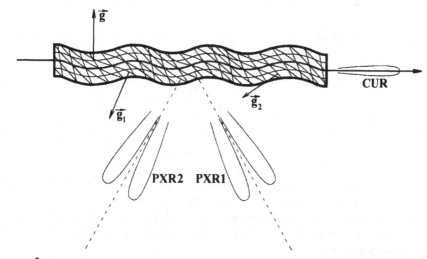

Figure 5. Generation of PXR reflections from the crystal undulator. The channeling particle emits crystal undulator radiation (CUR) in the forward direction. Besides, it emits PXR reflections in Bragg directions relative to different crystallographic planes of the crystal undulator. Only two sets of crystallographic planes denoted by the reciprocal lattice vectors $\vec{g}_1$ and $\vec{g}_2$ as well as the related Bragg directions and PXR reflections are shown in the figure.

7.4 Extraction of crystal undulator radiation

It is obvious from Fig. 5 that the crystal undulator radiation is going through the long crystal and may be attenuated significantly. Therefore it would be of interest to extract the CUR through the side of the undulator. This may be done provided that the CUR frequency is equal to the Bragg frequency for one of the crystallographic planes. In this case, the CUR will be diffracted and extracted from the undulator at a Bragg angle similarly to the channeling radiation case in Fig. 3. The exact tuning of CUR frequency may be provided by a correct choice of the incident particle energy. The angular positions of the PXR reflection, and of diffracted and extracted crystal undulator radiation (DCUR) are shown in Fig. 6. Note that the frequency band of extracted CUR may be proportional to the amplitude of the periodical crystal bend, as the Bragg condition should vary periodically, according to the crystal lattice curvature.

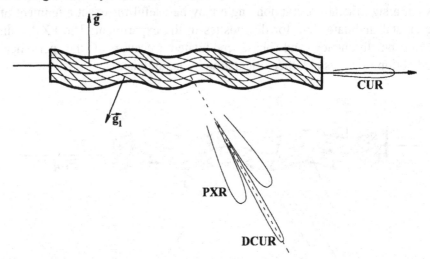

Figure 6. Extraction of undulator radiation from the crystal undulator. The diffracted and extracted crystal undulator radiation (DCUR) is going in the Bragg direction, and the PXR reflection is around the Bragg direction for the crystallographic planes denoted by the reciprocal lattice vector $\vec{g}_1$. The frequency of CUR should satisfy to Bragg condition for these crystallographic planes.

7.5   Focusing of crystal undulator radiation

In some cases, the extracted CUR may be focused similarly to the above-described focusing of channeling radiation. The focusing of extracted CUR is possible from a smoothly bent crystal undulator. To exemplify, Fig. 7 shows the focusing of CUR extracted at a right angle from a cylindrically bent crystal undulator. Positrons are channeling along the crystallographic planes denoted by the reciprocal lattice vector $\vec{g}$. A part of the CUR with frequency ω_{CUR} equal to the Bragg frequency ω_B and with polarization in the plane of Fig. 7b is diffracted in the crystallographic planes denoted by the reciprocal lattice vector $\vec{g}_1$ and is focused at the point F_3 on the axis f of crystal curvature. Besides, the CUR tail of frequency somewhat below ω_B may appear on the left of the axis f in Fig. 7a due to of CUR diffraction at some distance from the particle trajectory. The ordinary PXR is focused at points F_1 and F_2, just as in Fig. 1. The focused PXR train will possess the periodical frequency and phase modulation with the period $T = l / V$, where l the crystal undulator period and V is the particle velocity.

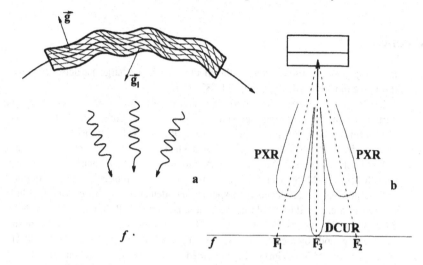

Figure 7. Diffraction, extraction and focusing of crystal undulator radiation. Positrons or other positive particles are channeling along a cylindrically bent crystal undulator in the crystallographic planes denoted by the reciprocal lattice vector $\vec{g}$. They emit the CUR of frequency $\omega_{CUR} = \omega_B$ along of particles trajectory (ω_B is the Bragg frequency relative to the crystallographic planes at $45°$ to the particle trajectory, that are denoted by the reciprocal lattice vector $\vec{g}_1$). The diffracted and extracted part of the crystal undulator radiation (DCUR) is focused at the point F_3 at axis f. The ordinary PXR reflection is focused at points F_1 and F_2 just as in Fig. 1. The focused PXR train will possess the periodical frequency and phase modulation with the period $T = \dfrac{l}{V}$, where l the crystal undulator period and V is the particle velocity.

7.6   About efficiency of the ChR and CUR focusing

Note that the focusing of ChR and CUR by the schemes shown in Figs. 4, 7 may be not very effective, because ChR and CUR should be polarized mainly in the plane of Fig. 4a and 7a that is parallel to the direction of the transverse motion of particles in the channel or undulator. But only the radiation polarized in the plane of Figs. 4b and 7b may be diffracted at $90°$ and focused due to properties of Bragg diffraction. Thus, schemes shown in Figs. 4, 7 only illustrate the idea but they are not optimal for focusing of ChR and CUR. The effectiveness of the focusing may be increased if we will choice another alignment of the set of crystallographic planes denoted by the reciprocal lattice vector $\vec{g}_1$

ACKNOWLEDGEMENTS

The paper became possible partially due to STCU grant 1030.

References

1. E.N. Tsyganov, Some aspects of the mechanism of a charge particle penetration through a monocrystal, Fermilab, TM-682 (1976).

2. V.M. Biryukov, Yu.A. Chesnokov, V.I. Kotov, Steering of high-energy charged particle beams by bent single crystals, Uspekhi Fizicheskikh Nauk,164, 1017-140 (1994) [in Russian]. Physics-Uspekhi 37, 937-961 (1994) [in English].

3. V.A. Andreev, V.V. Baublis, E.A. Damaskinskij, A.G. Krivshich, L.G. Kudin, V.V. Marchenkov, V.F. Morozov, V.V. Neljubin, E.M. Orishchin, G.E. Petrov, G.A. Rjabov, V.M. Samsonov, L.E. Samsonov, E.M. Spiridenkov, V.V. Sulimov, O.I. Sumbaev, V.A. Schegelsky, Experimental detection of the effect of volume capture in a channelling mode in the bent single crystal, Pis'ma ZhETF, 36, No 9, 340-343 (1982); Focussing of a beam 1 GeV protons during volume capture in a channelling mode by the bent monocrystal, Pis'ma ZhETF, 39, No 2, 58-61 (1984).

4. A.A. Asseev, M.Yu. Gorin, Crystal-aided non-resonant extraction of 70 GeV protons from the IHEP accelerator, Nucl. Instr. and Meth. B 119, 210-214 (1996).

5. Yu.A. Chesnokov, Review of IHEP experiments for focusing and deflection 70 GeV proton beam with bent crystals, Nucl. Instr. and Meth. B 119, 163-171 (1996).

6. K. Elsener, G. Fidecaro, M. Gyr, W. Herr, J. Klem, U. Mikkelsen, S.P. Møller, E. Uggerhøj, G. Vuagnin, E. Weisse, Proton extraction from the CERN SPS using bent silicon crystals, Nucl. Instr. and Meth. B 119, 215-230 (1996).

7. A. Baurichter, K. Kirsebom, R. Medenwaldt, S.P. Møller, T. Worm, E. Uggerhøj, C. Biino, M. Clément, N. Doble, K. Elsener, L. Gatignon, P. Grafström, U. Mikkelsen, A. Freund, Z. Vilakazi, P. Siffert, M. Hage-Ali, New results from the CERN-SPS beam deflection experiments with bent crystals, Nucl. Instr. and Meth. B 119, 172-180 (1996).

8. N. Doble, L. Gatigton, P. Grafstrom, A novel application of bent crystal channeling to the production of simultaneous particle beams, Nucl. Instr. and Meth. B 119, 181-190 (1996).

9. C.T. Murphy, R. Carrigan, D. Chen, G. Jackson, N. Mokhov, H.-J. Shih, B. Cox, V. Golovatyuk, A. McManus, A. Bogacz, D. Cline, S. Ramachandran, J. Rhoades, J. Rosenzweig, B. Newberger, J.A. Ellison, S. Baker, C.R. Sun, W. Gabella, E. Tsyganov, A. Taratin, A. Asseev, V. Biryukov, A. Khanzadeev, T. Prokofieva, V. Samsonov, G. Solodov, First results from bent crystal extraction at the Fermilab Tevatron, Nucl. Instr. and Meth. B 119, 231-238 (1996).

10. M.L. Ter-Mikaelian, High-Energy Electromagnetic Processes in Condensed Media, Edition of Armenian Academy of Sciences, Erevan, 1969 [in Russian]. Wiley-Interscience, New York, 1972 [in English].

11. M.L. Ter-Mikaelian, Electromagnetic radiation processes in periodic media at high energies, Uspekhi Fizicheskikh Nauk, 171, 597-624 (2001) [in Russian]. Physics-Uspekhi 44, 571-596 (2001) [in English].

12. A.V. Shchagin, X.K. Maruyama, Parametric X-rays, in: Accelerator-based atomic physics technique and applications, eds. S.M. Shafroth, J.C. Austin (AIP Press, New York, 1997) pp. 279-307.

13. P. Rullhusen, A. Artru, P. Dhez, Novel radiation sources using relativistic electrons, World Scientific Publishers, Singapore, 1998.

14. A.V. Shchagin, Investigations and properties of PXR, in: Electron-photon interactions in dense media, ed. by H. Wiedemann, NATO Science Series, II Mathematics, Physics and Chemistry, Vol. 49, Kluwer Academic Publishers, Dordrecht/Boston/London, 2002, pp. 133-151.

15. V.P. Afanasenko, V.G. Baryshevsky, R.F. Zuevsky, A.S. Lobko, A.A. Moskatelnikov, S.B. Nurushev, V.V. Panov, V.P. Potsilujko, V.V. Rykalin, S.V. Skorokhod, D.S. Shvarkov, Detection of proton parametric x-ray radiation in silicon, Phys. Lett. A 170, 315-318 (1992).

16. A.V. Shchagin, Focusing of Parametric X-ray Radiation from a Bent Crystal, Communication at Workshop "Relativistic Channeling and Related Coherent Phenomena", March 23 – 26, 2004, INFL-Laboratori Nazionali di Frascati, Frascati, Italy. E-Preprint: physics/0404137 (April 2004). Correction: Number of quanta for PXR(12 00) in Table 1 must be $3.98 \cdot 10^{-8}$ instead of $1.28 \cdot 10^{-7}$. Submitted to NIM B.

17. R.B. Fiorito, D.W. Rule, M.A. Piestrup, X.K. Maruyama, R.M. Silzer, D.M. Skopik, A.V. Shchagin, Polarized angular distributions of parametric x radiation and vacuum-ultraviolet transition radiation from relativistic electrons, Phys. Rev. E 51, R2759-R2762 (1995).

18. A.V. Shchagin, Linear polarization of parametric X-rays, Phys. Lett. A 247, 27-36 (1998).

19. A.V. Shchagin, Parametric X-rays at the right angle to the particle beam, Phys. Lett. A 262, 383-388 (1999).

20. A.V. Shchagin, V.I. Pristupa, N.A. Khizhnyak, A fine structure of parametric X-ray radiation from relativistic electrons in a crystal, Phys. Lett. A 148, 485-488 (1990).

21. K.-H. Brenzinger., B. Limburg., H. Backe., S. Dambach., H. Euteneuer., F. Hagenbuck., C. Herberg, K.H. Kaiser., O. Kettig, G. Kube, W. Lauth, H. Schope, Th. Walcher, How Narrow is the Linewidth of Parametric X-Ray Radiation?, Phys. Rev. Lett. 79, 2462-2465 (1997).

22. A.V. Shchagin, N.A. Khizhnyak, Differential properties of parametric X-ray radiation from a thin crystal, Nucl. Instr. and Meth. B 119, 115-122 (1996).

23. T. Ikeda, Y. Matsuda, H. Nitta, Y.H. Ohtsuki, Parametric X-ray radiation by relativistic channeled particles, Nucl. Instr. and Meth. B 115, 380-383 (1996).

24. A.V. Shchagin, V.I. Pristupa, N.A. Khizhnyak, K-shell ionization cross section of Si atoms by relativistic electrons, Nucl. Instr. and Meth. B 84, 9-13 (1994).

25. A.V. Shchagin, N.A. Khizhnyak, R.B. Fiorito, D.W. Rule, X. Artru, Parametric X-ray radiation for calibration of X-ray space telescopes and generation of several X-ray beams, Nucl. Instr. and Meth B 173, 154-159 (2001).

26. A.V. Shchagin, Focusing of parametric X-ray radiation, JETP Letters, 80, 535-540 (2004) [in Russian].

27. S. Bellucci, S. Bini, V. M. Biryukov, Yu. A. Chesnokov, S. Dabagov, G. Giannini, V. Guidi, Yu. M. Ivanov, V. I. Kotov, V. A. Maisheev, C. Malagu, G. Martinelli, A A. Petrunin, V. V. Skorobogatov, M. Stefancich, and D. Vincenzi, Experimental Study for the Feasibility of a Crystalline Undulator, Phys. Rev. Lett. 90, 034801 (2003).

CHOICE OF OPTIMAL TARGET FOR MONOCHROMATIC TUNABLE X-RAY SOURCE BASED ON LOW-ENERGY ACCELERATOR

Alexander Lobko and Olga Lugovskaya
Institute for Nuclear Problems, Belarus State University, 11 Bobruiskaya Str., Minsk 220050, Belarus

Abstract: Parametric x-rays (PXR) and other than PXR radiation mechanisms resulting from periodical structure of a crystal target can be used for development of quasi-monochromatic tunable x-ray source for medical imaging. We present numerical calculations of spectral-angular density and angular distributions of x-rays producing by relatively low energy electron beam of a medical accelerator. Calculations were performed for silicon crystals depending on target thickness and diffraction geometry for the purpose of optimal choice of x-rays generation conditions. Production technology and parameters of thin (down to sub-micron thickness) Si targets are presented.

Key words: Parametric x-rays; PXR; low-energy medical accelerators; thin crystal target.

1.   INTRODUCTION

Parametric x-rays (PXR) are very prospective radiation mechanism for the development of quasi-monochromatic tunable x-ray source for x-ray medical imaging. It was established [1] that such source based on existing relatively low-energy medical accelerators is feasible and can provide x-ray flux needed for high quality medical images obtaining. Now, it is important to find optimal generation parameters for maximal x-rays yield, i.e. in the first place target medium, its thickness, and geometry.

H. Wiedemann (ed.), Advanced Radiation Sources and Applications, 47–54.
© 2006 *Springer. Printed in the Netherlands.*

As was shown in [1,2], at low energies below threshold energy $E<<E_{tr}$ ($E_{tr}\approx mc^2\omega_B/\omega_L$, where ω_B – frequency of emitted quanta, ω_L – Langmuir frequency of a crystal) and/or relatively thick targets $L\geq L_{Br}$ (L_{Br} – *Bremsstrahlung* coherent length) considerable amounts of diffracted transition radiation and *Bremsstrahlung* connected with multiple scattering are emitted to addition of PXR almost to the same ranges of angles and frequencies. As all these contributions are comparable in such cases, calculations must take all radiation mechanisms into consideration. Evaluations of [1] have shown, that despite a decrease of radiation quantum yield at low energies, the integral amount of quanta do not decrease very drastically and the number of x-ray quanta needful for quality image obtaining can still be achieved at ~0.1 A beam current. More accurate calculations will be presented below.

2.   X-RAYS PARAMETERS CALCULATIONS

List of target materials, which is experimentally confirmed by PXR generation, is quite long. It is include diamond, pyrolytic graphite, Si, Ge, GaAs, LiF, W, quartz and some others. As main criterion for an imaging application is a highest PXR yield, diamond must be a best choice. On the other hand, microelectronics provides crystals of a very high degree of structure perfection together with sophisticated technology of their processing. Using this technology one can produce targets of exactly optimal thickness and orientation. It is especially important for low-energy beam x-ray generation, because a target thickness going to be as low as some microns and such targets cannot be produced without a well-developed technology. From these considerations we have chosen silicon among other probable materials taking also into account its minimal PXR absorption.

Accordingly, calculations of PXR spectral-angular and angular distributions were performed for following conditions: target of silicon single crystal; 25 MeV electron beam that is typical for medical accelerators; strong (111) reflection; ~33 keV x-rays with $\Delta\omega/\omega\leq10^{-2}$ that are required for subtractive digital angiography; image angular dimensions ~70 mrad.

Spectral-angular distributions of x-radiation generated in symmetrical Laue geometry at 0.01 cm target are shown for three polar angles, namely 3.5, 7.4, and 6.3 mrad. Azimuth angle is $\pi/2$, i.e. radiation has σ-polarization. Calculations were performed on general expressions taking into account radiation at both dispersion branches, diffracted transition radiation contribution, PXR and diffracted transition radiation interference. On account of electron multiple scattering results in the radiation integral intensity increase ~1.4 times due to diffracted *Bremsstrahlung* contribution and in change of differential (spectral-angular and angular) characteristics. Radiation strictly in diffraction maximum direction, i.e. in polar angle=0 rad appears. As one can see from the Fig.1, spectral-angular distribution half-width for a specific polar angle is about $5 \cdot 10^{-5} \omega_B$. While radiation angle increase, central frequency of distribution ω_0 moves along frequency axis away from ω_B (see Fig. 2). Thus, total frequency width of x-ray reflex going to be $\sim 5 \cdot 10^{-1} \omega_B$.

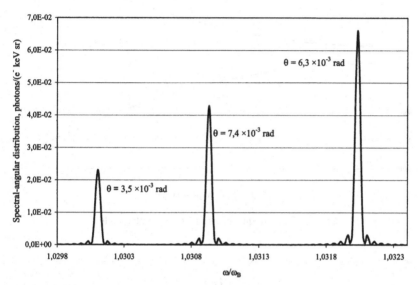

Figure 1. Spectral-angular distributions of PXR generated in symmetrical Laue (111) geometry at 0.01 cm Si target.

PXR spectral-angular distributions for crystal targets of various thicknesses from $1 \cdot 10^{-4}$ through $5 \cdot 10^{-2}$ cm are shown in Fig. 3. Calculations are performed for polar angle θ = 20.3 mrad, where radiation angular distributions reach maximal value. Spectral-angular density rapidly increases with crystal target increase, but saturation cannot be achieved in above mentioned thickness range because absorption length is equal about 0.5 cm. Angular distributions that was obtained by frequency integration of spectral-angular distributions from Fig. 3 within integral limits $\Delta\omega/\omega_B = 5 \cdot 10^{-2}$ are presented in Fig. 4.

For comparison, radiation angular distributions for crystal targets of four thicknesses in symmetrical Bragg geometry are displayed in Fig. 5. As effective target thickness (i.e. radiative length of a particle trajectory inside crystal) for Bragg geometry is $L_0/\cos\theta_B$, where L_0 is target thickness, so radiation intensity increases as function of a crystal thickness. This process goes until effective thickness reaches an absorption length. As one can see from the Fig. 5, maximal value in angular distribution for 0.1 cm target becomes less than that of 0.05 cm target distribution. Frequency ω_0 in the Bragg geometry decreases with polar angle increase (Fig. 2).

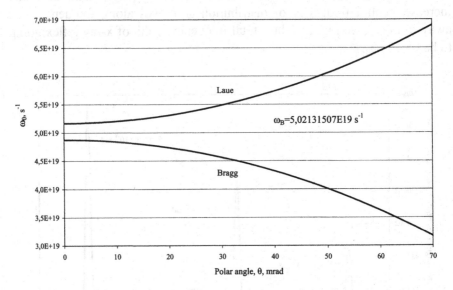

Figure 2. X-ray frequency as function of polar angle.

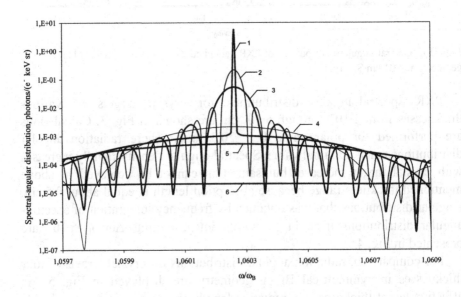

Figure 3. PXR spectral-angular distributions for targets of various thicknesses: 1 - $5 \cdot 10^{-2}$ cm; 2 - $1 \cdot 10^{-2}$ cm; 3 - $5 \cdot 10^{-3}$ cm; 4 - $1 \cdot 10^{-3}$ cm; 5 - $5 \cdot 10^{-4}$ cm; 6 - $1 \cdot 10^{-4}$ cm.

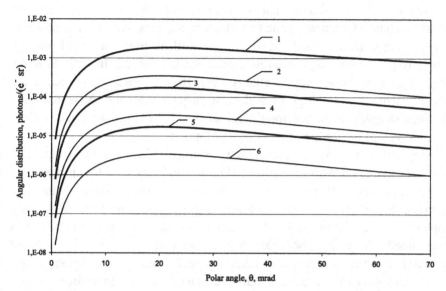

Figure 4. PXR angular distributions for targets of various thicknesses: 1 - $5 \cdot 10^{-2}$ cm; 2 - $1 \cdot 10^{-2}$ cm; 3 - $5 \cdot 10^{-3}$ cm; 4 - $1 \cdot 10^{-3}$ cm; 5 - $5 \cdot 10^{-4}$ cm; 6 - $1 \cdot 10^{-4}$ cm.

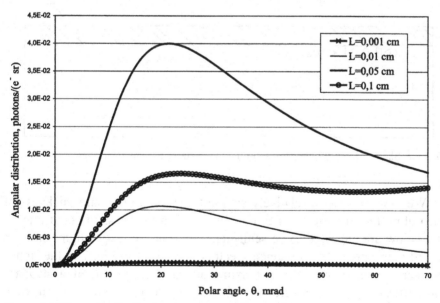

Figure 5. PXR angular distributions in symmetrical Bragg geometry.

3.   THIN SILICON TARGETS

Analysis above shows that optimal target thickness for examined situation should be between one to some tens microns. Application of lower energy beams than considered above, as well as usage in multi-passage source (e.g. [3]) challenge even thinner crystalline targets. In order to make preparations to low beam energy experiments, we have requested development of technology of a thin Si target production. Now we have obtained samples of sub-micron thickness, detailed technology route of their production will be published in a devoted journal.

Target (Fig. 6) is the silicon crystal substrate of 2x2 mm dimensions and ~200 µm thickness with ~0.5 µm thickness membrane of 1.0 mm diameter. Basic plane has (100) orientation. Membrane material is layer of undoped epitaxial Si of ~0.9–1.0 µm thickness deposited on substrate of heavily doped p^+ Si of KDB 0.01 <100> grade. Choice of such structure was determined by electrochemical etching technique, in which undoped epitaxial Si serves as termination layer. For membrane of other thickness one should take structures with epitaxial layer thickness close to desired. Precise membrane thickness adjustment can be performed by ion-beam technology with ~10-15 nm/min rate.

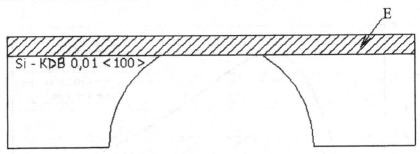

Figure 6. Sketch of Si (100) target of sub-micron thickness, E = epitaxial layer.

We have developed relatively simple technique to measure thickness of such ultra-thin Si targets. As Si membranes with thickness of about micron and below are appeared to be semi-transparent in the visible light, one can record their optical transmittance spectra (example is in Fig. 7 in the right). We can see here interference maxima against background of a standard transmittance. After background subtraction accurate values of maxima locations can be obtained (Fig. 7 in the left).

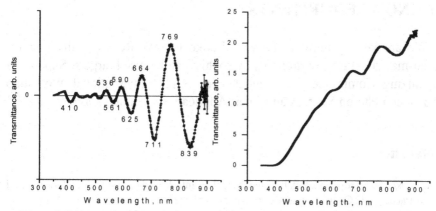

Figure 7. Target optical transmittance (right) and interference maxima locations (left)

Combining interference conditions for neighbor maxima one can obtain following expression for thickness of a parallel-sided plate.

$$d = \frac{\lambda_N \lambda_{N+1}}{2(n_{N+1}\lambda_N - n_N\lambda_{N+1})},$$

where n - refraction index, λ - wavelength in a maximum, N and $N+1$ - represent values belonging to adjoined interference maxima. Data on some measured samples took randomly of a last production lot are presented in Table 1.

Table 1. Results of target thickness measurements

Sample No	1	2	3	4	5
Thickness, nm	520±2	514±3	505±25	495±17	467±25

4.   CONCLUSION

Numerical calculations of parameters of x-rays that produce by electrons with energies below threshold value of parametric x-rays show, that despite of comparable contributions to x-ray reflex of other than PXR radiation mechanisms, main features of the x-radiation remain acceptable for development of a medical tunable quasi-monochromatic source. Value of optimal thickness of a crystal target is contained between one through some tens microns.

Devoted technology of an ultra-thin Si crystal targets was developed. Dispersion of measured thickness values of five targets randomly took from one lot is below 5%.

ACKNOWLEDGEMENTS

We are very grateful to Edward Lobko (Minsk Research Institute for Radio-materials) for providing us by sophisticated crystal targets. Support of Organizing Committee for attendance of NATO International Workshop "Advanced Photon Sources and Their Application" is much appreciated.

References

1. A. Lobko, O. Lugovskaya Relativistic Channeling Int. Workshop (Frascati, 2004), Book of Abstracts, p. 16. Complete manuscript will be published in conference issue of NIM.
2. V.P. Afanasenko, V.G. Baryshevsky, A.S. Lobko et al NIM A334 (1993) 631.
3. V.V. Kaplin, S.R. Uglov, O.F. Bulaev et al NIM B173 (2001) 3.

POLARIZED RADIATION FROM ELECTRONS AT OFF-AXIS CRYSTAL ORIENTATION

V.Strakhovenko
Budker Institute of Nuclear Physics, Novosibirsk, Russia,
v.m.strakhovenko@inp.nsk.su

Abstract   A complete description of spectral and polarization characteristics of the radiation emitted by arbitrarily polarized high-energy electrons is derived for the case when the electron velocity makes an angle with respect to some major crystal axis that is appreciably larger than the axial-channeling angle. A nonperturbative contribution to the probability of the process caused by electric fields of the crystal planes is exactly taken into account. A possibility of the use of oriented crystals in a polarized hard-photon source is evaluated basing on the results of numerical calculations and on the elaborated qualitative picture of the phenomenon.

Keywords: electron / photon / crystal / polarization

1.   Introduction

When relativistic electrons pass through single crystals, the specific radiation is emitted. Its spectrum, intensity and polarization depend on the crystal orientation and on the energy and polarization of an electron. Pronounced enhancement (as compared to the corresponding amorphous medium) of the radiation takes place when the electron velocity is almost aligned with some major crystal axis. Basing on these properties of the radiation, tunable sources of polarized high-energy photons may be developed for various applications.

We start from the well known formulas describing QED processes in external fields, which were obtained by means of the so-called quasi-classical operator method (QCOM, see e.g., [1]). In [2] and [3], the approach was developed which is valid in a wide range of the incidence angles and allows one to complete the calculations. The case of unpolarized electrons, when only linearly polarized photons may be obtained

H. Wiedemann (ed.), Advanced Radiation Sources and Applications, 55–62.
© *2006 Springer. Printed in the Netherlands.*

using crystals, was considered in [3]. In the present paper we apply this approach to investigate the general case, where all the involved particles are polarized. Then circularly polarized photons may be obtained from longitudinally polarized electrons.

Let us remind one the main features of our approach. Remember that for a wide class of external fields, satisfying some conditions specified in [4], the probabilities of QED processes are expressed within QCOM by way of classical trajectories of charged particles. The crystal electric field, which is responsible for the coherent processes, meets these conditions as well. So, to calculate the probability, we must solve a two-dimensional mechanical problem, what, generally, can not be done analytically. In crystals, as shown in [2], this problem becomes essentially one-dimensional for the off-axis orientation. More precisely, the angle ϑ_0 between the electron velocity and the nearest major crystal axis (incidence angle) should satisfy the condition $\theta_{sp} \leq \vartheta_0 \ll 1$. In this region of incidence angles, only a finite set of the strongest planes containing the axis is important. Some of these planes are schematically shown in Fig.1 along with adjacent angular domains, where the influence of each plane on the motion should be taken into account exactly. From Fig.1, the meaning of the characteristic angle θ_{sp} is that the different domains do not overlap for $\theta_0 \geq \vartheta_{sp}$. The magnitude of the angle θ_{sp} depends on the electron energy and the axis chosen, being appreciably larger than the characteristic angle for axial channeling, θ_{ax}. So, according to [2], the transverse (with respect to the axis) velocity of an electron reads for $\vartheta_0 \geq \theta_{sp}$ as $\mathbf{v}_\perp \simeq \mathbf{v}_\perp^{slow} + \mathbf{v}_\perp^{fast}$. Here, the term $\mathbf{v}_\perp^{slow}$ represents the exact solution for the transverse velocity in the field of the only one plane of the set. The term $\mathbf{v}_\perp^{fast}$ gives the contribution of other planes, which can be taken into account using a perturbation theory and, additionally, the rectilinear trajectory approximation. The component $\mathbf{v}_\perp^{slow}$ is present when the transverse velocity points to one of the blank areas in Fig.1, being almost aligned with one of the planes. It disappears when the velocity of a particle forms sufficiently large angles with each plane of the set (hatched areas in Fig.1). In the latter case, the perturbation theory in the whole crystal potential is applicable being the essence of the so-called coherent bremsstrahlung (CB) theory (see, e.g. [5]). It is clear that the transverse motion of a particle is two-dimensional but, for $\vartheta_0 \geq \theta_{sp}$, the exact calculation is necessary only for its one-dimensional component. To find the latter, the new approximate form of the actual planar potential was proposed in [3], which provides the precise fit for any crystal plane and allows one to find an analytic expression for the trajectory. That was done in [3], where the explicit form of the velocity Fourier-transform was also obtained for both electrons and positrons. In

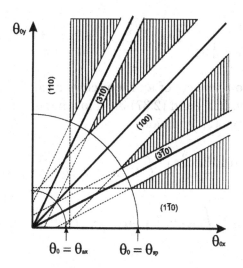

Figure 1. Characteristic angular regions around the $< 001 >$-axis (z-axis on the diagram) of diamond at electron energy of 150 GeV, when $\theta_{sp} \simeq 1.1 \cdot 10^{-4}$ and $\theta_{ax} \simeq 3.3 \cdot 10^{-5}$. Three strongest planes (thick solid lines) are shown along with domains (blank areas bounded by dashed lines) where the one-dimensional motion problem should be solved exactly.

the present paper, we consider characteristics of the radiation at given electron energy. They can be directly used to describe the radiation from thin crystals. In a practical use, sufficiently thick crystals are needed to get a noticeable yield. In this case, the obtained characteristics of the radiation-emission process are used to describe the development of an electromagnetic shower. For unpolarized initial electrons this was done in [3].

2.   Formulas and Discussion

The following vectors enter general formulas describing spectral, angular, and polarization distributions of photons emitted by electrons (positrons): the photon momentum $\mathbf{k}$, the photon polarization $\mathbf{e}$, and the particle velocity $\mathbf{v}$. We assume that the angular divergence of the electron beam is small enough. Then the same is true for the resulting photon beam provided that electron energy is sufficiently large ($\gamma = \varepsilon/m \gg 1$, m is the electron mass, ε is the electron energy), as the emission angle of a photon with respect to the particle velocity typically is of the order of γ^{-1}. In this case, we can choose some axis (z-axis), such that the momenta of all particles make small angles with respect to it. Let $\mathbf{a}_\perp$ denotes the component of an arbitrary vector $\mathbf{a}$ transverse to the axis chosen. Using relations:

$$v_z \simeq 1 - \tfrac{1}{2}(\gamma^{-2} + \mathbf{v}_\perp^2), \quad e_z = -(\mathbf{k}_\perp \mathbf{e}_\perp)/k_z \simeq (\mathbf{n}_\perp \mathbf{e}_\perp),$$
$$n_z \simeq 1 - \tfrac{1}{2}\mathbf{n}_\perp^2, \qquad\qquad \mathbf{n} = \mathbf{k}/\omega,$$

we can remove z-components of all the vectors from the problem. Then we obtain from Eqs.(2.27) and (2.37) in [1] for the completely differential probability of the photon emission

$$dw \quad = \frac{\alpha\,\omega\,d\omega\,d^2 n_\perp}{32\pi^2} \int dt_1 dt_2\, L(t_1, t_2)\, \exp(-iD)\ ,$$

$$D \quad = \frac{\omega\varepsilon}{2\varepsilon'} \int\limits_{t_1}^{t_2} dt\,[\,\gamma^{-2} + (\mathbf{n}_\perp - \mathbf{v}_\perp(t))^2\,]\,, \tag{1}$$

where $\alpha = 1/137$, $\varepsilon' = \varepsilon - \omega$, $\mathbf{v}_\perp(t)$ is the transverse electron velocity.

$$
\begin{aligned}
L(t_1, t_2) &= a_1 a_2^* + b_1^l b_2^{*l} + i(a_2^* b_1^l - a_1 b_2^{*l})(\zeta_{in}^l + \zeta_f^l) \\
&\quad + i\epsilon^{lkm} b_1^k b_2^{*m}(\zeta_f^l - \zeta_{in}^l) + \zeta_{in}^k \zeta_f^m \left[b_1^k b_2^{*m} + b_1^m b_2^{*k} \right. \\
&\quad \left. + \delta^{km}(a_1 a_2^* - b_1^l b_2^{*l}) - \epsilon^{lkm}(a_2^* b_1^l + + a_1 b_2^{*l}) \right], \tag{2}
\end{aligned}
$$
$$b_{1,2}^l = u\,\epsilon^{lkm} e_\perp^{*k}\,(n_\perp^m - v_\perp^m(t_{1,2}) + v^m/\gamma)\,,$$
$$a_{1,2} = (2+u)(\mathbf{e}_\perp^*, \mathbf{v}_\perp(t_{1,2}) - \mathbf{n}_\perp)\,,$$
$$u = \omega/\varepsilon'\,.$$

Here ϵ^{lkm} is the antisymmetric tensor, ζ_{in} and ζ_f are the polarization vectors of the initial and final electron, respectively. The unit vector ν is directed along the nearest major crystal axis (z-axis). In what follows, we use the Cartesian basis $\mathbf{e}_x, \mathbf{e}_y, \mathbf{e}_z = \nu$; $\mathbf{e}_x$ is within some crystal plane containing the axis and $\mathbf{e}_y$ is perpendicular to this plane. The polarization of radiation is described by the matrix, dw_{ki} in the contraction $dw = e_i e_k^* dw_{ki}$. Then we have for the Stokes vector of the radiation, $\eta = \mathbf{B}/A$, where

$$A \quad = dw_{11} + dw_{22}, \quad B_3 = dw_{11} - dw_{22},$$
$$B_1 \quad = dw_{12} + dw_{21}, \quad B_2 = i(dw_{12} - dw_{21})\,. \tag{3}$$

The quantity A defined in this equation gives the probability summed up over photon polarizations. Performing the integration over $d^2 n_\perp$ in (1), we obtain the matrix, w_{ij}^{sp}, describing the spectral characteristics of the radiation:

$$dw_{ij}^{\mathrm{sp}} = \frac{i\alpha d\omega}{16\pi\gamma^2} \int\limits_{-\infty}^{\infty} dt \int\limits_{-\infty}^{\infty} \frac{d\tau}{\tau - i0} \exp[-i\lambda\tau(1 + \rho(t,\tau))] \, L_{ij}, \qquad (4)$$

$$L_{ij} = P_{ij} + i\varphi_1(u)(\zeta_{in}^l + \zeta_f^l)Q_{ij}^l + \frac{i}{2}\varphi_2(u)(\zeta_f^l - \zeta_{in}^l)R_{ij}^l + \zeta_{in}^k\zeta_f^m S_{ij}^{km},$$

where $\varphi_1(u) = u(2+u)/(1+u)$, $\varphi_2(u) = u^2/(1+u)$, $t = (t_1 + t_2)/2$, $\tau = t_2 - t_1$, $\lambda = \omega m^2/(2\varepsilon\varepsilon')$. The tensors P_{ij}, Q_{ij}^l, and R_{ij}^l read

$$P_{ij} = \delta_{ij}^{\perp}\left[\varphi_2(u)\frac{(\mathbf{g}_2 - \mathbf{g}_1)^2}{2} - \frac{2}{i\lambda\tau_-}\right] + \varphi_2(u)g_{2i}g_{1j} - [4 + \varphi_2(u)]\, g_{2j}g_{1i},$$

$$Q_{ij}^l = \nu_k\left[\epsilon^{ljk}g_{1i} - \epsilon^{lik}g_{2j} + \nu_l\left(\epsilon^{imk}g_{2j}g_{1m} - \epsilon^{jmk}g_{2m}g_{1i} + \frac{\epsilon^{ijk}}{i\lambda\tau_-}\right)\right],$$

$$R_{ij}^l = \nu_k\left[\epsilon^{ijk}(g_{2l} + g_{1l} - 2\nu_l) + (\epsilon^{imk}\delta_{jl}^{\perp} + \epsilon^{jmk}\delta_{il}^{\perp})(g_{1m} - g_{2m})\right]. \qquad (5)$$

The tensor S_{ij}^{km} describing the spin-spin correlation effects reads

$$
\begin{aligned}
S_{ij}^{km} = {} & -\delta_{km}\left\{\delta_{ij}^{\perp}\left[\tfrac{1}{2}\varphi_2(u)(\mathbf{g}_2 - \mathbf{g}_1)^2 + \frac{2 + \varphi_2(u)}{i\lambda\tau_-}\right]\right. \\
& \left. + [4 + \varphi_2(u)]\, g_{2j}g_{1i} + \varphi_2(u)g_{2i}g_{1j}\right\} \\
& + \varphi_1(u)\epsilon^{kml}\left[\epsilon^{ljn}g_{1i}(\nu_n - g_{2n}) + \epsilon^{lin}g_{2j}(\nu_n - g_{1n})\right] \\
& - \varphi_2(u)\left[(\nu_l - g_{2l})(\nu_n - g_{1n}) + \frac{\delta_{ln}^{\perp}}{2i\lambda\tau_-}\right]\left(\epsilon^{kin}\epsilon^{mjl} + \epsilon^{kjl}\epsilon^{min}\right).
\end{aligned} \qquad (6)
$$

In above formulas $\mathbf{g}_{1,2} \equiv \mathbf{g}(t_{1,2})$, $\tau_- = \tau - i0$ and

$$\mathbf{g}(s) = \gamma\left[\mathbf{v}_{\perp}(s) - \frac{1}{\tau}\int\limits_{-\tau/2}^{\tau/2} dx\, \mathbf{v}_{\perp}(t+x)\right], \qquad \rho(t,\tau) = \frac{1}{\tau}\int\limits_{-\tau/2}^{\tau/2} dx\, \mathbf{g}^2(t+x).$$

We emphasize that formulas (1) and (4) are rather general. In particular, they describe spectral-angular distributions and all the polarization effects of radiation emitted from relativistic electrons moving in an undulator, in a laser wave or in an arbitrarily oriented crystal. If polarization of the final electron is not detected, we must sum up over its polarization states. Corresponding formulas are obtained from Eqs. (1),(4) by omitting the terms, which are proportional to ζ_f, and doubling all others. For example, in this case L_{ij} in Eq.(4) goes over into

$\tilde{L}_{ij} = 2\left\{P_{ij} + i\zeta_{in}^l \left[\varphi_1(u)Q_{ij}^l - \frac{1}{2}\varphi_2(u)R_{ij}^l\right]\right\}$. For the unpolarized initial electron, $\tilde{L}_{ij}$ reduces to $2P_{ij}$ and Eq.(5) in [3] is reproduced.

For crystals at off-axis orientation, we can advance in calculations using the approach developed in [2],[3]. Recollect that the transverse velocity can be represented as the sum: $\mathbf{v}_\perp \simeq \mathbf{v}_\perp^{slow} + \mathbf{v}_\perp^{fast}$, where $\mathbf{v}_\perp^{fast}$ (see Eq.(3) in [2]) is characterized by relatively small amplitudes and large frequencies. On the contrary, large amplitudes and small frequencies are inherent in the term $\mathbf{v}_\perp^{slow}$, which corresponds to the one-dimensional motion in the field of a plane. Its explicit form reads

$$\mathbf{v}_\perp^{slow}(t) = \mathbf{e}_y \sum_n v_n \exp(in\omega_0 t).$$

Here ω_0 is the frequency of this motion, and v_n is the velocity Fourier transform calculated in [3]. Note that the trajectories of electrons and positrons, moving in the electric field of the same crystal plane, are very different. This is true for the quantities v_n as well. Only high above the plane potential barrier, where v_n can be calculated by means of the perturbation theory, they coincide for electrons and positrons.

So, the quantity $\mathbf{g}(s)$ in (4) turns into the sum $\mathbf{g}(s) = \mathbf{l}(s) + \mathbf{w}(s)$ where

$$\mathbf{l}(s) = \mathbf{e}_y \sum_n (\gamma v_n)\left[e^{in\omega_0 s} - \frac{\sin(n\omega_0\tau/2)}{n\omega_0\tau/2}e^{in\omega_0 t}\right],$$

$$\mathbf{w}(s) = -\widetilde{\sum_{\mathbf{q}_\perp}}\frac{G(\mathbf{q}_\perp)\mathbf{q}_\perp}{mq_\parallel}\left[e^{iq_\parallel s} - \frac{\sin(q_\parallel\tau/2)}{q_\parallel\tau/2}e^{iq_\parallel t}\right]e^{i(\mathbf{q}_\perp\rho_0)}. \qquad (7)$$

Here $\mathbf{q}$ are discrete reciprocal lattice vectors, $G(\mathbf{q})$ are coefficients in the Fourier series describing the crystal potential (see Chapter 9 of [1] for the explicit form of $\mathbf{q}$ and $G(\mathbf{q})$), and $q_\parallel = (\mathbf{q}_\perp\mathbf{v})$. The tilde in the expression for $\mathbf{w}(s)$ means that the sum does not contain $\mathbf{q}_\perp \parallel \mathbf{e}_y$ which just form the plane potential.

In principle, using (7), one can perform the integration over t and τ straight in (4). However, as we have already obtained the quantity $\mathbf{w}(s)$ by means of some perturbation procedure, it is more convenient to continue consequently in the same way. So, as was done in [2] and [3], we can expand the exponential function in (4) in $\mathbf{w}(s)$. Keeping quadratic terms, we obtain the probability dw as a sum of three terms (cf. Eq.(7) in [3]):

$$dw = dw_{slow} + dw_{int} + dw_{fast}.$$

The term dw_{slow} describes the radiation in the electric field of the plane. Using explicit form of v_n, we can calculate dw_{slow} without additional approximations like CFA (the constant field approximation).

Remember that CFA becomes valid at rather high electron energy (see [3] for details). The spectrum provided by dw_{slow} is peaked in the soft photon region (at $x = \omega/\varepsilon \ll 1$) and radiation has a high degree of the linear polarization. The latter is due to the one-dimensional character of the motion. Examples of the power spectra and of the polarization distributions are given in Fig.2 of [3] for the unpolarized initial electron. Let us remind one that the contribution dw_{slow}, as well as the interference term dw_{int}, is present when the transverse velocity points to one of the blank areas in Fig.1. At such orientation, called SOS, the term dw_{fast} may also be slightly affected by the "slow" motion (see criteria for disregard of such an influence in [3]), [6]. Essentially, the term dw_{fast} describes Compton scattering of equivalent photons, which may represent the crystal electric field in the corresponding frame (see [7]). Disregarding the influence of the "slow" motion on Compton scattering and going over to the probability per unite time, we obtain dw_{fast} expressed via the quantities $A, \mathbf{B}$, defined in Eq.(3). Summing up over the final electron states, we have whith the step function $\Theta(1 - \beta)$ for electrons with a longitudinal polarization ζ(cf. Eq.(13) in [3])

$$(A, \mathbf{B}) = \frac{\alpha}{\gamma^2} \sum_{\mathbf{q}_\perp} \left| \frac{G(\mathbf{q}_\perp)\mathbf{q}_\perp}{mq_\parallel} \right|^2 (a, \mathbf{b}) \Theta(1 - \beta); \quad \beta = \frac{\omega m^2}{2\varepsilon\varepsilon'|q_\parallel|},$$

$$a = \frac{1}{2} + \frac{u^2}{4(1 + u)} - \beta(1 - \beta), \, b_2 = \zeta x \left(a + \frac{x}{4} \right), \, x = \frac{\omega}{\varepsilon},$$

$$b_1 = \beta^2 h_1 h_2, \quad b_3 = \frac{1}{2}\beta^2(h_1^2 - h_2^2), \quad \mathbf{h} = \frac{\mathbf{q}_\perp}{|\mathbf{q}_\perp|}. \tag{8}$$

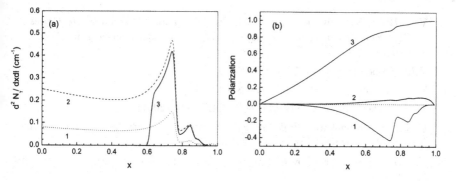

Figure 2. (a) non-collimated spectra at CB (1) and at SOS (2); collimated ($\vartheta_{col} = 4 \cdot 10^{-5}$) spectrum at SOS (3); (b) polarization: η_3 at CB (1), η_3 at SOS (2), and η_2/ζ (3).

Shown in Fig.2 are some examples of the spectral and polarization distributions calculated according to Eq.(8) at $\varepsilon = 12$ GeV for two different orientations (CB and SOS) near the (110)-plane and the $< 001 >$-axis of a thin diamond crystal. The angles of incidence are chosen in such a way that the main spectral peak has the same position in both cases. From Fig.2, the collimation of emitted photons within $\vartheta_{col} \sim 1/\gamma$ may substantially monochromate the spectrum. For the chosen basis e_x, e_y, the positive (negative) values of the parameter η_3 correspond to the linear polarization, which is parallel (perpendicular) to the plane. The parameter η_1 (not shown) typically is very small. The linear polarization depends on the incident electron energy and on the crystal orientation (cf. curves 1 and 2 in Fig.2(b)). The latter is due to the orientational effects in the linear polarization of the equivalent photon flux (see [3]).

In contrast to the linear polarization, the circular one does not reveal any significant dependence on the parameters of the problem. Really, from Eq.(8), the quantity η_2, which describes the circular polarization of the radiation, reduces to $\eta_2 \simeq \zeta x$ for $x \ll 1$ being, thereby, independent of any crystal characteristics, including the orientation. Such independence also takes place at growing x, as seen in Fig.2(b), where the single curve (3) represents η_2/ζ for both orientations. It is interesting that the dependence of the circular polarization on x obtained in crystals is very similar to that obtained for longitudinally polarized electrons in amorphous media (see Eq.(8.10) and Fig.5 in [8]). Moreover, according to [8], the shape of the corresponding curve is independent of the atomic number and the incident electron energy (above 20 MeV). Such a similarity hints at a universal character of the physical mechanism providing the helicity transfer from longitudinally polarized electrons to photons.

References

[1] V.N. Baier, V.M. Katkov, and V.M. Strakhovenko, *Electromagnetic Processes at High Energies in Oriented Single Crystals* , World Scientific Publishing Co, Singapore, 1998.

[2] V.M. Strakhovenko, NIM **B 145** (1998) 120.

[3] V.M. Strakhovenko, Phys.Rev.**A 68**, 042901 (2003)

[4] V.M. Katkov and V.M. Strakhovenko, JETP **92** (2001) 561.

[5] L.M. Ter-Mikhaelyan, *High-Energy Electromagnetic Processes in Condenced Media* , Wiley-Interscience , New-York, 1972.

[6] V.N. Baier, V.M. Katkov and V.M. Strakhovenko, Nucl. Instr. Methods **B69** (1992) 258.

[7] V.N. Baier, V.M. Katkov and V.M. Strakhovenko, Sov. Phys. Usp.**32** (1989) 972.

[8] H.Olsen and L.C.Maximon, Phys. Rev. **114** (1959) 887.

RADIATION EMITTED FROM RELATIVISTIC PLANAR CHANNELED POSITRONS

Valentin Boldyshev and Michail Shatnev
Akhiezer Institute for Theoretical Physics of NSC KIPT1, Akademicheskaya St., Kharkov 61108, Ukraine

Abstract: We present a detailed calculation of channeling radiation of planar-channeled positrons from crystal targets in the framework of our approach, which was proposed earlier The development stemmed from the idea that the amplitude for a given process is the sum of the transition amplitudes for each transition to lower state of transverse energy with the same energy differences between bound-bound transitions. It seems that a consistent interpretation is only possible if positrons move in a nearly harmonic planar potential with equidistant energy levels.

Key words: radiation from positrons; quantum theory of channeling; harmonic potential.

1.   INTRODUCTION

Charged particles directed into a crystal approximately parallel to one of the crystal planes will be planar channeled. between the crystal planes. In classical terms, the particle's momentum forms some For positively charged particles, such as positrons, the channel is small angle θ with respect to the crystal plane. This angle must be less than Lindhard's critical angle θ_L for planar channeling to occur. The motion of the particles then consists of a periodic back-and-forth reflection of the boundaries of the planar channel. An accelerated periodic motion of this kind will lead to the emission of radiation.

From a quantum-mechanical viewpoint, the channel is the source of a one-dimensional potential well for planar channeling in the direction transverse to the particle's motion, which gives rise to transversely bound states for the particle. Its transitions to lower levels are accompanied by the

H. Wiedemann (ed.), Advanced Radiation Sources and Applications, 63–70.

emission of radiation with frequencies related to the energy differences of the levels.

Theoretical studies of the radiation from planar-channeled electrons and positrons have been done by M.A. Kumakhov [1,5], N.K. Zhevago [2,5], A.I. Akhiezer, I.A. Akhiezer and N.F. Shul`ga [3,4].

Experimentally, the channeling radiation of positrons was observed by the different groups[6,7], demonstrating strong and sharp peaks in the spectrum. The purpose of the present work is to calculate the spectral-angular distribution of the channeling radiation intensity emitted from positrons in the framework of approach, which was proposed recently [8].

2.   TRANSVERSE POTENTIAL AND RADIATION AMPLITUDE

We consider a relativistic charged particle incident onto a crystal at a small angle to a crystal planes. In the planar channeling case for positively charged positrons, the channel is between the crystal planes. This channel is the source of a potential well in the direction transverse to particle motion giving rise to transversely bound states for the particle. Transitions to lower-energy states lead to the phenomenon known as channeling radiation (CR). Calculations of the CR process are carried out by using the rules of quantum electrodynamics. The Doppler formula for the photon energy $\omega_{nn'}$ corresponding to the transition $n \to n'$ is derived using the energy and momentum conservation laws. For this energy one gets

$$\omega_{nn'} = \frac{\tilde{\omega}_{nn'}}{1 - \beta_{//} \cos\theta} , \tag{1}$$

where $\tilde{\omega}_{nn'} = \varepsilon_n - \varepsilon_{n'}$, ε_n and $\varepsilon_{n'}$ are the discrete energy levels of the transverse oscillations of the positron in the channel before and after radiation, respectively; $\beta_{//} = \dfrac{|\vec{p}_{//}|}{E}$, E and $\vec{p}_{//}$ are the energy and the longitudinal momentum of the positron, and θ is the angle of radiation emission relative to the direction of motion of the channeled positron. For positrons of not too high transverse energies, a good approximation [1,2] is the harmonic potential leading to equidistant energy levels

$$\varepsilon_n = \Omega(n + 1/2) \tag{2}$$

where $\Omega = \frac{2}{d_p}\sqrt{\frac{2U_0}{E}}$, d_p is the distance between planes in the corresponding units, and U_0 is the depth of the potential well. Since $\frac{|\vec{p}_{//}|}{E} \approx (1 - \frac{1}{2}\gamma^{-2})$, where $\gamma = \frac{E}{m}$, $\cos\theta \approx (1 - \frac{1}{2}\theta^2)$ and taking into account Eq. (2), Eq. (1) can be expressed as

$$\omega_{nn'} = 2\gamma^2 \frac{\Omega(n-n')}{(1+\theta^2\gamma^2)} \tag{3}$$

It follows from Eq. (3) that the radiation of a maximum frequency

$$\omega_{nn'} = 2\gamma^2\Omega(n-n') \tag{4}$$

is emitted in the forward direction (at $\theta = 0$). The case $n - n' = 1$ corresponds to the peak values of the experimental channeling radiation spectra[7], being the first harmonic with the photon energy $\omega = 2\gamma^2\Omega$. As it follows from Eq. (3), photons emitted via positron transition from any initial level n to the final level $n-1$ are identical (i.e., have the same energies for the same emission angles). This means that the resulting amplitude should be given by an additive superposition of amplitudes of all such transitions. The positron state outside the crystal ($z < 0$) is a plane wave, whereas inside the crystal ($z > 0$), the part of its wave function corresponding to the transverse motion is a superposition of the harmonic oscillator eigenvectors. Factors c_n describing transitions from the initial state to states with the transverse energy levels n can be found using boundary conditions set upon the wave function at the crystal boundary ($z = 0$). Then, a transition to the closest lower level $n-1$ occurs with emission of a photon having energy ω. One may expect that the total amplitude of the transition from the initial to final state accompanied by the photon emission is determined by products of the amplitudes c_n and $M_{n,n-1}$. Following the rules of the quantum mechanics, we express this amplitude as

$$A \propto \sum_n c_n M_{n,n-1}, \tag{5}$$

were summation is performed over all the harmonic oscillator levels. Also we must find an additive superposition of amplitudes of all transitions $n \to n-2$, $n \to n-3,...$ etc.

3.   WAVE FUNCTIONS AND INTENSITY OF CHANNELING RADIATION

Taking into account above considerations, we can write the transition matrix element in the form

$$\left|M_{if}\right|^2 = \sum_j \left|\sum_n c_n M_{n,n-j}\right|^2, \tag{6}$$

where $j = n - n'$. It is well known from the quantum electrodynamics that the transition matrix element is given by [9]

$$\left|M_{if}\right|^2 = \frac{2\pi \cdot e^2}{\omega \cdot V}\left|J_\lambda\right|^2, \tag{7}$$

where $e^2 = \dfrac{1}{137}$, $V = L^3$ is the normalization volume, $\lambda = 1, 2$ indicates the linear photon polarization, and

$$J_\lambda = \int \Psi'^{\dagger} \alpha_\lambda e^{-i\vec{k} \cdot \vec{r}} \Psi d^3 r, \tag{8}$$

with $\alpha_\lambda = \vec{\alpha} \cdot \vec{\varepsilon}_\lambda^*$. The prime indicates the final state. The wave function is the solution of the time-independent Dirac equation for a relativistic particle moving with momentum $\vec{p}_{//} = (0, p_y, p_z)$ in a one-dimensional planar potential $U(x)$ periodic in the x-direction (which is normal to the channeling planes)

$$(i\vec{\alpha} \cdot \vec{\nabla} + E - \beta m)\Psi = U(x)\Psi, \tag{9}$$

where m and E are the particle's mass and energy, $\vec{\alpha}$ and β are the Dirac matrices. Separating the wave function Ψ into large and small components,

$$\Psi = \begin{pmatrix} \Psi_a \\ \Psi_b \end{pmatrix} \tag{10}$$

$$\vec{\sigma} \cdot \vec{\nabla}(E - U(x) + m)^{-1}\vec{\sigma} \cdot \vec{\nabla}\Psi_a + (E - U(x) - m)\Psi_a = 0 \tag{11}$$

Since a potential $U(x)$ is independent of y and z, the solution of last equation is a plane wave in the yz plane

$$\Psi_a \propto \exp[i(p_z z + p_y y)]\varphi(x)\chi \tag{12}$$

This allows us to transform a Pauli-type equation into a one-dimensional, relativistic Schrödinger equation for the transverse motion

$$-\frac{1}{2E}\frac{d^2\varphi(x)}{dx^2}+U(x)\varphi(x)=\varepsilon\varphi(x),\tag{13}$$

where

$$\varepsilon=\frac{E^2-m^2-p_z^2-p_y^2}{2E}.\tag{14}$$

In a given potential, the latter will assume certain bound-state eigenvalues $\varepsilon_n<0$ $(n=0,1,2,...)$, with corresponding eigenfunctions $\varphi_n(x)$. The wave function of Eq. (10) is finally obtained in the form

$$\Psi=\frac{N}{L}\left(\begin{array}{c}\chi\\-i\vec{\sigma}\cdot\vec{\nabla}\\\frac{}{E+m}\chi\end{array}\right)\exp(i\vec{p}_{//}\vec{r}_{//})\varphi_n(x),\tag{15}$$

where L^2 being the two-dimensional normalization volume for the plane waves, $N=\sqrt{\dfrac{E+m}{2E}}$, and χ being a two-component spinor which is $\begin{pmatrix}1\\0\end{pmatrix}$ or $\begin{pmatrix}0\\1\end{pmatrix}$ when the particle spin points in the $+z$ or the $-z$ direction in the rest frame, respectively. For the harmonic potential $U(x)=U_0x^2$, it is well known that the corresponding eigenfunctions being given by

$$\varphi_n(x)=\sqrt[4]{\frac{E\Omega}{\pi}}\frac{1}{\sqrt{2^n n!}}\exp(-E\Omega x^2/2)H_n(\sqrt{E\Omega}\,x),\tag{16}$$

where H_n are the Hermite polynomials. According to Eq. (8), we find, using Eq. (15), the first-order matrix element corresponding to the $n\to n-j$ transverse transition

$$J_\lambda=(2\pi)^2\delta(\vec{p}_{//}-\vec{p}_{//}'-\vec{k}_{//})NN'\chi'^*\vec{\varepsilon}_\lambda^*(\vec{A}+i[\vec{B}\vec{\sigma}])\chi,\tag{17}$$

where with

$$I_{n,n-j}^{(1)}=\int\exp(-ik_x x)\varphi_{n-j}^*(x)\varphi_n(x)dx,$$

$$I_{n,n-j}^{(2)}=\int\exp(-ik_x x)\varphi_{n-j}^*(x)\frac{d\varphi_n(x)}{dx}dx,\tag{18}$$

$$A_x = -iI^{(2)}_{n,n-j}(\frac{1}{E+m} + \frac{1}{E-\omega+m}),$$

$$\vec{A}_{/\!/} = I^{(1)}_{n,n-j}\vec{p}_{/\!/}(\frac{1}{E+m} + \frac{1}{E-\omega+m}),$$

$$B_x = iI^{(2)}_{n,n-j}(\frac{1}{E-\omega+m} - \frac{1}{E+m)}) + \frac{k_x}{E-\omega+m}I^{(1)}_{n,n-j},$$ (19)

$$\vec{B}_{/\!/} = I^{(1)}_{n,n-j}(\frac{\vec{p}_{/\!/}}{E+m} - \frac{\vec{p}'_{/\!/}}{E-\omega+m}),$$

Taking into account that the transverse eigenfunctions for the parabolic potential being given by Eq. (16), we find

$$I^{(1)}_{n,n-j} = i^j \sqrt{\frac{(n-j)!}{n!}} e^{-y/2} y^{j/2} L^j_{n-j}(y)$$

$$I^{(2)}_{n,n-j} = (-i)^j \sqrt{\frac{(n-j)!}{n!}} \sqrt{\frac{\Omega}{2E}} (j+y)e^{-y/2} y^{(j-1)/2} L^j_{n-j}(y)$$ (20)

where $y = \frac{k_x^2}{2E\Omega}$ and $L^j_{n-j}(y)$ are the Laguerre polynomials. Then we calculate the matrix element and differential intensity, and after summation over polarization of emitted photon and the positron using the formulae

$$\varepsilon_i\varepsilon_k^* \to \delta_{ik} - n_i n_k,$$

$$\chi*\chi = \frac{1+\vec{\sigma}\vec{\varsigma}}{2} \to \frac{1}{2},$$ (21)

$$\chi'*\chi' = \frac{1+\vec{\sigma}\vec{\varsigma}'}{2} \to 1,$$

where $\vec{\varsigma}$ and $\vec{\varsigma}'$ are the unit spin vectors of the initial and final positron (in the rest frame of the positron), respectively, using Fermi's golden rule for the differential intensity,

$$d^2 I_\lambda = 2\pi\omega \left|M_{if}\right|^2 \delta(\omega - \omega\beta_{/\!/}\cos\theta - \tilde{\omega}_{nn'})\frac{Vd^3k}{(2\pi)^3}\frac{L^2 d^2 p'_{/\!/}}{(2\pi)^2},$$ (22)

and taking into account that $\tilde{\omega}_{n,n-j} = j\Omega$ according to Eq. (2), we find by integrating over $d^2 p'_{/\!/}$

$$\frac{d^2 I}{d\omega do} = \frac{e^2 \omega^2}{2\pi} \sum_j \left| \sum_n c_n M_{n,n-j} \right|^2 \delta(\omega - \beta_{//} \omega \cos\theta - j\Omega), \tag{23}$$

where

$$\sum_j \left| \sum_n c_n M_{n,n-j} \right|^2 \approx \sum_j \left(1 + \frac{\omega}{E - \omega} + \frac{1}{2} \frac{\omega^2}{(E-\omega)^2} \right) *$$

$$\left(\left| \sum_n c_n I^{(1)}_{n,n-j} \right|^2 \theta^2 + \left| \sum_n c_n I^{(2)}_{n,n-j} \right|^2 - 2\theta \cos\varphi \, \mathrm{Re} \sum_n c_n^2 I^{(1)}_{n,n-j} I^{(2)*}_{n,n-j} \right) \tag{24}$$

$$+ \sum_j \frac{1}{2} \frac{\omega^2 \gamma^{-2}}{(E-\omega)^2} \left| \sum_n c_n I^{(1)}_{n,n-j} \right|^2,$$

in the approximation $\theta \ll 1$. The factors c_n are, in the case of the parabolic potential and when the initial positron is a plane wave, given by

$$c_n = \sqrt[4]{\frac{\pi}{E\Omega}} \frac{i^n}{\sqrt{2^{n-1} n!}} \exp\left(-\frac{p_x^2}{2E\Omega}\right) H_n\left(\frac{p_x}{\sqrt{E\Omega}}\right). \tag{25}$$

4.   CONCLUSION

Following N.K. Zhevago [2], the spectral-angular distribution of emitted photons is represented as

$$\frac{d^2 I}{d\omega do} \propto \sum_f |M_{if}|^2 \tag{26}$$

The sum entering Eq. (26) is the one over the quantum numbers f of the transverse motion of the particle. Then, the probability of having a definite transverse energy is taken into account by multiplying each term of this sum by a corresponding factor. In our consideration, discrete levels of the transverse motion in the harmonic oscillator potential refer to the intermediate state of the particle. Accordingly, the contribution to the intensity due to transitions, e.g., between the closest levels is determined by the square of the absolute value of Eq. (5)

$$\frac{d^2 I}{d\omega do} \propto \left| \sum_n c_n M_{n,n-1} \right|^2. \tag{27}$$

In other words, unlike in [2], we get an expression that contains interference terms mixing amplitudes of photon emission from different equidistant levels. We would like to note that dynamics of channeling electron in a crystal differs from that of the positron case. The transverse potential well for the electron does not give rise to equidistant energy levels for transverse particle motion. Therefore, there are no interference contributions to the photon emission intensity similar to those present in Eq. (27). In our opinion, this could explain the greater intensity in case of channeling positron compared to that for the electron observed in experiment [7]. Corresponding numerical calculations will be given in a subsequent publication.

ACKNOWLEDGEMENTS

One of the authors (M. Sh.) is grateful to Nikolai Shul`ga, Vladimir Baier, Vladimir Strakhovenko, Xavier Artru and Dick Carrigan for useful discussions. He also wishes to thank very much Robert Avagyan and Helmut Wiedemann for inviting him to Nor Hamberd and supporting his stay there.

REFERENCES

1. M.A. Kumakhov and R. Wedell, Theory of radiation of relativistic channellled particles, Phys.Stat.Sol. B**84**, 581-593 (1977).
2. N.K. Zhevago, Emission of γ quanta by channeled particles, Zh. Eksp. Teor. Fiz. **75**, 1389-1401 (1978).
3. A.I. Akhiezer, I.A. Akhiezer and N.F. Shul`ga, Theory of relativistic electron and positron bremsstrahlung in crystals, Zh. Eksp. Teor. Fiz. **76**, 1244-1253 (1979).
4. A.I. Akhiezer and N.F. Shul`ga, Radiation of relativistic particles in single crystals, Sov.Phys.Usp. **25**(8), 541-564 (1982).
5. V.A. Bazylev, V.V. Beloshitsky, V.I. Glebov, N.K. Zhevago, M.A. Kumakhov and Ch. Trikalinos, Radiation emitted by channeling positrons in a continuous potential of crystal planes, Zh. Eksp. Teor. Fiz. **80**, 608-626 (1981).
6. R.O. Avakian, I.I. Miroshnichenko, J. Murrey and T. Fieguth, Radiation emission by ultrarelativistic positrons moving in a single crystal near the crystallographic axes and planes, Zh. Eksp. Teor. Fiz. **82**, 1825-1832 (1982).
7. J. Bak, J.A. Ellison, B. Marsh, F.E. Meyer, O. Pedersen, J.B.B. Petersen, E. Uggerhøj and K. Østergaard, Nucl.Phys. Channeling radiation from 2-55 GeV/c electrons and positrons, Nucl.Phys. B**254**, 491-527 (1985).
8. V.F. Boldyshev, M.G. Shatnev, On the question of interference in radiation produced by relativistic channeled particles, Proc. of the First Feynman Festival (August, 2002, University of Maryland, College Park, Maryland, USA), quant-ph/0210203 (30 Oct 2002).
9. W. Heitler, The quantum theory of radiation (Clarendon Press, Oxford, 1954).

RADIATION PRODUCED BY FAST PARTICLES IN LEFT-HANDED MATERIALS (LHM) AND PHOTONIC CRYSTALS (PHC)

K.A. Ispirian
Yerevan Physics Institute, Br. Alikhanians 2, 375036 Yerevan, Armenia

Abstract: After a short review of some predicted and observed properties and perspectives of LHM and photonic crystals (PhC) it is discussed the published theoretical and experimental works devoted to the Cherenkov and Smith-Purcell radiation processes produced by high energy particles in them.

Keywords: Cherenkov radiation/ Smith-Purcell radiation/ Left handed materials/ Photonic crystals

1.   Introduction

The name LHM was given [1] to hypothetic materials the electric permittivity, ε, and magnetic permeability, μ, of which are negative because according to the Maxwell's equations for a plane monochromatic wave

$$\left[\vec{k}\vec{E}\right] = \frac{\omega\mu}{c}\vec{H},$$ (1)

$$\left[\vec{k}\vec{H}\right] = -\frac{\omega\varepsilon}{c}\vec{E}$$ (2)

the vectors of the electric field, $\vec{E}$, of magnetic field, $\vec{H}$, and wave vector, $\vec{k}$, give a left-handed system in contrast to the usual right-handed materials (RHM) with $\varepsilon > 0$ and $\mu > 0$.

71

H. Wiedemann (ed.), Advanced Radiation Sources and Applications, 71–96.
© *2006 Springer. Printed in the Netherlands.*

Since

$$n = \pm\sqrt{\varepsilon\mu} \qquad (3)$$

such LHM can be amphoteric with negative index (NIM) and positive n (PIM).

Furthermore, since the Poynting vector does not depend on ε and μ and is equal to

$$\vec{S} = \frac{c}{4\pi}\left[\vec{E}\vec{H}\right], \qquad (4)$$

we find in contrast to RHM for LHM that $\vec{S}$ and $\vec{k}$ as well as the phase $\vec{v}_{ph} = \omega/\vec{k}$, and the group velocities $\vec{v}_{gr} = d\omega/d\vec{k}$ are anti-parallel to each other.

These properties of LHM results in new, sometimes paradoxical physical phenomena with useful applications. Here we shall only mention some of which and begin less or more detailed discussion of the radiation produced in LHM. The refraction of electromagnetic waves passing through the boundary between RHM and LHM which takes place according to Snell's law $\sin\theta_{LHM} = n_{RHM}\sin\theta_{RHM}/n_{LHM}$ in LHM will results in negative refraction with angle $\theta_{LHM} < 0$ since in the right hand side $n_{RHM} > 0$, $\theta_{RHM} > 0$, while $n_{LHM} < 0$. This is experimentally observed for the first time in [2]. Perhaps, the most challenging is the considered in [1] possibility of construction of flat lenses with perfect imaging without diffraction limit of the order of wavelength as it has been shown in [3]. After many theoretical and numerical works such an effect of negative refraction has been observed experimentally in [4].

Though LHM have been postulated in 1967 only recently it has become possible to have LHM artificially for the following reason. The fact that some materials have negative values of ε and μ was well known long ago. The problem was that there was no material with negative ε and μ in the same frequency region. For instance, silver, gold and aluminum thin layers have negative ε in optical regions, while some semiconductors and insulators only in Teraherz and IR regions only. Meantime resonance phenomena and negative values of μ of magnetic systems take place at only much lower frequency regions. Composite LHM can be obtained combining metallic split ring resonators with regular lattice of thin metallic wires providing, respectively, negative μ and ε in the same frequency regions (see [5]) near resonances.

However the negative index (-n) is not a prerequisite for obtaining negative refraction. Recently the amphoteric refraction, i.e. negative and positive refraction index has been observed [6] for the boundary between twin structures of uniaxial crystals YVO$_4$ with $n_0 = 2.01768$ and $n_e = 2.25081$ at $\lambda = 532$ nm under various incident angles. These crystals are not LHM, however, the facts that this experiment is the single one carried out in the optical frequency region and no reflection is visible in this case are of great interest.

PhC are another alternative of obtaining negative refraction without having $\varepsilon < 0$ and $\mu < 0$. PhC are artificial composites with periodically varying $\varepsilon > 1$ in 1D, 2D or 3D ($\mu \approx 1$). For instance, a layer of dielectric beads periodically arrayed makes a 2D PhC. As it has been shown in [7] in the region of wavelengths of the order of PhC period the photons behaves in PhC just as electrons in crystals, i.e. there are photonic band gaps and the distinction of refraction and diffraction vanishes. New dispersion relations can take place when the frequency disperses negatively with wave vector, and negative refraction takes place with $n_{eff} < 0$. This has been confirmed by numerical calculations in [8] and further experimental consequences shown in Fig. 1.

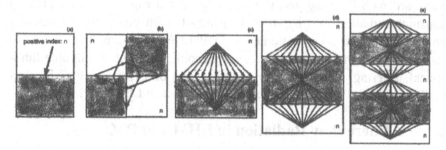

Figure 1. Light propagation in PhC with $n_{eff} < 0$: (a) negative refraction, (b) formation of an open cavity, (c) mirror-inverted imaging, (d) imaging by a slab and (e) by a stack [8].

All these materials show the above-discussed properties only in narrow frequency regions and are dispersive. It seems very difficult to prepare composite LHM and PhC for short, say optical or X-ray region, due to very small sizes of components, and the production and observation of LHM and PhC properties in materials in these region is an important challenge of the XXI century. Generally speaking, from the beginning up today there are many controversies in the understanding of the predicted and observed properties of LHM. The discussion of the validity of many LHM properties and their orthodox explanation has begun from the work [9], and the possibility of the alternative interpretation of the facts is going on.

At present there are a few published works devoted to the numerical and analytical consideration of the radiation produced by fast charged particles passing through or close to LHM and PhC. Smith-Purcell type radiation has been considered in [10,11], while in [12,13] the Cherenkov radiation has been considered theoretically and it has been shown that in

agreement with [1] the Cherenkov radiation can be directed backward. The first indirect experimental results on Cherenkov radiation have been obtained in [14], The SP radiation produced in PhC has been directly observed in [15].

Before the publication of the above mentioned works, such and many above discussed properties of LHM as well as of Cherenkov radiation have been theoretically and experimentally studied for anisotropic materials (see [16] and references therein) the detailed consideration of which comes out of this review. Let us remind that in anisotropic media there are two refraction indices, ordinary, n_0 and extraordinary, n_e and as it has been mentioned in the work [5] it has been observed negative refraction for a uniaxial crystals. Near the resonance regions the corresponding ε_0 and ε_e may become negative, and the Cherenkov radiation [16] can be directed backward with Pointing vector antiparallel to the wave vector (in [17] the transition radiation also has been considered) with many other interesting properties, though these materials are not LHM with simultaneous $\varepsilon < 0$ and $\mu < 0$. Without discussing these phenomena as well as possible future optical, filtering, lasing switching applications of LHM and PhC here we shall discuss only radiation processes taking place in LHM and PhC.

2.   Cherenkov Radiation in LHM and PhC

Still in the work [1] it has been shown that in LHM the Doppler Effect is reversed, i.e. in contrast to the usual Doppler Effect the frequency ω_0 emitted by a source and detected by the detector decreases if the source is moving with velocity V towards the detector. For an observer under observation angle θ the frequency is given by

$$\omega = \frac{\omega_0}{\gamma(1 - p\beta\cos\theta)}, \tag{5}$$

where p $=$ +1 and -1 for RHM and LHM respectively, $\gamma = E/mc^2 = \left(1 - \beta^2\right)^{-1/2}$ and $\beta = V/c$.

It has been intuitively, without mathematical consideration, mentioned that Cherenkov radiation produced in LHM by fast moving particles also will be reversed. Since in contrast to RHM in LHM the Pointing and wave vector are antiparallel then for Cherenkov radiation of a particle moving with velocity v the angle is determined by

$$\cos\theta = p\left|\sqrt{\frac{c^2}{v^2 n^2}}\right| \tag{6}$$

and can be greater than 90^0 (see Fig. 2).

Before considering the mathematical solution of the Cherenkov radiation given in [12] for the case when the wavelength is larger than the inherent inhomogeneities of LHM let us discuss the problem of momentum-energy conservation. The momentum of the electromagnetic waves (Cherenkov photons) is equal to

$$\vec{p} = \vec{D}x\vec{B} = \varepsilon\mu\vec{E}\vec{H} = \varepsilon\mu\vec{S}, \tag{7}$$

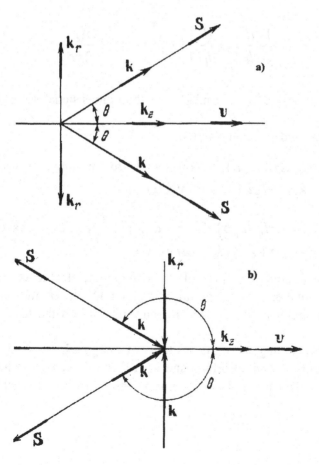

Figure 2. Cherenkov radiation for RHM a) and LHM b) [1].

and is parallel to the Pointing vector and directed backward since $\varepsilon < 0$ and $\mu < 0$. This implies that momentum of the charged particle increases which is in contradiction with the third law of thermodynamics: the particle losing energy cannot increase its energy. This paradox vanishes in quantum theory of Cherenkov radiation (see [16]) according to which the momentum of a photon is equal to $\vec{p} = \hbar\vec{k}$. Since the r-component k_r is cancelled due to the axial symmetry of the radiation the wave vector is directed along the particle velocity, and the particle momentum is decreased.

As in [18] and [12] one can present the current of the charged particle in the form

$$\vec{J}(\vec{r},t) = \hat{z}qv\delta(z - vt)\delta(x)\delta(y).$$ (8)

Using the Lorentz gauge one obtains the solution of the Maxwell's equation in the Poisson form in the cylinder coordinate frame

$$\left[\frac{1}{\rho}\frac{d}{d\rho}\left(\rho\frac{d}{d\rho}\right) + k_\rho^2\right]g(\rho) = -\frac{\delta(\rho)}{2\pi\rho},$$ (9)

where $k_\rho = (\omega/c)\sqrt{\beta^2 n^2 - 1}$ and $g(\rho)$ is two-dimensional scalar Green's function.

Eq. (9) has 2 independent solutions:

Case1: $g(\rho) = iH_0^{(1)}(k_\rho\rho)/4$ corresponds to an outgoing wave with $\vec{k} = k_\rho\hat{\rho} + k_z\hat{z}$ ($k_z = \omega/v > 0$).

Case2: $g(\rho) = -iH_0^{(1)}(k_\rho\rho)/4$ to an outgoing wave with $\vec{k} = -k_\rho\hat{\rho} + k_z\hat{z}$ ($k_z = \omega/v > 0$).

Then the authors of [12] calculate accurately the electric and magnetic fields and the total energy per unit area $W_i(\vec{\rho}) = \int S_i(\vec{r},t)dt$ radiated out in $i = \hat{\rho}$ and $\hat{z}$ directions in far field region, and confirm the results [1] of Fig. 2.

Still in [1] coming out of the positivity of energy it has been shown that LHM are dispersive and absorbing and in [12] the Cherenkov radiation for such LHM is calculated. Following Drude-Lorentz model and [19] one can write

$$\varepsilon(\omega) = 1 - \frac{\omega_{ep}^2 - \omega_{e0}^2}{\omega^2 - \omega_{e0}^2 + i\Gamma_e\omega}, \tag{10}$$

$$\mu(\omega) = 1 - \frac{\omega_{\mu p}^2 - \omega_{\mu 0}^2}{\omega^2 - \omega_{\mu 0}^2 + i\Gamma_\mu\omega}, \tag{10a}$$

where ω_{ip} and ω_{i0} ($i = e, \mu$) are the plasma and resonance frequency of LHM, which are determined only by the geometry of the LHM lattice and not by electron charge, mass and density as in RHM, while $\Gamma_{e,\mu}$ describe the attenuation.

Following [12] let us in the beginning consider nonabsorbing LHM $(\Gamma_{e,\mu} = 0)$. The following points are critical.

$$\omega_{ec} = \sqrt{\frac{\omega_{ep}^2 + \omega_{e0}^2}{2}}, \qquad \text{for which } \varepsilon(\omega_{ec}) = -1 \tag{11a}$$

$$\omega_{\mu c} = \sqrt{\frac{\omega_{\mu p}^2 + \omega_{\mu 0}^2}{2}}, \qquad \text{with } \mu(\omega_{\mu c}) = -1 \tag{11b}\backslash$$

$$\omega_c = \sqrt{\frac{\omega_{ep}^2\omega_{\mu p}^2 - \omega_{e0}^2\omega_{\mu o}^2}{\omega_{ep}^2 + \omega_{\mu p}^2 - \omega_{e0}^2 - \omega_{\mu 0}^2}},$$

for which $\varepsilon(\omega_c)\mu(\omega_c) = 1$. Fig. 3 shows the regions in which Cherenkov radiation can be produced for $\beta = 1$.

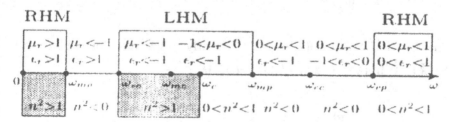

Figure 3. Frequency bands for RHM and LHM within the models (6a) and (6b) [12].

Then the authors calculate $E_z, E_\rho, H_\phi, S_z, S_\rho$ and the total energy per unit area radiated in $\hat{z}$ and $\hat{\rho}$ directions

$$W_z(\vec{\rho}) = \frac{e^2}{8\pi^2\rho v}\left[\int_0^{\omega_{\mu 0}} d\omega \frac{k_\rho}{\varepsilon(\omega)} + \int_{\omega_{e0}}^{\omega_c} d\omega \frac{k_\rho}{\varepsilon(\omega)}\right], \tag{12a}$$

$$W_\rho(\vec{\rho}) = \frac{e^2}{8\pi^2\rho}\left[\int_0^{\omega_i} d\omega \frac{k^2_\rho}{\omega\varepsilon(\omega)} - \int_{\omega_.}^{\omega} d\omega \frac{k^2_\rho}{\omega\varepsilon(\omega)}\right]. \qquad (12b)$$

Using (12a) and (12b) they come to the following conclusions for $\beta \to 1$:

In $\hat{z}$ direction: From (12a) it is seen that the first integral gives CR in RHM band with $\varepsilon > 0$ and $\mu > 0$. The energy flows along the positive $\hat{z}$ direction. However the second integral gives CR in LHM band with $\varepsilon < 0$ and $\mu < 0$. The energy flows negative $\hat{z}$ direction, backward. The total energy trough x-y plane is radiated in two bands, and the net result on which one dominates. If one detects only one frequency the energy can be forward and backward.

In $\hat{\rho}$ direction: From (12a) it is seen that the first integral gives CR in RHM band, therefore the energy flows out of $\hat{\rho}$ direction. The second integral gives CR in LHM band with $\varepsilon < 0$. However since there is − before the integral the whole second term is positive. Therefore, the energy flow in LHM band goes out in the $\hat{\rho}$ direction.

Finally following [12] let us take into account the absorption, $\Gamma \neq 0$. This means

$$\varepsilon(-\omega) = \varepsilon^*(\omega)$$
$$\mu(-\omega) = \mu^*(\omega)$$

with

$$\operatorname{Im}(\varepsilon(\omega)) > 0$$
$$\operatorname{Im}(\mu(\omega)) > 0$$

For RHM with $g(\rho) = iH_0^{(1)}(k_\rho\rho)/4$ one has

$$k_\rho = \sqrt{\frac{\omega^2}{c^2}\mu\varepsilon - \frac{\omega^2}{v^2}} = \operatorname{Re}(k) + i\operatorname{Im}[k],$$

where $\operatorname{Re}(k) > 0, \operatorname{Im}(k) > 0$. Similarly, one has for LHM with $\operatorname{Re}(k) > 0, \operatorname{Im}(k) < 0$ and $g(\rho) = -iH_0^{(1)}(k_\rho\rho)/4$

$$k_\rho = \sqrt{\frac{\omega^2}{c^2}\mu\varepsilon - \frac{\omega^2}{v^2}} = \operatorname{Re}(k) + i\operatorname{Im}[k].$$

Writing the complex magnitudes in the forms $k_\rho = \frac{\omega}{v}\eta\exp(i\theta)$ and $\varepsilon = |\varepsilon|\exp(i\theta_\varepsilon)$ the authors as before calculate $E_z, E_\rho, H_\phi, S_z, S_\rho$ and

$$W_z(\vec{\rho}) = \frac{e^2}{8\pi^2\rho v}\int_{LH}\exp[2\operatorname{Im}(k_\rho)]\frac{|k_\rho|}{|\varepsilon(\omega)|}\cos\theta_\varepsilon d\omega, \qquad (13a)$$

$$W_\rho(\vec{\rho}) = \frac{e^2}{8\pi^2 \rho} \int_{LH} \exp\left[2\,\mathrm{Im}(k_\rho)\right] \frac{-|k_\rho|^2}{|\varepsilon(\omega)|} \cos(\theta - \theta_\varepsilon)\,d\omega. \tag{13b}$$

As it is seen the direction of the radiated power is determined by θ and θ_ε. The real and imaginary parts of the complex n as well as the real part of n^2 are shown in Fig. 4. Note that in the model used, the imaginary parts of $\varepsilon(\omega)$ and $\mu(\omega)$ are always positive. These properties are summarized in Table 1:

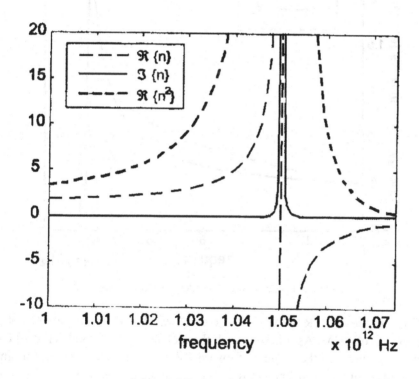

Figure 4. $\mathrm{Re}\,n(\omega), \mathrm{Im}\,n(\omega), \mathrm{Re}\,n^2(\omega)$ near the resonance frequency.

Table 1.

Magnitude	RHM	LHM
$\mathrm{Re}\,\varepsilon(\omega)$	>0	<0
θ_ε	$[0, \pi/2]$	$[\pi/2, \pi]$
$\mathrm{Im}(k)$	>0	<0
θ	$[0, \pi/2]$	$[3\pi/2, 2\pi]$

In the case of $\Gamma = 0$ (no losses) for both RHM and LHM $\theta = 0$, while for RHM $\theta_\varepsilon = 0$ and for LHM $\theta_\varepsilon = \pi$. Of course, in this case (13a) and (13b) become (12a) and (12b).

For illustration Fig. 5 shows the spectral distribution of CR energy calculated with the help of (13a) and (13b) for $\omega_{\mu p} = \omega_{ep} = 2\pi 1.09 x 10^{12}$ rad/s, $\omega_{\mu 0} = \omega_{e0} = 2\pi 1.05 x 10^{12}$ rad/s and $\Gamma_\mu = \Gamma_e = \Gamma = 1 x 10^8$ rad/s

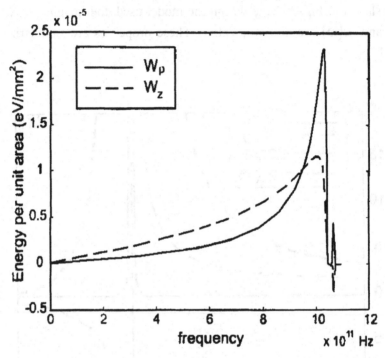

Figure 5. Energy density dependence upon frequency [12].

Fig. 6 shows CR pattern for several of Γ and above given other parameters of LHM. As it is seen with the increase of Γ 1) the backward CR power increases, 2) the angle of forward CR increases approaching 90^0 and 3) the angle of backward power decreases approaching 90^0. With such large angles one expects an increases of CR intensity and a possibility to improve the CR detectors.

Similar backward CR has been obtained [13] by numerical computations for the radiation produced in a 2D PhC. However in PhC the radiation is connected with transition and SP radiations and is produced, including in backward direction, even when the particle velocity is less than the phase velocity in PhC $v < v_{ph}$. However, the radiation for $v < v_{ph}$ depends on the particle velocity and according to the author of [13] can be used for particle identification when the usual CR is not useful. For such an application as well as for the above-mentioned application of CR in LHM many new calculation and R&D works are necessary.

As it has been mentioned CR produced in LHM and PhC by passing particles has not been observed directly. In [4] a new method of obtaining LHM has been developed by replacing the C and L parts of electrical transmission lines. Then the piece of LHM as a receiving antenna was rotated in 1^0 steps with respect to an illuminating (15 GHz) transmitting horn antenna ~5 m away. The measured pattern of the absorbed radiation shows 14.6 GHz intensities at negative angles $\theta = 38.5^0$ and in backward directions, which according to the authors [4] (without charged particle) witnesses that $n = -0.623$ and that there can be indeed backward CR. In the other indirect experiment [14] on CR in PhC the authors have observed a characteristic energy loss at 8 eV of electrons passing through PhC, which they ascribe to CR.

Concluding this part one can say that very interesting properties of CR produced in LHM are predicted theoretically and their confirmation in the microwave and optical regions is a challenge. The numerical results obtained in [13] for Cherenkov radiation (mixed with Transition radiation) produced in PhC need in more detailed analytical and numerical consideration for discussion and experimental investigations.

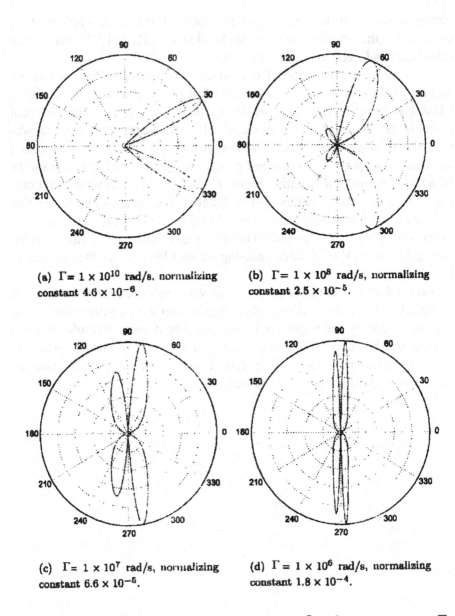

(a) $\Gamma = 1 \times 10^{10}$ rad/s, normalizing constant 4.6×10^{-6}.

(b) $\Gamma = 1 \times 10^{8}$ rad/s, normalizing constant 2.5×10^{-5}.

(c) $\Gamma = 1 \times 10^{7}$ rad/s, normalizing constant 6.6×10^{-5}.

(d) $\Gamma = 1 \times 10^{6}$ rad/s, normalizing constant 1.8×10^{-4}.

Figure 6. CR pattern for a LHM described in the text for $\beta \rightarrow 1$ at various values of Γ.

3. Smith-Purcell Radiation in PhC

In the case of the above considered CR when the wavelength was larger than the inhomogeneities period of LHM or PhC it was reasonable to assume that the LHM or PhC media are characterized by average permittivity and permeability. Now let us consider radiation with wavelength of the order of inhomogeneities of PhC, namely the relatively simplest SP type or, in brief, SPR arising when a fast charged particle flies near the surface of PhC. The case when the particle passes through the PhC is much more complicated because it is necessary to take into account the transition radiation, the scattering and diffraction of the already produced radiation in the remained part of PhC, etc.

Let the PhC (see Fig. 7) is composed of N identical layers of 2D infinite arrays of dielectric spheres of radius a with distance between the spheres or with cubic lattice of constant d. The center of a sphere in the upper layer is taken as the origin of Cartesian coordinate system, and the OX and OY or the [100] and [010] axes are directed along the corresponding centers of spheres in this layer. Let a charged particle flies with velocity v in parallel to the axis OX at a distance $D + a$, where D is the distance between the trajectory and the bottom tangent plane of PhC. We want to study in the beginning the SP type radiation with the help of a detector placed in the plane y = 0 above PhC and accepting photons under angle θ with respect to the particle momentum. Here we can describe only the physics of the process of generation of SR giving the references [10,11] for more details.

The current density and its Fourier transform can be written in the form [10]

$$\vec{j}(\vec{r},t) = -\hat{x}ev\delta(x - vt)\delta(y)\delta(z - D - a), \qquad (10a)$$

$$\vec{j}(\vec{r},\omega) = -\hat{x}e\exp(ik_x x)v\delta(x - vt)\delta(y)\delta(z - D - a), \qquad (10b)$$

where $\hat{x}$ is a unit vector along OX, and $k_x = \omega / v$.

Again using the Lorentz gauge and solving Maxwell equations, one can derive the following expression for the field from the charge

$$\vec{E}(\vec{r},\omega) = \frac{e}{2c^2}\omega\int\frac{dq_y}{2\pi}\left(-1 + \frac{1}{\beta^2}, \frac{q_y}{k\beta}, \frac{\Gamma}{k\beta}\right)\frac{\exp(i\vec{k}\vec{r})}{\Gamma}, \qquad (11)$$

where $k = \omega / c$,

$$\vec{k}_i = (k_x, q_y, \Gamma) \qquad (12)$$

and

$$\Gamma = \sqrt{\frac{\omega^2}{c^2} - k_x^2 - q_y^2} = \sqrt{\frac{\omega^2}{c^2} - \frac{\omega^2}{v^2} - q_y^2}. \qquad (13)$$

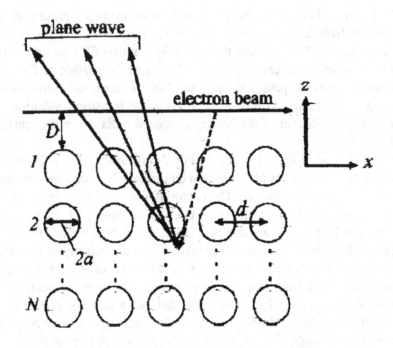

Figure 7. PhC parameters, geometry and electron beam.

Expressions (11)-(13) mean that the field of the particle near PhC is a superposition over q_y of plane waves with wave vector $\vec{k}_i$ (index i means incident) and with decay constant Γ.

Let us first consider Γ. If v is as usual smaller than c then Γ is imaginary. Therefore the field of the particle incident on PhC is evanescent and cannot be detected by the far detector. However, if these evanescent photons will undergo some scattering then, just as the pseudo photons, they can be detected as real photons after some umklapp processes. It is better to explain the physics with the help of Fig. 8 where the line of the light cone $\omega = \pm cq_x$ as well as the dispersion curve of the evanescent light $\omega = vq_x$ are given (it is suitable to put $q_y = 0$). Let us call these lines as c- and v-lines.

The point P_1 on the v-line corresponds to an evanescent photon with wave vector k_x and frequency ω. This photon is not observable since the point P_1 is out of the light cone. It arrives PhC with a decay constant Γ which is equal to the distance P_1-P_2. It can be observable if the v-line will be inside the light cone. For instance, in the case of Cherenkov radiation the light cone

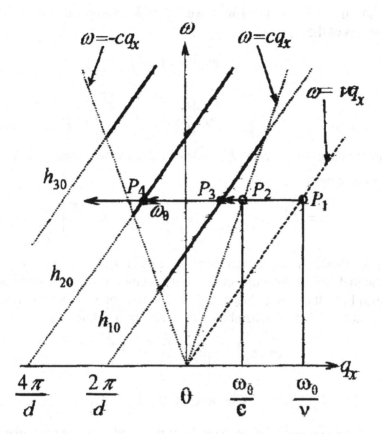

Figure 8. The dispersion curves, c- and v-lines, of the light cone, evanescent photons. The other lines, h_{10}, h_{20} and h_{30} are the new v-lines after the umklapp momentum shifts.

widens with a tangents angle $c/\sqrt{\varepsilon}$ instead of c, and Γ becomes real. The Bragg diffraction during which integer number of the reciprocal lattice vectors can be added to the photon momentum is another possibility of converting the evanescent photons into real ones. The incident photon momentum can obtain umklapp momentum shifts.

$$h_x = n_x \frac{2\pi}{d}, \quad h_y = n_y \frac{2\pi}{d}, \quad n_{x,y} = \pm 1, \pm 2..., \tag{14}$$

As it is seen from Fig. 8 parts of the shifted v-lines, the new h_{10}, h_{20} and h_{30}-lines with $n_x = -1$, -2 and -3 are in the light cone. Now SPR photon production is interpreted as the result of umklapp scattering of the evanescent photons determined by (11) on PhC. Because of larger momentum transfer or larger distance P_2P_4, etc the intensity of SP photons with larger n_x must be reduced. The scattering of q_y waves take place independently. Thus to incident waves with k_i correspond new, scattered plane waves of the type

$$\exp(i\vec{k}_s\vec{r}) \tag{15}$$

with

$$\vec{k}_s = \left(k_x - n_x\frac{2\pi}{d}, q_y - n_y\frac{2\pi}{d}, \Gamma_{n,n}\right). \tag{16}$$

Therefore,the wave vector $\vec{k}_S$ is determined by two integers n_x and n_y and a z-component

$$\Gamma_{mn} = \left[\left(\frac{\omega}{c}\right)^2 - \left(k_x - n_x\frac{2\pi}{d}\right)^2 - \left(q_y - n_y\frac{2\pi}{d}\right)^2\right]^{1/2}, \tag{17}$$

The calculation of the prefactors of the plane waves (15) is the lengthy and essential part of obtaining of the SPR spectra [10,11] which will not be described here. It can be reduced to the calculation of a 3x3 tensor reflection coefficient $\mathbf{R(k_S, K_i)}$. In general the SPR intensity is defined as

$$I(n_x, n_y)_q = \left|E_1^{(n,n)}(k_x, q_y, \Gamma)\right|^2, \tag{18}$$

where

$$E_1^{(n,n)}(k_x, q_y, \Gamma) = R(k_s, k_i)E_0(k_x.q_y, \Gamma). \tag{19}$$

In the last expression E_0 is the field of the incident wave (..) without the irrelevant prefactor [11].

The SPR intensity $I(n_x, 0)_0$ when $q_y = 0$ and $n_y = 0$ as well as $I(1,0)_0$ when $|n_x| = 1$ (the radiation around the line h_{10} is of most interest since $n_y \neq 0$ gives small contribution, while the photons with $|n_x| > 1$ have higher energy and lower intensity. In these cases the radiation with frequency ω emitted in the plane y = 0 makes an angle θ with the axis x which according to (16) is given by

$$\cos\theta = |k_S|_x / |k_S| = \frac{k_x - 2\pi/d}{\omega/c} \tag{19}$$

Consider the relation between the PhC band structure or photonic bands (PB) and SPR spectra. As it is well known just as in the case of electrons in crystals, the photons in PhC have certain density of states (DOS) within the light cone. Again there are a few methods for calculation of these DOS or PhC band structures which we cannot consider. The position of the peaks of DOS gives the frequency of the modes and their width, the lifetime due to leakage. The calculated PB structures can be experimentally verified, in particular by measuring the incident angle dependent transmission spectra. In particular, a good agreement between theoretical and experimental DOS has been obtained [19] for a PhC which was further used in a SPR experiment (see below).

Figure 9 shows the band structure and calculated $I(n_x,0)_0$ for a PhC with a/d =0.42, D=d/2, N=4 layers of spheres with $\varepsilon = 3.2^2$, As it is seen SPR peaks are radiated at the crossing of the line h_{10} with the band structure curves which are the curves of maximal DOSes (for details and polarization see [11]).

To confirm the above said and to show the D-dependence in Fig. 10 it is shown a part of Fig. 9 with calculated $I(1,1)_1$ for D/d=0.5 and D/d=0. The calculated few result show that, as expected, the SPR intensity increases with the increase of N and decreases when the electron beam is under angle to the axis OX.

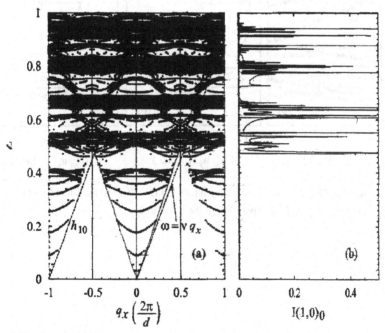

Figure 9. Band structure a) and SPR spectrum b) on the line h_{10}. The vertical axis in a) and b) is the frequency ω in units $2\pi c/d$.

So far the intensities (18) were given in arbitrary units. In the work [11] it is given the so-called absolute differential cross section for SPR. It has been shown that the SPR intensity emitted into a solid angle per period d of PhC is given by

$$\frac{dW}{d\Omega} = \sum_{\vec{h}} \frac{dW(\vec{h})}{d\Omega} = \sum_{\vec{h}} \Xi(\vec{h}) I(n_x, n_y)_q \quad (20)$$

where $\vec{h}$ is the reciprocal lattice vector,

$$\Xi(\vec{h}) = \pi^2 \frac{e^2}{\varepsilon_0} \left(\frac{d}{\lambda^3 h_x} \right) \sin^2 \theta(\vec{h}) \cos^2 \phi(\vec{h}). \quad (21)$$

The SPR polar $\theta(\vec{h})$ and azimuth $\phi(\vec{h})$ angles are defined in Fig. 11.

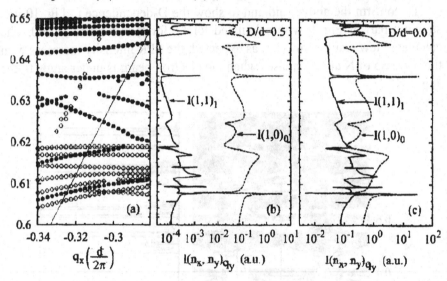

Figure 10. A part of Figure 9 with results for I(1,1) and D/d=0. In a) the filled (open) circles are for polarization perpendicular (parallel) to the plane y=0.

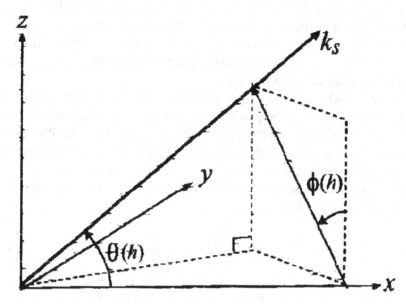

Figure 11. Polar and azimuth angles definition.

Using (18) (20) and (21) in the work [11] it has been calculated SPR produced by PhC and made a comparison with that produced with the help of a metallic grating. Figure 12 shows the parameters of the grating and PhC with $N = 1$, $D/d = 0.5$.

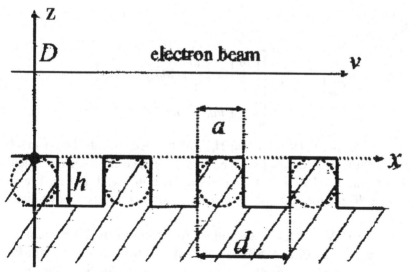

Figure 12. The diffraction grating with rectangular groove and PhC.

Figure 13 shows the comparison of SPR produced with the help of these radiators. The advantage of PhC is evident though PhC results are not optimized. In particular, it is assumed $q_y = 0$ and there is a "leak" from one layer PhC which can be compensated by a metallic mirror or increasing N the latter resulting in sharper peaks.

To our knowledge there is only one published work [15] devoted to the experimental study of SPR produced in PhC the discussion of which we begin. The experimental arrangement is shown in Fig. 14d. The PhC consist of a single layer of polytetrafluoroethylene (PTFE) beads with diameter $d = 3.2$ mm and refractive index $n = 1.437+ \text{i}0.0002$ arrayed in contrast to the above considered cubic in a 2D triangular lattice. Therefore, the basis vectors are (see Fig. 14b)

$$\vec{k}_1 = \frac{4\pi}{d\sqrt{3}}(1,0) \equiv (k_{1x}, k_{1y})$$

$$\vec{k}_2 = \frac{4\pi}{d\sqrt{3}}\left(\frac{1}{2}, \frac{\sqrt{3}}{2}\right) \equiv (k_{2x}, k_{2y})$$

(19)

while instead of (16) and (17) we have

$$\vec{k}_S = \left[k_x + mk_{1x} + nk_{2x}, q_y + mk_{1y} + nk_{2y}, \Gamma_{mn}(k_x q_y)\right]$$

(20)

with

$$\Gamma_{mn}(k_x, q_y) = \left[\left(\frac{\omega}{c}\right)^2 - \left(\frac{\omega}{v} + mk_{1x} + nk_{2x}\right)^2 - \left(q_y + mk_{1y} + nk_{2y}\right)^2\right]^{1/2}$$

(21)

Therefore, the dispersion line $\mathbf{H}_{m,n}$ of the evanescent photon in (k,ω) space is shifted by $m\vec{k}_1 + n\vec{k}_2$ (see Fig. 14c). The theoretically calculated [20] photonic band is shown later in this note in Fig. 17a. Fig. 14c shows how a SPR photon can be produced according to the above-discussed mechanism.

The 150 MeV electron beam with a cross section 10x12 mm^2 passes over the PhC at a height 10 mm. SPR is detected with the help of a Martin-Puplett type Fourier-transform spectrometer having a He-cooled Si bolometer under $\theta = 60^0$-112^0 with a resolution 3.75 GHz.

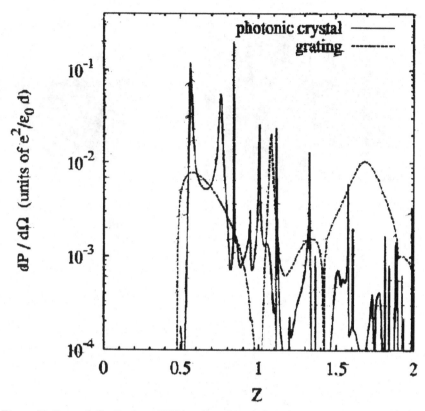

Figure 13. Spectral distribution of SPR produced by perfect grating and PhC with N=1.

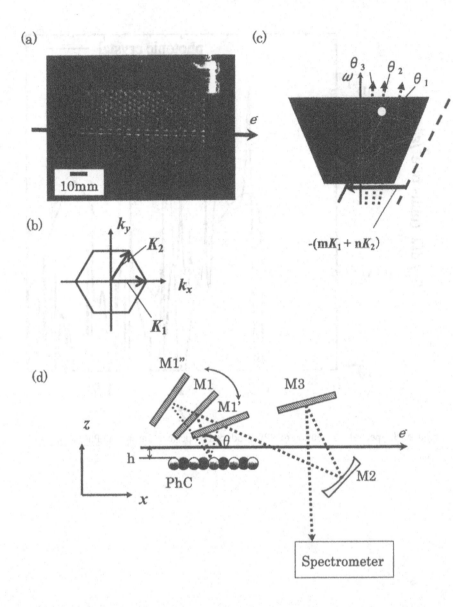

Figure 14. a) the top view of PhC and electron beam, b) The two basis vectors and the first Brillouin zone, c) SPR production, the spot is the cross point of three curves, d) Side view of the arrangement. M_i are mirrors.

The spectral distributions of the SPR detected under various angles without a) and with PhC b) and c) are shown in Fig. 15. A series of lines with FWHM about 5 GHz are observed.

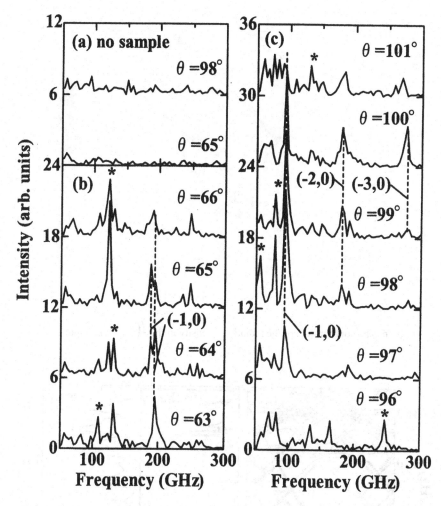

Figure 15. The measured spectral distribution of the radiation under various angles.

Figure 16 shows the intensity of the detected radiation and how the radiation is concentrated around the shifted lines (-1,0), (-2,0) and (-3,0). Figure 17 shows the calculated PB structure of the used PhC and the expected and measured spectral distribution of the radiation.

Some peaks indicated by crosses can not be explained by the discussed SPR mechanism and are generated, probably, due to the deformation of PhC. By the damping effect due to distortions are explained the difference between the intensity ratios observed and expected peaks at various frequencies.

The satisfactory interpretation of the results of this first and single experiment on SPR produced in PhC witnesses that PhC can serve as good transformer of the particle energy into monochromatic SPR lines.

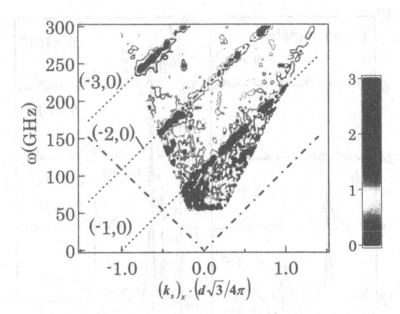

Figure 16. The contour map of the radiation intensity in the $\left((k_S)_x, \omega\right)$ plane.

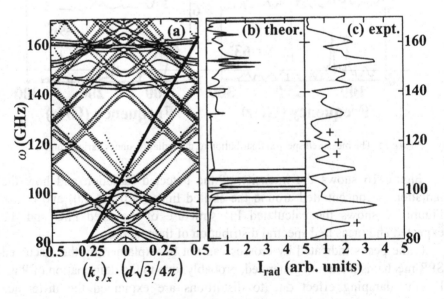

Figure 17. The calculated PB structure of the PhC a) and the theoretical b) and experimental emission spectra along the line **H(-1,0)**.

4. Conclusions and Perspectives

Due to their specific properties it is expected that LHM and PhC will find wide application in optics, drastically changing the components of optical devices, in high energy physics, improving the possibility of the existing Cherenkov detectors, in microwave technology, filtering, lasing, switching at long wavelengths, in opto-electronic technology, replacing in some regions the semiconductor silicon devices, etc.

In difference from the usual materials LHM has negative refraction index in certain frequency regions and the vectors of velocity, electric and magnetic fields make a left handed system. As a result one cane expect that the photons radiated in LHM are emitted in backward directions. In difference from single crystals and carbon nanotubes and fullerites the PhC have larger lattice constants, and this fact can result in specific properties for the radiation produced in them when the radiated wavelength are of the order or less than PhC inhomogeneity periods.

At present there are a few published works devoted to the theoretical and experimental study of the radiation properties of these materials. The first theoretical results on Cherenkov radiation in LHM and PhC developed in analogy of the Cherenkov radiation in ordinary media has shown that the corresponding radiation acquires interesting, sometimes even paradoxical properties. For instance, despite to the usual cases the Cherenkov photons are emitted in backward directions. Such predictions have been experimentally observed in indirect measurements and in microwave frequency region only. Theoretical results carried out for the region showed that the intensity of the emitted by the Smith-Purcell type mechanism radiation when the wavelength is of the order of PhC inhomogeneities is much richer and higher than in the case of Smith-Purcell radiation from metallic gratings. The last predictions have been experimentally verified only in a single experiment.

The search of LHC and PhC with similar properties in the optical and higher frequency including X-ray regions is a challenge for the century. Nevertheless, it is reasonable to study the radiation produced in various frequency region when a charged particle passes through or crosses the surface of LHM and PhC. First using the well known classical and quantum electro-dynamical methods it is necessary to consider the general case of radiation when a charged particle passes through LHM and PhC, while the virtual and produced real photons undergo various processes; scattering, Bragg scattering, absorption, etc. Taking into account the peculiarities of LHM and the photonic band (PB) structures of PhC it is necessary to develop the theories 1) of coherent bremsstrahlung (CB), 2) of transition radiation produced at a single interface between LHM (or PhC) and usual material as well as at many such interfaces taking into account the interference phenomena, 3) of parametric or more correctly diffracted radiation in various frequency regions.

References

1. V.G. Veselago, Electrodynamics of substances with simultaneous negative values of permittivity and permeability, Soviet Physics Uspekhi, **10**, 505 (1968).
2. R.A. Shelby, D.R. Smith and S. Schultz, *Science*, **292** 77 (2001).
3. J.B. Pendry, Some problems with negative refraction, *Phys. Rev. Lett.* **87** 249702 and 249704 (2001).
4. A. Grbich and G.V. Eleftheriades, Experimental verification of backward-wave radiation from a negative refractive index material *J. Appl. Phys.* **92**, 5930 (2002).
5. P. Markos and C.M. Soukoulis, Structures with negative index of refraction, *Phys. Stat. Sol. (a)* **197**, 595 (2003).
6. Y.Zhang, B. Fluegel and A. Mascarenhas, Total negative refraction in real crystals for ballistic electrons and light, *Phys. Rev. Lett.* **91**, 157404 (2003).
7. E. Yablonovich, Inhibited spontaneous emission in solis-state physics and electronics, *Phys. Rev. Lett.* **58**, 2059 (1987).
8. M. Notomi, Negative refraction in photonic crystals *Optical and Quantum Electronics,* **34**, 133 (2002).
9. P.M. Valanju, R.M. Walser and A.P. Valanju, Wave refraction in negative-index media: Always positive and very inhomogeneous, *Phys. Rev. Lett.* **88**, 187401 (2002).
10. K.Ohtaka and S. Yamaguti, Smith-Purcell radiation from a charge running near the surface od a photonic crystal, *Opt. and Quantum Electronics,* **34**, 235 (2002).
11. S. Yamagouti, J. Inoue, O. Haeberle and K. Ohtaka, *Phys.Rev. B66*, 195202, (2002).
12. J.Liu, T.M. Grzegorzyk et al, *Optics Express*, **1**. 723 (2003).
13. C. Luo, M.Ibanescu, S.G. Johnson and J.D. Joannopoulos, Cerenkov radiation in photonic crystals, *Science*, **299**, 368 (2003).
14. F.J. Garcia de Abajo et al *Phys. Rev. Lett.* **91**, 143902 (2003).
15. K. Yamamoto et al, Phys. Rev. E, Accepted for rapid publication in 2004.
16. V.P. Zrelov, Cherenkov Radiation In high energy Physics (Moscow, Atomizdat, 1968).
17. V.E. Pafomov, *Zh. Eksper. Teor. Fiz.* **36**, 1853 (1959).
18. I.M. Frank and I.G. Tamm, *Dokladi Akade Nauk SSSR,* **14**, 109 (1937).
19. T. Kondo et al, Transmission characteristics of 2D PhC. *Phys. Rev. B66*, 033111 (2002).
20. K.Ohtaka and Y. Tanabe, *J. Phys. Soc. Jpn.* **65**, 2265 (1996).

POLARIZATION EFFECTS ON ELECTROMAGNETIC PROCESSES IN ORIENTED CRYSTALS AT HIGH ENERGY

V. N. Baier and V. M. Katkov

Budker Institute of Nuclear Physics
630090 Novosibirsk, Russia

Abstract   Under quite generic assumptions the general expression is derived for the probability of circularly polarized photon emission from the longitudinally polarized electron and for the probability of pair creation of longitudinally polarized electron (positron) by circularly polarized photon in oriented crystal in a frame of the quasiclassical operator method. For small angle of incidence the expression turns into constant field limit with corrections due to inhomogeneous character of field in crystal. For relatively large angle of incidence the expression gets over into the theory of coherent radiation or pair creation. It is shown that the crystal is a very effective device for helicity transfer from an electron to photon and back from a photon to electron or positron.

Keywords: Radiation, pair creation, longitudinal polarization, oriented crystals

1.   Introduction

The study of processes with participation of polarized electrons and photons permits to obtain important physical information. Because of this experiments with use of polarized particles are performed and are planning in many laboratories (BINP, CERN, SLAC, Jefferson Natl Accl Fac, etc). In this paper it is shown that oriented crystal is a unique tool for work with polarized electrons and photons.

The quasiclassical operator method developed by authors [1]-[3] is adequate for consideration of the electromagnetic processes at high energy. The probability of photon emission has a form (see [4], p.63, Eq.(2.27); the method is given also in [5],[6])

$$
\mathrm{d}w = \frac{e^2}{(2\pi)^2} \frac{\mathrm{d}^3 k}{\omega} \int \mathrm{d}t_2 \int \mathrm{d}t_1 R^*(t_2) R(t_1) \exp\left[-\frac{\mathrm{i}\varepsilon}{\varepsilon'}(kx_2 - kx_1)\right], \quad (1)
$$

H. Wiedemann (ed.), Advanced Radiation Sources and Applications, 97–108.

where $k^\mu = (\omega, \mathbf{k})$ is the 4-momentum of the emitted photon, $k^2 = 0$, $x^\mu(t) = (t, \mathbf{r}(t))$, $x_{1,2} \equiv x(t_{1,2})$, t is the time, and $\mathbf{r}(t)$ is the particle location on a classical trajectory, $kx(t) = \omega t - \mathbf{k}\mathbf{r}(t)$, ε is the energy of initial electron, $\varepsilon' = \varepsilon - \omega$, we employ units such that $\hbar = c = 1$. The matrix element $R(t)$ is defined by the structure of a current. For an electron (spin 1/2 particle) one has

$$R(t) = \frac{m}{\sqrt{\varepsilon\varepsilon'}}\overline{u}_{s_f}(\mathbf{p}')\hat{e}^* u_{s_i}(\mathbf{p}) = \varphi^+_{\zeta_f}\left(A(t) + i\sigma\mathbf{B}(t)\right)\varphi_{\zeta_i},$$

$$A(t) = \frac{1}{2}\left(1 + \frac{\varepsilon}{\varepsilon'}\right)\mathbf{e}^*\vartheta(t),$$

$$\mathbf{B(t)} = \frac{\omega}{2\varepsilon'}\left[\mathbf{e}^* \times \left(\frac{\mathbf{n}}{\gamma} - \vartheta(t)\right)\right], \tag{2}$$

here $\mathbf{e}$ is the vector of the polarization of a photon (the Coulomb gauge is used),the four-component spinors u_{s_f}, u_{s_i} describe the initial (s_i) and final (s_f) polarization of the electron and we use for the description of electron polarization the vector ζ describing the polarization of the electron (in its rest frame), ζ_i is the spin vector of initial electron, ζ_f is the spin vector of final electron, the two-component spinors $\varphi_{\zeta_i}, \varphi_{\zeta_f}$ describe the initial and final polarization of the electron, $\mathbf{v}=\mathbf{v}(t)$ is the electron velocity, $\vartheta(t) = (\mathbf{v} - \mathbf{n}) \simeq \mathbf{v}_\perp(t)$, $\mathbf{v}_\perp$ is the component of particle velocity perpendicular to the vector $\mathbf{n} = \mathbf{k}/|\mathbf{k}|$, $\gamma = \varepsilon/m$ is the Lorentz factor. The expressions in Eq.(2) are given for radiation of ultrarelativistic electrons, they are written down with relativistic accuracy (terms $\sim 1/\gamma$ are neglected) and in the small angle approximation.

The important parameter χ characterizes the quantum effects in an external field, when $\chi \ll 1$ we are in the classical domain and with $\chi \geq 1$ we are already well inside the quantum domain while for pair creation the corresponding parameter is κ

$$\chi = \frac{|\mathbf{F}|\varepsilon}{F_0 m}, \quad \kappa = \frac{|\mathbf{F}|\omega}{F_0 m}, \quad \mathbf{F} = \mathbf{E}_\perp + (\mathbf{v} \times \mathbf{H}), \quad \mathbf{E}_\perp = \mathbf{E} - \mathbf{v}(\mathbf{v}\mathbf{E}), \tag{3}$$

where $\mathbf{E}(\mathbf{H})$ is an electric (magnetic) field, $F_0 = m^2/e = (m^2 c^2/e\hbar)$ is the quantum boundary (Schwinger) field: $H_0 = 4.41 \cdot 10^{13}$Oe, $E_0 = 1.32 \cdot 10^{16}$V/cm.

The quasiclassical operator method is applicable when $H \ll H_0$, $E \ll E_0$ and $\gamma \gg 1$.

Summing the combination $R^*(t_2)R(t_1) = R_2^*R_1$ over final spin states we have

$$\sum_{\zeta_f} R_2^*R_1 = A_2^*A_1 + \mathbf{B}_2^*\mathbf{B}_1 + i\left[A_2^*(\zeta_i\mathbf{B}_1) - A_1(\zeta_i\mathbf{B}_2^*) + \zeta_i\left(\mathbf{B}_2^* \times \mathbf{B}_1\right)\right],$$

$$\tag{4}$$

where the two first terms describe the radiation of unpolarized electrons and the last terms is an addition dependent on the initial spin.

For the longitudinally polarized initial electron and for circular polarization of emitted photon we have [10]

$$\sum_{\zeta_f} R_2^* R_1 = \frac{1}{4\varepsilon'^2} \left\{ \frac{\omega^2}{\gamma^2}(1+\xi) + \left[(1+\xi)\varepsilon^2 + (1-\xi)\varepsilon'^2\right] \vartheta_1 \vartheta_2 \right\}, \quad (5)$$

where $\xi = \lambda\zeta$, $\lambda = \pm 1$ is the helicity of emitted photon, $\zeta = \pm 1$ is the helicity of the initial electron. In this expression we omit the terms which vanish after integration over angles of emitted photon.

The probability of polarized pair creation by a circularly polarized photon can be found from Eqs.(1), (2) using standard substitutions:

$$\varepsilon' \to \varepsilon', \quad \varepsilon \to -\varepsilon, \quad \omega \to -\omega, \quad \lambda \to -\lambda,$$

$$\zeta_i \equiv \zeta \to -\zeta', \quad \zeta_f \equiv \zeta' \to \zeta, \quad \xi \to \xi, \quad \xi' \to -\xi'. \quad (6)$$

Performing the substitutions Eq.(6) in Eq.(5) the combination for creation of pair with polarized positron by circularly polarized photon

$$\sum_{\xi'} R_2 R_1^* = \frac{m^2}{8\varepsilon^2\varepsilon'^2} \left\{ \omega^2(1+\xi) + \gamma^2\vartheta_2\vartheta_1 \left[\varepsilon^2(1+\xi) + \varepsilon'^2(1-\xi)\right] \right\}. \quad (7)$$

It should be noted that a few different spin correlations are known in an external field. But after averaging over directions of crystal field only the longitudinal polarization considered here survives.

2. General Approach to Electromagnetic Processes in Oriented Crystals

The theory of high-energy electron radiation and electron-positron pair creation in oriented crystals was developed in [7]-[8], and given in [4]. In these publications the radiation from unpolarized electrons was considered including the polarization density matrix of emitted photons. Since Eqs.(5), (7) have the same structure as for unpolarized particles, below we use systematically the methods of mentioned papers to obtain the characteristics of radiation and pair creation for longitudinally polarized particles.

Let us remind that along with the parameter χ which characterizes the quantum properties of radiation there is another parameter

$$\varrho = 2\gamma^2 \left\langle (\Delta \mathbf{v})^2 \right\rangle, \quad (8)$$

where $\left\langle (\Delta \mathbf{v})^2 \right\rangle = \left\langle \mathbf{v}^2 \right\rangle - \left\langle \mathbf{v} \right\rangle^2$ and $\langle \ldots \rangle$ denotes averaging over time. In the case $\varrho \ll 1$ the radiation is of a dipole nature and it is formed during the time of the order of the period of motion. In the case $\varrho \gg 1$ the radiation is of magnetic bremsstrahlung nature and it is emitted from a small part of the trajectory.

In a crystal the parameter ϱ depends on the angle of incidence ϑ_0 which is the angle between an axis (a plane) of crystal and the momentum of a particle. If $\vartheta_0 \leq \vartheta_c$ (where $\vartheta_c \equiv (2V_0/\varepsilon)^{1/2}$, V_0 is the scale of continuous potential of an axis or a plane relative to which the angle ϑ_0 is defined) electrons falling on a crystal are captured into channels or low above-barrier states, whereas for $\vartheta_0 \gg \vartheta_c$ the incident particles move high above the barrier. In later case we can describe the motion using the approximation of the rectilinear trajectory, for which we find from Eq.(8) the following estimate $\varrho(\vartheta_0) = (2V_0/m\vartheta_0)^2$. For angles of incidence in the range $\vartheta_0 \leq \vartheta_c$ the transverse (relative to an axis or a plane) velocity of particle is $v_\perp \sim \vartheta_c$ and the parameter obeys $\varrho \sim \varrho_c$ where $\varrho_c = 2V_0\varepsilon/m^2$. This means that side by side with the Lindhard angle ϑ_c the problem under consideration has another characteristic angle $\vartheta_v = V_0/m$ and $\varrho_c = (2\vartheta_v/\vartheta_c)^2$.

We consider here the photon emission (or pair creation) in a thin crystal when the condition $\varrho_c \gg 1$ is satisfied. In this case the extremely difficult task of averaging of Eqs.(5),(7), derived for a given trajectory, over all possible trajectories of electrons in a crystal simplifies radically. In fact, if $\varrho_c \gg 1$ then in the range where trajectories are essentially non-rectilinear ($\vartheta_0 \leq \vartheta_c$, $v_\perp \sim \vartheta_c$) the mechanism of photon emission is of the magnetic bremsstrahlung nature and the characteristics of radiation can be expressed in terms of local parameters of motion. Then the averaging procedure can be carried out simply if one knows the distribution function in the transverse phase space $dN(\varrho, \mathbf{v}_\perp)$, which for a thin crystal is defined directly by the initial conditions of incidence of particle on a crystal.

Substituting Eq.(5) into Eq.(1) we find after integration by parts of terms $\mathbf{nv}_{1,2}$ ($\mathbf{nv}_{1,2} \to 1$) the general expression for photon emission probability (or probability of creation of longitudinally polarized positron)

$$dw_\xi = \sigma \frac{\alpha m^2}{8\pi^2} \frac{d\Gamma}{\varepsilon\varepsilon'} \int \frac{dN}{N} \int e^{i\sigma A} \left[\varphi_1(\xi) - \frac{\sigma}{4}\varphi_2(\xi)\gamma^2 (\mathbf{v}_1 - \mathbf{v}_2)^2 \right] dt_1 dt_2,$$

$$A = \frac{\omega\varepsilon}{2\varepsilon'} \int_{t_1}^{t_2} \left[\frac{1}{\gamma^2} + [\mathbf{n} - \mathbf{v}(t)]^2 \right] dt,$$

$$\varphi_1(\xi) = 1 + \xi\frac{\omega}{\varepsilon}, \quad \varphi_2(\xi) = (1+\xi)\frac{\varepsilon}{\varepsilon'} + (1-\xi)\frac{\varepsilon'}{\varepsilon}. \tag{9}$$

where $\alpha = e^2 = 1/137$, the vector $\mathbf{n}$ is defined in Eq.(1), the helicity of emitted photon ξ is defined in Eq.(5), $\sigma = -1$, $d\Gamma = d^3k$ for radiation and $\sigma = 1$, $d\Gamma = d^3p$ for pair creation and for pair creation one have to put $\int dN/N = 1$.

The circular polarization of radiation is defined by Stoke's parameter $\xi^{(2)}$:

$$\xi^{(2)} = \Lambda(\zeta \mathbf{v}), \quad \Lambda = \frac{dw_+ - dw_-}{dw_+ + dw_-}, \tag{10}$$

where the quantity $(\zeta \mathbf{v})$ defines the longitudinal polarization of the initial electrons, dw_+ and dw_- is the probability of photon emission for $\xi=+1$ and $\xi=-1$ correspondingly. In the limiting case $w \ll \varepsilon$ one has $\varphi_2(\xi) \simeq 2(1 + \xi w/\varepsilon) = 2\varphi_1(\xi)$. So the expression for the probability dw_ξ contains the dependence on ξ as a common factor $\varphi_1(\xi)$ only. Substituting in Eq.(10) one obtains the universal result independent of a particular mechanism of radiation $\xi^{(2)} = w(\zeta \mathbf{v})/\varepsilon$.

The periodic crystal potential $U(\mathbf{r})$ can be presented as the Fourier series (see e.g.[4], Sec.8)

$$U(\mathbf{r}) = \sum_{\mathbf{q}} G(\mathbf{q}) e^{-i\mathbf{q}\mathbf{r}}, \tag{11}$$

where $\mathbf{q} = 2\pi(n_1, n_2, n_3)/l$; l is the lattice constant. The particle velocity can be presented in a form $\mathbf{v}(t) = \mathbf{v}_0 + \Delta\mathbf{v}(t)$, where $\mathbf{v}_0$ is the average velocity. If $\vartheta_0 \gg \vartheta_c$, we find $\Delta\mathbf{v}(t)$ using the rectilinear trajectory approximation

$$\Delta\mathbf{v}(t) = -\frac{1}{\varepsilon} \sum \frac{G(\mathbf{q})}{q_\parallel} \mathbf{q}_\perp \exp[-i(q_\parallel t + \mathbf{q}\mathbf{r})], \tag{12}$$

where $q_\parallel = (\mathbf{q}\mathbf{v}_0)$, $\mathbf{q}_\perp = \mathbf{q} - \mathbf{v}_0(\mathbf{q}\mathbf{v}_0)$. Substituting Eq.(12) into Eq.(9), performing the integration over $\mathbf{u} = \mathbf{n} - \mathbf{v}_0$ ($d^3k = \omega^2 d\omega\, d\mathbf{u}$, $d^3p = \varepsilon^2 d\varepsilon\, d\mathbf{u}$) and passing to the variables t, τ : $t_1 = t - \tau$, $t_2 = t + \tau$, we obtain after simple calculations the general expression for the probability of photon emission (or probability of pair creation) for polarized case valid for any angle of incidence ϑ_0

$$dw_\xi = \frac{i\alpha m^2 d\Gamma_1}{4\pi} \int \frac{dN}{N} \int \frac{d\tau}{\tau + i\sigma 0} [\varphi_1(\xi) + \sigma\varphi_2(\xi)]$$
$$\times \left(\sum_{\mathbf{q}} \frac{G(\mathbf{q})}{mq_\parallel} \mathbf{q}_\perp \sin(q_\parallel \tau) e^{-i\mathbf{q}\mathbf{r}} \right)^2 \Bigg] e^{i\sigma A_2}, \tag{13}$$

where

$$A_2 = \frac{m^2\omega\tau}{\varepsilon\varepsilon'}\left[1 + \sum_{\mathbf{q},\mathbf{q}'}\frac{G(\mathbf{q})G(\mathbf{q}')}{m^2 q_\| q'_\|}(\mathbf{q}_\perp \mathbf{q}'_\perp)\Psi(q_\|, q'_\|, \tau)\exp[-i(\mathbf{q}+\mathbf{q}')\mathbf{r}]\right]$$

$$\Psi(q_\|, q'_\|, \tau) = \frac{\sin(q_\| + q'_\|)\tau}{(q_\| + q'_\|)\tau} - \frac{\sin(q_\|\tau)}{q_\|\tau}\frac{\sin(q'_\|\tau)}{q'_\|\tau}, \tag{14}$$

where $d\Gamma_1 = \omega d\omega/\varepsilon^2(\varepsilon d\varepsilon/\omega^2)$ for radiation (pair creation).

3. Radiation and Pair Creation for $\vartheta_0 \ll V_0/m$ (constant field limit)

The behavior of probability dw_ξ Eq.(13) for various entry angles and energies is determined by the dependence of the phase A_2 on these parameters given Eq.(14). In the axial case for $\vartheta_0 \ll V_0/m \equiv \vartheta_v$ the main contribution to A_2 give vectors $\mathbf{q}$ lying in the plane transverse to the axis [4] and the problem becomes two-dimensional with the potential $U(\boldsymbol{\varrho})$. Performing the calculation for axially symmetric $U = U(\varrho^2)$ we obtain

$$dw_\xi^F(\omega) = \frac{\alpha m^2\omega d\Gamma_1}{2\sqrt{3}\pi}\int_0^{x_0}\frac{dx}{x_0}\left\{D\,R_0(\lambda) - \frac{1}{6}\left(\frac{m\vartheta_0}{V_0}\right)^2\left[\frac{xg'' + 2g'}{xg^3}\right.\right.$$

$$\left.\left.\times R_1(\lambda) - \frac{\lambda}{20g^4 x^2}\left(2x^2 g'^2 + g^2 + 14gg'x + 6x^2 gg''\right)R_2(\lambda)\right]\right\}, \tag{15}$$

where

$$R_0(\lambda) = \varphi_2(\xi)K_{2/3}(\lambda) + \varphi_1(\xi)\int_\lambda^\infty K_{1/3}(y)dy,$$

$$R_1(\lambda) = \varphi_2(\xi)\left[(K_{2/3}(\lambda) - \frac{2}{3\lambda}K_{1/3}(\lambda)\right],$$

$$R_2(\lambda) = \varphi_1(\xi)\left[K_{1/3}(\lambda) - \frac{4}{3\lambda}K_{2/3}(\lambda)\right]$$

$$- \varphi_2(\xi)\left[\frac{4}{\lambda}K_{2/3}(\lambda) - \left(1 + \frac{16}{9\lambda^2}\right)K_{1/3}(\lambda)\right], \tag{16}$$

and $D = \int dN/N(D = 1)$ for the radiation (pair creation), $K_\nu(\lambda)$ is the modified Bessel function (McDonald's function), we have adopted a new variable $x = \rho^2/a_s^2$, $x \leq x_0$, $x_0^{-1} = \pi a_s^2 dn_a = \pi a_s^2/s$, a_s is the effective screening radius of the potential of the string, n_a is the density of atoms in a crystal, d is the average distance between atoms of a chain forming the axis. The term in Eq.(15) with $R_0(\lambda)$ represent the spectral probability in the constant field limit. The other terms are the correction proportional ϑ_0^2 arising due to nongomogeneity of field in crystal. The notation $U'(x) = V_0 g(x)$ is used in Eq.(15) and

$$\lambda = \frac{m^3 a_s \omega}{3\varepsilon\varepsilon' V_0 g(x)\sqrt{x}} = \frac{u}{3\chi_s g(x)\sqrt{x}} = \frac{2\omega^2}{3\varepsilon\varepsilon'} \frac{\sqrt{\eta}}{\kappa_s \psi(x)}, \quad \psi(x) = 2\sqrt{\eta x} g(x),$$
(17)

here $\kappa_s = V_0 \omega/(m^3 a_s)$, $\chi_s = V_0 \varepsilon/(m^3 a_s)$ are a typical values of corresponding parameters in crystal. For specific calculation we use the following for the potential of axis:

$$U(x) = V_0 \left[\ln\left(1 + \frac{1}{x + \eta}\right) - \ln\left(1 + \frac{1}{x_0 + \eta}\right) \right].$$
(18)

For estimates one can put $V_0 \simeq Ze^2/d$, $\eta \simeq 2u_1^2/a_s^2$, where Z is the charge of the nucleus, u_1 is the amplitude of thermal vibrations, but actually the parameter of potential were determined by means of a fitting procedure using the potential Eq.(11) (table of parameters for different crystals is given in Sec.9 of [4]). For this potential

$$g(x) = \frac{1}{x + \eta} - \frac{1}{x + \eta + 1} = \frac{1}{(x + \eta)(x + \eta + 1)}$$
(19)

The results of numerical calculations for the spectral intensity ($dI_\xi = \omega dw_\xi$), $\varepsilon \, dI_\xi^F(\omega)/d\omega$ in a tungsten crystal, axis $< 111 >$, $T = 293\ K$ are given in Fig.1.

In Fig.2 the circular polarization $\xi^{(2)}$ of radiation is plotted versus ω/ε for the same crystal. This curve is true for both energies: $\varepsilon = 250$ GeV and $\varepsilon = 1$ TeV. Actually this means that it is valid for any energy in high-energy region. At $\omega/\varepsilon = 0.8$ one has $\xi^{(2)} = 0.94$ and at $\omega/\varepsilon = 0.9$ one has $\xi^{(2)} = 0.99$.

Now we pass to a pair creation by a photon. In the limit $\kappa_s \ll 1$ one can substitute the asymptotic of functions $K_\nu(\lambda)$ at $\lambda \gg 1$ in the term with $R_0(\lambda)$ in Eq.(15). After this one obtains using the Laplace method in integration over the coordinate x

$$\frac{dW_\xi^F}{dy} = \frac{\sqrt{3}\alpha V_0 \left(y(1 + \xi) + (1 - y)^2\right)}{2mx_0 a_s} \sqrt{\frac{\psi^3(x_m)}{4\eta|\psi''(x_m)|}} e^{-\frac{2}{3y(1-y)\kappa_m}}, \quad (20)$$

where $\kappa_m = \kappa_s \psi(x_m)/\sqrt{\eta}$. So, at low energies the pair creation probability is suppressed. With photon energy increase the probability increases also and at some energy $\omega \simeq \omega_t$ it becomes equal to the standard Bethe-Maximon W_{BM} probability in the considered medium, e.g. for W, T=293

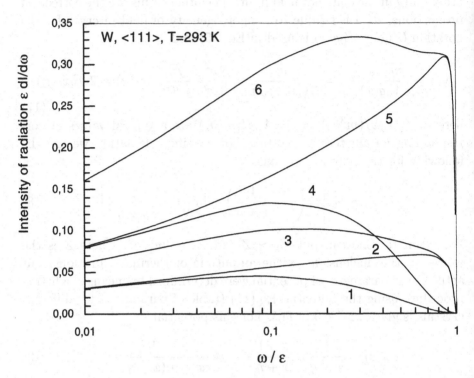

Figure 1. Spectral intensity of radiation in units αm^2 vs ω/ε. The curves 1, 4 are for $\xi = -1$, the curves 2, 5 are for $\xi = 1$, the curves 3, 6 are the sum of previous contributions (the probability for unpolarized particles). The curves 1, 2, 3 are for the initial electron energy ε=250 GeV and the curves 4, 5, 6 are for the initial electron energy ε=1 TeV.

K, axis $< 111 > \omega_t = 22$ GeV. For high energies the probability of pair creation may be much higher than W_{BM} (see [4]).

This is seen in Fig.3 where the spectral probability of pair creation dw_ξ/dy in tungsten, T=293 K, axis $< 111 >$ is given. Near the end of spectrum the process with $\xi = 1$ dominates. The sum of curves at the indicated energy gives unpolarized case. For the integral (over y) probability the longitudinal polarization of positron is $\zeta = 2/3$.

4.   Modified Theory of Coherent Bremsstrahlung and Pair Creation

The estimates of double sum in the phase A_2 made at the beginning of previous section: $\sim (\vartheta_v/\vartheta_0)^2 \Psi$ remain valid also for $\vartheta_0 \geq \vartheta_v$, except that now the factor in the double sum is $(\vartheta_v/\vartheta_0)^2 \leq 1$, so that the values $|q_{\parallel}\tau| \sim 1$ contribute. We consider first the limiting case $\vartheta_0 \gg \vartheta_v$, then

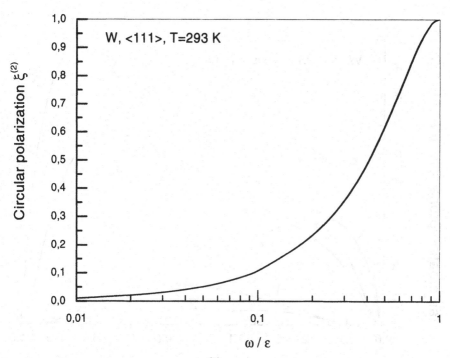

Figure 2. The circular polarization $\xi^{(2)}$ of radiation (for $(\boldsymbol{\zeta v})=1$) vs ω/ε for the tungsten crystal, axis $<111>$, $T = 293\ K$. The curve is valid for both energies: $\varepsilon=250$ GeV and $\varepsilon=1$ TeV.

this factor is small and $\exp(-iA_2)$ can be expanded accordingly. After integration over coordinate $\mathbf{r}$ we obtain

$$dw_\xi^{\mathrm{coh}} = \frac{\alpha d\Gamma_1}{8} \sum_{\mathbf{q}} |G(\mathbf{q})|^2 \frac{\mathbf{q}_\perp^2}{q_\parallel^2} \left[\varphi_2(\xi) + \sigma\varphi_1(\xi)\frac{2m^2\omega}{\varepsilon\varepsilon' q_\parallel^2} \left(|q_\parallel| - \frac{m^2\omega}{2\varepsilon\varepsilon'} \right) \right]$$
$$\times \vartheta \left(|q_\parallel| - \frac{m^2\omega}{2\varepsilon\varepsilon'} \right). \tag{21}$$

For unpolarized electrons (the sum of contributions with $\xi = 1$ and $\xi = -1$) Eq.(21) coincides with the result of standard theory of coherent bremsstrahlung (CBS), or coherent pair creation see e.g. [9].

In the case $\chi_s \gg 1$ (χ_s is defined in Eq.(17)), one can obtain a more general expression for the spectral probability:

$$dw_\xi^{\mathrm{coh}} = \frac{\alpha d\Gamma_1}{8} \sum_{\mathbf{q}} |G(\mathbf{q})|^2 \frac{\mathbf{q}_\perp^2}{q_\parallel^2} \left[\varphi_2(\xi) + \sigma\varphi_1(\xi)\frac{2m^2\omega}{\varepsilon\varepsilon' q_\parallel^2} \left(|q_\parallel| - \frac{m_*^2\omega}{2\varepsilon\varepsilon'} \right) \right]$$
$$\times \vartheta \left(|q_\parallel| - \frac{m_*^2\omega}{2\varepsilon\varepsilon'} \right), \quad m_*^2 = m^2(1+\frac{\rho}{2}), \quad \frac{\rho}{2} = \sum_{\mathbf{q}, q_\parallel \neq 0} \frac{|G(\mathbf{q})|^2}{m^2 q_\parallel^2} \mathbf{q}_\perp^2 \tag{22}$$

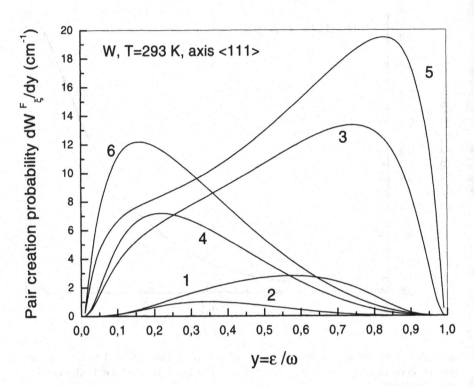

Figure 3. The spectral probability of pair creation dw_ξ^F/dy, the curves 1 and 2 are for energy $\varepsilon=22$ GeV, the curves 3 and 4 are for energy $\varepsilon=100$ GeV, the curves 5 and 6 are for energy $\varepsilon=250$ GeV. The curves 1, 3 and 5 are for $\xi=1$, and the curves 2, 4 and 6 are for $\xi=-1$.

The spectral probabilities Eqs.(21) and (22) can be much higher than the Bethe-Maximon bremsstrahlung probability W_{BM} for small angles of incidence ϑ_0 with respect to selected axis. For the case $\vartheta_0 \ll 1$ the quantity $q_\parallel$ can be represented as

$$q_\parallel \simeq \frac{2\pi}{d}n + \mathbf{q}_\perp\mathbf{v}_{0.\perp} \qquad (23)$$

In the extreme limit of $\lambda \equiv 2\varepsilon|q_\parallel|_{\min}/m^2 \sim \varepsilon\vartheta_0/m^2 a_s \gg 1$, the maximum of probability of coherent bremsstrahlung is attained at such values of ϑ_0 where the standard theory of coherent bremsstrahlung becomes invalid. Bearing in mind that if $\lambda \gg 1$ and $\vartheta_0 \sim V_0/m$, then $\chi_s \sim \lambda \gg 1$, we can conveniently use a modified theory of coherent bremsstrahlung.

The direction of transverse components of particle's velocity in Eq.(23) can be selected in a such way, that the spectral probability described by Eq.(22) has a sharp maximum near the end of spectrum at $\omega = 2\varepsilon\lambda(2 + 2\lambda + \varrho)^{-1} \simeq \varepsilon$ with relatively small (in terms of λ^{-1}) width

$\Delta\omega \sim \varepsilon(1+\varrho/2)/\lambda = m^2(1+\varrho/2)/2|q_{\parallel}|_{\min}$:

$$(dw_\xi)_{\max} = \frac{\alpha\varepsilon d\Gamma_1\varrho|q_{\parallel}|_{\min}}{4(2+\varrho)}\left(1+\xi+\frac{1-\xi}{(1+u_m)^2}\right), \quad u_m = \frac{2\lambda}{2+\varrho}. \quad (24)$$

It is seen that in the maximum of spectral distribution the radiation probability with opposite helicity ($\xi = -1$) is suppressed as $1/(1+u_m)^2$. At $u > u_m$ one have to take into account the next harmonics of particle acceleration. In this region of spectrum the suppression of radiation probability with opposite helicity is more strong, so the emitted photons have nearly complete circular polarization.

Comparing Eq.(24) with W_{BM} for $\varepsilon' \ll \varepsilon$ we find that for the same circular polarization ($\xi^{(2)} \simeq (\zeta\mathbf{v})$) in the particular case $\varrho = 1$ the magnitude of spectral probability in Eq.(24) is about $\chi_s(\varepsilon_e)$ times larger than W_{BM}. For tungsten $\chi_s(\varepsilon_e) = 78$. From above analysis follows that under mentioned conditions the considered mechanism of emission of photons with circular polarization is especially effective because there is gain both in monochromaticity of radiation and total yield of polarized photons near hard end of spectrum.

5. Conclusions

At high energy the radiation from longitudinally polarized electrons in oriented crystals is circularly polarized and $\xi^{(2)} \to 1$ near the end of spectrum. This is true in magnetic bremsstrahlung limit $\vartheta_0 \ll V_0/m$ as well as in coherent bremsstrahlung region $\vartheta_0 > V_0/m$. This is particular case of helicity transfer.

In crossing channel: production of electron-positron pair with longitudinally polarized particles by the circularly polarized photon in an oriented crystal the same phenomenon of helicity transfer takes place in the case when the final particle takes away nearly all energy of the photon.

So, the oriented crystal is a very effective device for helicity transfer from an electron to photon and back from a photon to electron or positron. Near the end of spectrum this is nearly 100% effect.

Acknowledgments

We would like to thank the Russian Foundation for Basic Research for support in part this research by Grant 03-02-16154.

References

[1] V. N. Baier, and V. M. Katkov, Phys. Lett.,A **25** (1967) 492.

[2] V. N. Baier, and V. M. Katkov, Sov.Phys.JETP **26** (1968) 854.

[3] V. N. Baier, and V. M. Katkov, Sov.Phys.JETP **28** (1969) 807.

[4] V. N. Baier, V. M. Katkov and V. M. Strakhovenko, *Electromagnetic Processes at High Energies in Oriented Single Crystals*, World Scientific Publishing Co, Singapore, 1998.

[5] V. B. Berestetskii, E. M. Lifshitz and L. P. Pitaevskii, *Quantum Electrodynamics* Pergamon Press, Oxford, 1982.

[6] V. N. Baier, V. M. Katkov and V. S. Fadin, *Radiation from Relativistic Electrons* (in Russian) Atomizdat, Moscow, 1973.

[7] V. N. Baier, V. M. Katkov and V. M. Strakhovenko, Sov.Phys.JETP **63** (1986) 467.

[8] V. N. Baier, V. M. Katkov and V. M. Strakhovenko, Sov.Phys.JETP **65** (1987) 686.

[9] M. L. Ter-Mikaelian, *High Energy Electromagnetic Processes in Condensed Media*, John Wiley & Sons, 1972.

[10] V. N. Baier and V. M. Katkov, *"Radiation from polarized electrons in oriented crystals at high energy"* hep-ph/0405046, 2004; Preprint BINP 2004-26, Novosibirsk, 2004, Nucl.Instr.and Meth B (presented for publication).

CALCULATION OF PLANAR CHANNELING RADIATION PRODUCED BY 20 MEV ELECTRONS IN QUARTZ

R.O. Avakian, K.A. Ispirian and V.J. Yaralov

Yerevan Physics Institute, Brothers Alikhanian 2, 375036, Yerevan, Armenia

Abtract: The potential, energy levels and the spectral distribution of the radiation produced by 20 MeV electrons channeled in the (100) plane of the piezo-crystal quartz (SiO_2) are calculated taking into account the broadening due to scattering of electrons on thermal fluctuations and atomic electrons.

Keywords: Channeling radiation/Many-beam method

In order to prepare an experiment devoted to the first observation of channeling radiation (ChR) produced in quartz with and without external ultrasonic waves we have calculated the expected ChR spectra in the case of 20 MeV electrons channeling in the (100) plane of quartz, SiO_2.

There are three atoms of Si and six atoms of O in the quartz elementary cell. The coordinates of the atoms in the elementary cell of α-quartz are taken from [1]. We choose the axis OZ along the beam direction and coinciding with one of the two-fold axis (the electric axis) and the axis OY along the three-fold axis (the optical axis). The root mean square of thermal shifts of all the Si atoms are identical, and using the data of the work [2] one can show that the root mean square shifts in the directions x,y and z are

H. Wiedemann (ed.), Advanced Radiation Sources and Applications, 109–114.
© 2006 *Springer. Printed in the Netherlands.*

equal to (in A^o) $\rho_x = 0.0723$ $\rho_y = 0.085$ and $\rho_z = 0.0493$. The 6 oxygen atoms O can be divided into 3 pairs with identical values of $\rho_x = 0.1159, \rho_y = 0.1026$ and $\rho_z = 0.0801; \rho_x = 0.0828$,

$\rho_y = 0.1026$ and $\rho_z = 0.1140; \rho_x = 0.0973; \rho_y = 0.1026$ and

$\rho_z = 0.1018$.

The crystalline planar potential is calculated by averaging the sum of the potentials of the atoms in the crystal over the thermal oscillations and along the chosen crystallographic planes. The obtained potential depends only on one coordinate, V(x), and are periodic, V(x+d$_p$)=V(x), where d$_p$ is the shortest period. The beginning of the coordinates is in the center of the potential well so that the x-coordinate in each pair of the oxygen atoms are placed symmetrically with respect to the beginning of the coordinates. Taking the Fourier expansion of the potential

$$V(x) = \sum_n V_n \exp(ingx), \tag{1}$$

where $g = 2\pi / d_p$ is the reciprocal lattice vector, and averaging one obtains the following expressions for the coefficients V_n [3,4] for elementary cell containing different atoms.

$$V_n = \frac{2\pi\hbar^2}{m} n_{cell} \sum_\alpha \exp(ingx_\alpha) \exp\left[\frac{\rho_x^2(\alpha)}{2}(ng)^2 \right]$$

$$\sum_{i=1}^4 a_i^\alpha \exp\left[-\frac{b_i^\alpha}{16\pi^2}(ng)^2 \right] \tag{2}$$

where a_i^α, b_i^α are the coefficients corresponding to the atom α which are calculated in [5] for various chemical elements, x_α is the equilibrium position of the atom α in the elementary cell, n_{cell} is the density of elementary cells. The summation $\sum$ is taken over all the atoms in the elementary cells. If the function V(x) is even, V(x) = V(-x), as in the case of quartz then the expansion coefficients V_n are real.

The solutions of the 1-D transversal Schrödinger equation are the Bloch functions

$$\psi_{kn}(x) = \frac{\exp(ikx)}{\sqrt{L}} u_{kn}(x), \tag{3}$$

where $u_{kn}(x) = u_{kn}(x + d_p)$, n is the number of the energy zone, k is the Bloch-momentum and L is the crystal size in direction x. For each arbitrary k, limiting by a finite number of the terms of the expansion

$$u_{kn} = \sum_m c_m^{kn} \exp(imgx). \qquad (4)$$

and substituting into the Schrödinger equation (the many-beam method) [3] one obtains real symmetric matrix of linear equations for the coefficients c_m^{kn}. The eigenvalues of this matrix equation give the energy levels, while the eigenvectors give the real coefficients c_m^{kn}.

Solving the matrix equations numerically we took 21 terms in the expansion of u_{kn}. The number of the sublevels in each energy zone corresponding to a Bloch momentum equal to 8. The obtained wave functions $\psi_{kn}(x)$ make complete basis system of the electron states in the crystal in transversal direction. In Fig. 1 we show the potential of the (100) plane of the single crystal α-quartz SiO₂.

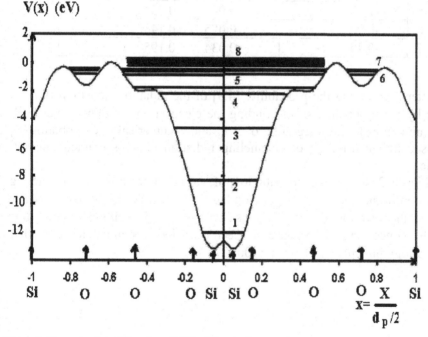

Figure1. The potential of the plane (100) of the single crystal α-quartz (SiO₂) and energy levels of the channeled electrons with E = 20 MeV.

The place of the Si and O planes are shown by thick arrows on the axis OX on which the distance are given in the units of $d_p/2$.

In this work the transition probabilities between the transversal levels (k,n) due to the scattering and widening of levels have been calculated by the methods described in [6-8] without taking into account the scattering on core electrons.

The transversal energy levels of the channeled 20 MeV electrons are shown in Fig.1 and given in Table 1 together with corresponding calculated widths due to scattering on thermal displacements and scattering on valence electrons.

Table 1. The transversal energy levels $E_n(eV)$ and their widths $\Delta E_n(eV)$ of 20 MeV electrons channeled in the (100) plane of SiO_2.

n	$E_n(eV)$	$\Delta E_n(eV)$	
		Therm.	Val.
1	-11.6	0.502	0.021
2	-8.23	0.309	0.044
3	-4.66	0.248	0.066
4	-2.23	0.139	0.099
5	-1.75	0.001	0.034
6	-0.96	0.055	0.148
7	-0.40	0.034	0.195

After multiplying the probability [4,9] of the radiation transition (k,i) → (k,f) by the probability of the finding the electron on the upper level (k,i), summing over k and integrating over k_y and z numerically, one obtains [6] the spectral distribution of channeling radiation photons emitted under 0 angle.

Figure 2 shows the contribution of various transitions of channeling radiation in the photon energy region ℏω = (9-14) KeV. In the given energy region the contribution of the transitions 3→2 is dominant. The same results concerning potentials and spectra we have obtained for the other crystallographic planes (101) and (001).

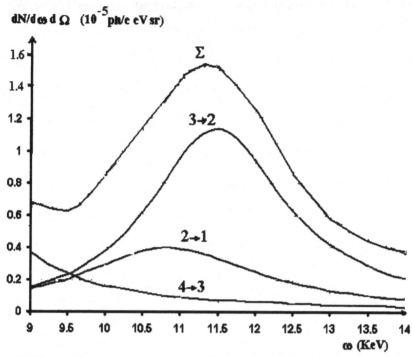

Figure 2. The spectral distribution of the channeling radiation due to transitions $2 \rightarrow 1$, $3 \rightarrow 2$ and $4 \rightarrow 3$ of 20 MeV electrons with angular divergence $\sigma_x = \sigma_y = 0.55$ mrad and channeled in the plane (100) of a 13 μm thick SiO_2 crystal.

References

1. W.A. Deer, R.A. Howie and J. Zussman, Rock-Forming Minerals, Vol. 4, Frame Silicates (Longnans, London, 1966).
2. R.A. Young and Ben Post, Electron density and thermal effects in alpha quartz, *Acta Cryst.* 15, 337 (1962).
3. J.U. Andersen, E. Bonderup, E. Laegsgaard, B.B. Marsh and A.H. Sorensern, Axial channeling radiation from MeV electrons, *Nucl. Intr. And Meth.* 194, 209 (1982).
4. J.U. Andersen, E. Bonderup, E. Laegsgaard and A.H. Sorensern, Incoherent scattering of electrons and linewidth of planar-channeling radiation, *Physica Scripta,* 28, 308 (1983).
5. P.A. Doyl and P.S. Turner, Relativistic Hartree-Fock X-ray and electron scattering factors, *Acta Cryst.* A24, 390 (1968).
6. H.Genz et al, Channeling radiation of electrons in natural diamond crystals and their coherence and occupation lengths, *Phys. Rev.* B53, 8922 (1996).
7. M. Weber, Dissertation, Channeling radiation of relativistic electrons, University of Erlangen, 1995.
8. V.A. Bazilev, V.I. Glebov and V.V. Goloviznin, Quantum Theory of inelastic scattering of negatively charged particles by oriented crystals, *Zh. Eksp. Teor. Fiz.91,* 25 (1986).

9. M.A. Kumakhov, Radiation of particles channeled in crystals, (Moscow, Energoatonizdat, 1986)

CHANNELING AT EXTREMELY HIGH BUNCH CHARGES AND SOLID STATE PLASMA ACCELERATION

Richard A. Carrigan, Jr.
Fermi National Accelerator Laboratory, Batavia, IL 60510, USA, Carrigan@fnal.gov

Abstract: Plasma and laser acceleration suggest possible ways to circumvent the limitations of conventional acceleration technology. The potentially high plasma densities possible in solids might produce extremely high acceleration gradients. The daunting challenges of solid state plasmas are discussed. Crystal channeling has been suggested as a mechanism to ameliorate these problems. However a high-density plasma in a crystal lattice could quench the channeling process. There is no experimental or theoretical guidance on channeling for intense charged particle beams. An experiment has been carried out at the Fermilab A0 photoinjector to observe electron channeling radiation at high bunch charges. An electron beam with up to 8 nC per electron bunch was used to investigate the electron-crystal interaction. No evidence was found of quenching of channeling at charge densities two orders of magnitude larger than in earlier experiments. Possible new channeling experiments are discussed for the much higher bunch charge densities and shorter times required to probe channeling breakdown and plasma behavior.

Keywords: Channeling, channeling radiation, accelerators

1. EXOTIC ACCELERATORS

Modern accelerators have been able to reach extremely high energies but at a cost. With available technology they must be very large and expensive to

H. Wiedemann (ed.), Advanced Radiation Sources and Applications, 115–127.
© 2006 *Springer. Printed in the Netherlands.*

reach the ever-rising energy frontier. Currently they are limited by available magnetic fields, roughly 10 Tesla, and by RF gradients, at best 100 MeV/m.

There are several possibilities for breaking the acceleration bottleneck. The first thought is to use the fantastic capabilities of lasers directly. A second possibility is to use a plasma wave to accelerate particles. This has been discussed in articles by Dawson [1] and his collaborators as well as the comprehensive review by Esarey et al. [2] All of these approaches use gas plasmas, not plasmas generated in a solid.

The good news about lasers is that they can give very high electric fields. The bad news is that the electric vectors point in the wrong direction (transverse to the laser beam direction) and optical frequencies make the construction of an electromagnetic cavity outside of the reach of conventional technology. Several ingenious ideas have been advanced to overcome this problem including the use of gratings, exotic boundary conditions at metallic surfaces, and other special modes. In the last several years there has been some real progress on this at the STELLA facility at Brookhaven [3]. A large-scale R&D program, LEAP, is now underway at Stanford/SLAC [4].

In STELLA a co-propagating CO_2 laser beam moving with the accelerating electron beam pumps an undulator field – electron beam system so that the electron beam either gains or loses energy. The first stage in STELLA is called a buncher and the second stage is an accelerator. This type of laser acceleration works. One problem is that practically it gives a smaller acceleration than a plasma system. It is necessary to cascade stages to get any substantial acceleration. The same is true for practical plasma acceleration. STELLA was built to investigate the challenges of cascading. A powerful laser system is required for STELLA with 24 MW instantaneous for the first stage and 300 MW for the second. Initially this device has been used to demonstrate rephasing but not acceleration.

Another exotic possibility is plasma acceleration. This is discussed in the next section.

2.   PLASMA WAKEFIELD ACCELERATION

In wakefield acceleration a laser or electron beam driver creates a moving wave in a plasma. In turn that wave accelerates a charged particle. Figure 1 illustrates a metaphor for the plasma wakefield acceleration process. In this illustration the wake driver is a powerful motorboat. The particle, here a surfboarder, is pushed along by the wave. Notice that the surfer is not connected to the boat by a rope. Phase stability is important. If the surfer drifts too far down the wave he no longer moves forward. He can accelerate for a while by moving up the wave. The plasma wakefield acceleration process is very similar to the metaphor.

Figure 1: Wake surfer metaphor for plasma wakefield acceleration (courtesy S. and L. Carrigan).

The gradient in a plasma is $G \approx 0.96\sqrt{n_0}$ where n_0 is the electron density (per cm^3) and G is in V/cm. The best wake is generated when the driver bunch is half the plasma wavelength. For a laser wakefield the acceleration length is the smallest of the dephasing length, the so-called pump depletion length, or the optical depth of field. For a good electron plasma in a gas the density might reach 10^{18}/cm^2 and produce a gradient of 1 GV/cm. For a solid the density could be 10,000 times higher corresponding to 100 GV/cm.

The difficulty with using a solid is that the driving and accelerating beams have very strong interactions with the medium resulting in a number of problems including high energy loss, multiple scattering, and radiation. Channeling has been suggested as a technique to ameliorate these problems. Channeling is the process whereby a charged particle moving along a crystal plane or axis is guided by the crystal fields.

Plasma acceleration has been demonstrated in several places including UCLA, SLAC, and Fermilab. In the last several years SLAC has run a series of increasingly sophisticated wakefield acceleration studies [5]. Typically a 30 GeV electron or positron beam impinges on a meter-long Li ion plasma produced by an ionizing laser. The bunched plasma wakefield is driven by the head of the electron beam itself rather than a laser. That wakefield accelerates the tail to O(100 MeV) depending on conditions. At Fermilab, Barov et al. [6] have produced similar results with the A0 photoinjector.

A different approach to plasma wakefield acceleration is to use a laser pulse some tens of femtoseconds long to produce a plasma in a millimeter thick gas stream without the requirement of an initiating electron beam system. Using this approach several teams [7] have recently produced small divergence 100 MeV electron beams with energies spreads of only several percent.

3.   SOLID STATE ACCELERATORS

For many years crystal channeling researchers have dreamed of using aligned crystals to accelerate charged particles. Robert Hofstadter [8] mused on the limitations of conventional accelerators and speculated on an early version of a channeling accelerator. In his words *"To anyone who has carried out experiments with a large modern accelerator there always comes a moment when he wishes that a powerful spatial compression of his equipment could take place. If only the very large and massive pieces could fit in a small room!"* What Hofstadter imagined was a tabletop accelerator he called **Miniac**. The device would consist of a single crystal driven by an x-ray laser. Channeling would be used to focus the beam. Hofstadter realized the device might be an after-burner to boost a conventional accelerator beam that was already up into the relativistic regime somewhat in the spirit of the recent SLAC studies. At the time channeling was a new subject and x-ray lasers were distant dreams.

In the last years there has been progress on something that might at first glance be considered solid-state acceleration. Groups from Livermore [9], Michigan [10], Rutherford [11] and LULI [12] have all seen energetic ions and electrons emanating from thin foils struck by extremely powerful picosecond laser pulses. At Livermore they observe "beams" of 10^{13} protons downstream of a foil irradiated with a 1000 TW, $3*10^{20}$ W/cm^2 laser pulse. A beam can be focused by curving the foils. The high-energy ions originate from deposits or contaminants on the downstream side of the foil. The accelerating electrostatic field at the downstream surface is produced by ponderomotively accelerated hot electrons generated by the laser pulse. While this is an interesting process it should probably not be considered a solid-state accelerator since the electrostatic field is outside the solid.

In the late 1990s Helen Edwards' group at Fermilab built a prototype photoinjector at A0 to work on development of the Tesla injector [13]. The Tesla photoinjector is basically a gigantic photocathode powered by a laser and followed by warm and cold RF stations. The accelerator can deliver very large 14 MeV picosecond electron pulses on the order of 10 nanocoulombs or 10^5 A/cm^2.

The Fermilab A0 photoinjector offered a means to probe in the direction of channeling conditions characteristic of those needed for solid-state acceleration and do observations of channeling under conditions never studied before. With the facility it was possible to study channeling behavior as the bunch charge increased and go several orders of magnitudes beyond earlier measurements. One could even wonder if it would be possible to see a hint of acceleration. In any case it was possible to do some unique studies of channeling and how solids behaved under unusual conditions. The channeling radiation experiment discussed below [14] was designed to initiate this program.

4. THE CHALLENGES OF CRYSTAL ACCELERATION

The basic solid-state acceleration paradigm is to excite a plasma wakefield in a crystal with a density thousands of time higher than a gas plasma. This possibility has been explored in some detail by Chen and Noble [15]. Recent developments in femtosecond laser and electron beam technology make this possibility thinkable. Channeling would be used to reduce energy loss, to focus the particles, and perhaps even cool the beam. However there are significant problems. The required electron or laser driver beam is so powerful that the crystal would probably be blown away. In addition there is the classical problem of dechanneling where the channeled particle is scattered out of the channel.

To consider a crystal accelerator one must first understand what happens to a crystal exposed to the intense radiation needed to create a plasma wakefield. Chen and Noble [16] have developed sketches of the process that are useful guides. Initially, an electronic plasma is excited by so-called tunnel ionization. The electron plasma frequency is

$$\omega_p = \left(4\pi n_0 e^2 / m_e\right)^{1/2} \qquad (1)$$

The electronic plasma decays via electronic interband transitions with plasma lifetimes in the femtosecond range. That decay to a hot electron gas excites phonons in the lattice. That can lead to crystal disorder, fracture, or vaporization. The ionic plasma lifetime goes as the square root of the ratio of ionic mass divided by the electron mass so that serious lattice damage may occur within 10 to 100 fs. At Livermore [17] hydrodynamic heating is seen to occur in the 1 to 10 ps range. These times are short but not so short that one can discount the possibility of plasma acceleration.

These comments have to be understood in the context of what is required for acceleration and some other observations of materials exposed to intense laser and particle beams. While the accelerating gradient in a plasma scales as the square root of the plasma density, the drive power required varies linearly with n and destruction problems can easily get out of hand. For a plasma gradient of 1 GeV/cm the required laser power density is of order 10^{19} W/cm^3. Laser crystal destruction occurs for power densities of order 10^{13} W/cm^3. The lattice is ionized with power fluences of 10^{15} to 10^{16} W/cm^2. Notice that these power levels are well below the requirements for crystal acceleration. Chen and Noble state that charge densities in a particle drive beam of order 10^{20} e/cm^2 are required to get near the acceleration regime. Current densities of 10^{11} A/cm^2 can fracture a crystal. Interestingly, charged particle beams are not quite as destructive and do not suffer as much from the equivalent of skin depth problems as laser drivers do.

No evidence of electron beam damage or short-term crystal disorder was seen over the course of the A0 channeling radiation experiment for instantaneous beam current densities of 10^5 A/cm^2 and an estimated electron beam fluence of 10^{17} electrons. On the other hand several adverse effects were observed in the laser system that powers the A0 photoinjector. The laser slab ruptured under continuous operation at 10 W/cm^3 corresponding to 10^{15} W/cm^3 for 10 fs. There was lens damage at 10^9 W/cm^2. In principle the A0 laser might be able to operate to deliver 1 Joule to a 10 micron spot in 1 ps or 10^{18} W/cm^2. This is already in the regime where it could be an interesting driver for a plasma accelerator.

In any case one can destroy single crystals. There are literally billions of them embedded in TVs and computers. Most of our knowledge about how solids behave when struck by femtosecond long packets of light or particles appears to be schematic. This is particularly true for the understanding of channeling properties. There are some interesting theoretical processes to study here.

How would channeling be affected as a crystal vaporized? I call this process *dynamic channeling*. There appears to be no well-developed theory of dynamic channeling. Andersen [18] has suggested a treatment based on a screening length that increases when the electrons are removed due to ionization. In the Andersen picture the channeling critical angle as a function of temperature and screening length is

$$\Psi_{1/2} = \frac{\Psi_L}{\sqrt{2}} \sqrt{\ln\left(\frac{r_0^2}{u_2^2 \ln 2}\right) + \ln\left(\frac{\left(\sqrt{3}a_{TF}\right)^2 + u_2^2 \ln 2}{\left(\sqrt{3}a_{TF}\right)^2 + r_0^2}\right)} \quad , \qquad (2)$$

where Ψ_L is the Lindhard angle. Here r_0 is some channel radius and u_2 is the rms two-dimensional lattice vibration amplitude equal to $0.006\sqrt{T}$ at high temperature for u_2 in Å and T in °K. Removing most of the electrons is equivalent to a large screening length or letting the Thomas-Fermi screening length, a_{TF}, become large. For practical purposes the screening length reaches its limiting value when $a_{TF} = r_0$. The behavior of the critical angle is shown in Fig. 2 as a function of temperature for two different screening lengths.

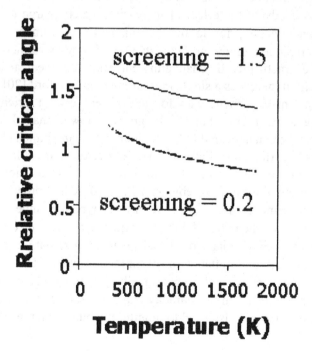

Figure 2: Axial critical angle for silicon as a function of temperature.

The critical angle for a fixed screening length does drop as the temperature increases but actually increases for larger screening lengths. The point is that the changes with temperature and screening length are not so large. Channeling would seem to survive until the nuclear centers in the lattice have displacements in the Å range, that is, until the crystal ceases to be a crystal. It would be useful to have a well-developed treatment of dynamic channeling as a framework for future studies of channeling under conditions similar to those needed for solid-state acceleration. Lacking a concrete theory, the dotted line in Fig. 5 suggests how the channeling radiation yield might diminish as the beam intensity increased to the regime where the crystal is vaporized.

5.   THE FERMILAB A0 STUDIES

The Fermilab channeling radiation experiment was undertaken to extend channeling studies toward the solid-state acceleration regime. Channeling radiation was investigated at A0 in part because earlier experiments had looked at this and in part because it was practical to do a channeling radiation experiment at the high bunch charges at A0.

Only one study of the effect of increasing bunch charge on channeling had been undertaken prior to the Fermilab A0 experiment. This was a channeling radiation study at Darmstadt at 5.4 MeV aimed in part at studying the practicality of channeling radiation for medical applications [19]. In addition there was a single measurement at Stanford [20] at 30 MeV. The Stanford measurement was for a different crystal orientation. The Stanford value at higher bunch charge was lower that the Darmstadt measurements which appeared to be flat with increasing bunch charge.

As noted earlier the Fermilab photoinjector was able to produce extremely large bunch charges. In addition it was also possible to operate with dark current alone and thereby get a $10^5 - 10^6$ reduction in bunch charge so that the experiment was able to cover a wide dynamic range.

Figure 3 is a schematic of the A0 experiment. The electrons moving along a crystal axis or plane produced channeling radiation. The electron beam typically had an emittance of 10 mm*mrad with a 10 ps long bunch. The beam spot size was characteristically 0.5 mm (σ). A spectrometer magnet swept the beam into a Faraday cup and beam dump. An integrating current transformer was also used to monitor the bunch charge. The channeling radiation showered in a calcium tungstate sheet and produced visible photons that were detected by a photosensitive device.

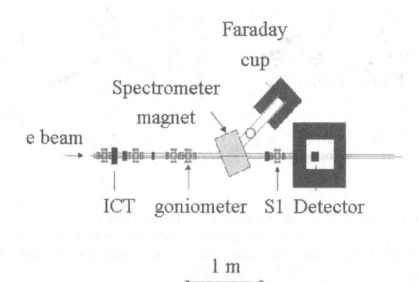

Figure 3: Fermilab A0 channeling radiation apparatus. The detectors and the Faraday cup were surrounded by lead shields. Here ICT stands for integrating current transformer. The S1 port housed one of the view screens for AberX-lite.

The design of the crystal goniometer was predicated on the extremely tight requirements on both vacuum and the need for a dust-free environment for the photoinjector and the superconducting cavity. In addition a large diameter, thin silicon crystal was used to eliminate background from the crystal holder due to beam halo. It was mounted with the <100> axis along the beam line. The crystal was aligned by looking at the x-ray signal as a function of the goniometer angles. Typical alignment curves are shown in Fig. 4. The experiment consisted of determining the axial and planar yields as a function of bunch charge and dark current.

Because of the extremely high rate it was not possible to count individual particles or photons. Two different detection approaches were used, both employing CaW screens viewed by either a photomultiplier (AberX lite) or a CCD camera (AberX). The photo-detectors were calibrated by placing them in a variable mono-energetic x-ray beam at the Argonne Advanced Photon Source.

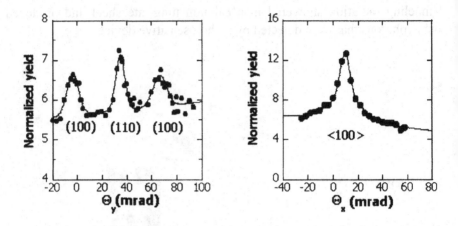

Figure 4: X-ray yields for planar (left panel) and axial (right panel) scans obtained from a Si crystal in the 14.4 MeV electron beam. The quantities Θ_x and Θ_y describe the rotation of the crystal around a horizontal and a vertical axis perpendicular to the beam axis, respectively. The yields were measured using the AberX-Lite detector.

Figure 5 shows the channeling radiation yield as x-rays/bunch in a 10% band for a twelve decade span of bunch charge.

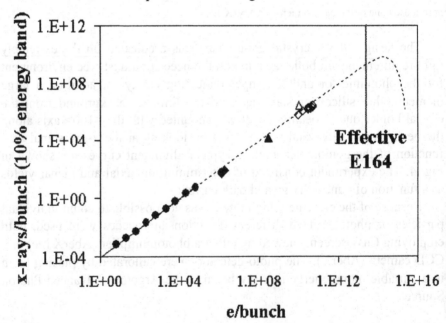

Figure 5: Channeling radiation as a function of bunch charge.

This unit was used to facilitate comparison to the earlier Darmstadt data. The A0 phototube data are represented by open and filled diamonds obtained for axial channeling from Si with the laser on and with dark current, respectively, and for the AberX detector as indicated by the open triangle for planar channeling. The filled circles result from the earlier Darmstadt measurement of axial channeling in a diamond crystal with electrons of 5.4 MeV and the filled triangle from the Stanford measurement of planar channeling in a Si crystal at 30 MeV. All points have been scaled to T = 5.4 MeV. Differences of order two are expected between the various data sets because of different materials, orientations, and crystal thickness. Over the 12 decades of the measurements the data trend for the yield per electron is flat.

6. THE FUTURE OF HIGH BUNCH CHARGE CHANNELING MEASUREMENTS BEYOND A0

Two advances are required to move channeling studies fully into the plasma acceleration regime. One is to increase bunch charge per unit area by a factor on the order of 10^7. Part of this can be accomplished by using more focused beams. The second is to use pulse lengths in the 10 fs regime. This is challenging but there have recently been significant developments in femtosecond laser technology that could help. Higher beam energies might reduce beam size and perhaps also help for channeling studies. However high energy channeling experiments are different and often harder to arrange than the A0 study.

An example of a potential facility for investigating channeling radiation is E164 at the SLAC 30 GeV FFTB facility [21]. This is being used for continuing plasma acceleration studies. By adding a crystal and a high energy gamma ray detector it might be possible to do a channeling radiation study there along the lines of experiments carried out at Serpukhov [22] and CERN [23]. The relative beam charge at SLAC is less than at A0 but the beam cross section is substantially smaller so that the bunch charge per unit area would be 500 times larger. The potential reach of SLAC E164 relative to A0 is shown schematically in Fig. 5 (schematically because the graph is expressed per bunch while the relevant factor is bunch charge per unit area). The 300 femtosecond pulse length at the facility is a step forward but not all the way to the plasma acceleration regime.

A second possibility is to use the 100 TW laser facilities at Livermore to get extremely high beam currents. A 100 fs laser capable of producing a 50 micron spot with a beam power density of $5*10^{18}$ W/cm^2 is used at Livermore to generate protons by a pseudo solid state acceleration process in the first foil. The proton beam produces 4 eV plasma conditions in a second foil. Could one do channeling studies with this geometry by replacing the second aluminum foil with a crystal? One possibility might be to try Rutherford back scattering although it is not obvious how the backscattering detector could be incorporated in the geometry. Another possibility might be a blocking experiment. Lattice behavior with time could be studied using pump and probe and streak camera techniques. "Available" petawatt lasers could get up into the 10^{14} protons/bunch regime.

There have also been recent studies in other fields that suggest different paths to follow for investigations of crystal lattices in dynamic situations. An intriguing example is a study of laser melting at Toronto using sub-picosecond electron diffraction [24]. They use a special 500 fs electron gun to study electron diffraction as a very thin polycrystalline aluminum foil is heated by a laser that is weak by the standards discussed above. The transition from the electronic plasma stage to phonon melting is clearly indicated in the successive Bragg diffraction pictures.

These are three very recent illustrations of potential experimental approaches from a rapidly emerging field studying behavior of solids under dynamic conditions. Many of the possibilities are driven by developments in laser technology. Channeling may have something to offer these studies. Conversely, one might learn interesting things about channeling. And maybe, one may be able to take a step toward solid-state acceleration employing oriented crystals.

7.   ACKNOWLEDGEMENTS

The help of J. Freudenberger, S. Fritzler, H. Genz, A. Richter, and A. Ushakov (Darmstadt), H. Edwards, W. Muranyi, J. Santucci (Fermilab), D. Haeffner, P. Lee, A. Mashayekhi, A. McPherson (Argonne), W. Hartung (Michigan State), R. Noble (Stanford), J. Carneiro (DESY), M. Fitch (Johns Hopkins), and N. Barov (NIU) is gratefully acknowledged.

*Operated by Universities Research Association, Inc. under contract No. DE-AC02-76CHO3000 with the United States Department of Energy

References

[1] T. Tajima and J. M. Dawson, *Phys. Rev. Lett.* **43**, 267 (1979).

[2] E. Esarey, et al., *IEEE Trans. On Plasma Sci.* **24**, 252 (1996).

[3] W. D. Kimura et al., PRL **92**, 054801 (2004).

[4] C. D. Barnes, E. R. Colby, and T. Plettner, p. 294 in *Advanced Accelerator Concepts: Tenth Workshop*, eds. C. E. Clayton and P. Muggli, Amer. Inst. of Physics Press CP647, New York (2002).

[5] M. J. Hogan, et al. *Phys. Plasmas* **7**, 2241 (2000).

[6] N. Barov et al. Particle Accelerator Conference Proceedings (2001), Fermilab-Conf-01/365.

[7] S. Mangles, et al. *Nature* **431**, 535 (2004). C. Geddes, et al. *Nature* **431**, 538 (2004). J. Faure, et al. *Nature* **431**, 541 (2004).

[8] R. Hofstadter, Stanford HEPL Report 560 (1968).

[9] R. A. Snavely et al., *Phys. Rev. Lett.* **85** 2945 (2000).

[10] K. Nemoto, et al., *Applied Physics Lett.* **78**, 595 (2001).

[11] M. Zepf, et al., *Phys. Rev. Lett.* **90**, 064801 (2003).

[12] M. Roth, et. al., *Phys. Rev. Special Topics – Accelerators and Beams* **5**, 061301 (2002).

[13] J. P. Carneiro *et al.*, FERMILAB-Conf-99/271, 1999.

[14] R. A. Carrigan, Jr. et al., *Phys. Rev.* **A68**, 062901 (2003).

[15] P. Chen and R. J. Noble, p. 273 in *Advanced Accelerator Concepts*, eds. S. Chattopadhyay, J. McCullough, and P. Dahl, Amer. Inst. of Physics Press C398, New York (1997).

[16] P. Chen and R. J. Noble, p. 517 in *Relativistic Channeling*, eds. R. A. Carrigan, Jr. and J. A. Ellison (Plenum, 1987).

[17] P. K. Patel, et al., *Phys. Rev. Lett.* **91**, 125004 (2003).

[18] J. U. Andersen, private communication.

[19] H. Genz, H.-D. et al., *App. Phys. Lett.* **57**, 2956 (1990). W. Lotz, et al. *Nucl. Instr. and Meth.* **B48**, 256 (1990).

[20] C. K. Gary, et al. *Nucl. Instr. and Meth.* **B51**, 458 (1990). C. K. Gary, et al., *Phys. Rev.* **B42**, 7 (1990).

[21] C. D. Barnes et al., *Proc. 2003 Particle Acc. Conf.* 1530 (2003).

[22] N. A. Filatova, et al., *Phys. Rev. Lett.* **48**, 488 (1982).

[23] K. Kirsebom, et al., *Nucl. Instr. and Meth.* **119**, 79 (1996).

[24] B. J. Siwick, et al., *Science* **302**, 1382 (2003).

TRANSITION RADIATION BY RELATIVISTIC ELECTRONS IN INHOMOGENEOUS SUBSTANCE

N.F.Shul'ga[1], V.V.Syshchenko[2]

[1]National Scientific CenterKharkov Institute of Physics and Technology, Kharkov, Ukraine;
[2]Belgorod State University, Belgorod, Russia

Abstract: The process of transition radiation of relativistic electrons in non-uniform media is considered. The method of description of this process based on the equivalent photons method and the eikonal approximation of the wave mechanics is proposed. The formulae for the spectral-angular density of the transition radiation that permit to examine the radiation in the case when the dielectric permittivity depends on more than one coordinate are obtained in this approximation. The comparison of the basic results obtained in Born and eikonal approximations of the transition radiation theory is carried out. The ranges of validity of these results are determined. The formulae obtained are applied to the analysis of the transition radiation process on the uniform plate and on the fiber-like target.

Keywords: transition radiation, Born approximation, eikonal approximation, fiber-like target, nanotube

1.   INTRODUCTION

Transition radiation (TR) arises when a charged particle crosses the boundary between two media with different dielectric properties [1-3]. Commonly the description of this process is carried out via "sewing" the fields generated by the particle in substances on their boundary. However, such approach to TR description could be developed only for the media with the simplest shape of the boundary between them (plane, spherical or cylindrical boundary [1-4]). In addition, it is assumed commonly that the dielectric permittivity of each medium is constant. For the cases of

H. Wiedemann (ed.), Advanced Radiation Sources and Applications, 129–148.
© 2006 Springer. Printed in the Netherlands.

complicated boundary configuration and fuzzy boundaries another approaches to the TR process description are to develop.

One of such approaches to description of TR of ultrarelativistic particles in the range of high frequencies is method based on the expansion of the radiation amplitude by small deviation of the dielectric permittivity from the unit, analogous to Born expansion in quantum theory of scattering [2,3]. However, the condition of validity of that expansion rapidly violates with the radiated photon frequency decrease. So, the development of methods that permit to work out of the frames of Born perturbation theory in the problem under consideration is necessary.

In the present paper the possibility of use of the eikonal approximation to description of the process of transition radiation by relativistic electrons in the medium with non-uniform dielectric permittivity is studied. The approximated method of TR description is based on the presentation of the particle's field in the form of the packet of free electromagnetic waves and application of the eikonal approximation for the description of the scattering of this wave packet on non-uniformities of the dielectric permittivity of the medium.

2.   SPECTRAL-ANGULAR DENSITY OF TR

Consider the particle with electric charge e moving with the constant velocity $\vec{v}$ in non-uniform medium with dielectric permittivity $\varepsilon_\omega(\vec{r})$. In this case Fourier component of the electric field

$$\vec{E}_\omega(\vec{r}) = \int_{-\infty}^{\infty} \vec{E}(\vec{r},t)\,e^{i\omega t}\,dt$$

created in the target under passage of the particle satisfies the equations [2]

$$\left(\Delta + \omega^2 \varepsilon_\omega\right)\vec{E}_\omega = grad\,div\,\vec{E}_\omega - 4\pi e i\omega \frac{\vec{v}}{v}\delta(\vec{\rho})e^{i\omega\frac{z}{v}}, \qquad (2.1)$$

$$div\,\varepsilon_\omega \vec{E}_\omega(\vec{r}) = 4\pi e\delta(\vec{\rho})e^{i\omega\frac{z}{v}}. \qquad (2.2)$$

For $\varepsilon_\omega = 1$ the solution of Eqs. (2.1), (2.2) is Coulomb field of the particle

$$\vec{E}_\omega^{(C)}(\vec{r}) = \frac{2e\omega}{v^2\gamma}e^{i\frac{\omega}{v}z}\left\{\frac{\vec{\rho}}{\rho}K_1\left(\frac{\omega\rho}{v\gamma}\right) - i\frac{\vec{v}}{v}\frac{1}{\gamma}K_0\left(\frac{\omega\rho}{v\gamma}\right)\right\} =$$

$$= \frac{ie}{\pi v} e^{i\frac{\omega}{v}z} \int \frac{\vec{\kappa} + \frac{\omega}{v^2\gamma^2}\vec{v}}{\omega^2 - \frac{\omega^2}{v^2} - \vec{\kappa}^2} e^{i\vec{\kappa}\vec{\rho}} d^2\kappa, \quad \vec{\kappa} \perp \vec{v}, \tag{2.3}$$

where $K_n(x)$ is the modified Bessel function of the third kind, $\gamma = (1 - v^2)^{-1/2}$ is Lorentz-factor of the particle. Using (2.2) and (2.3) we can transform Eq. (2.1) to the form

$$\left(\Delta + \omega^2\right)\left(\vec{E}_\omega - \vec{E}_\omega^{(C)}\right) = \omega^2\left(1 - \varepsilon_\omega\right)\vec{E}_\omega + grad\ div\left[\left(1 - \varepsilon_\omega\right)\vec{B}_\omega\right] \tag{2.4}$$

The last equation could be written in the integral form:

$$\vec{E}_\omega - \vec{E}_\omega^{(C)} = \int d^3r'\, G(\vec{r} - \vec{r}') \times$$

$$\times \left\{\omega^2\left(1 - \varepsilon_\omega(\vec{r}')\right)\vec{E}_\omega(\vec{r}') + grad\ div\left[\left(1 - \varepsilon_\omega(\vec{r}')\right)\vec{B}_\omega(\vec{r}')\right]\right\}, \tag{2.5}$$

where $G(\vec{r} - \vec{r}')$ is Green function for Eq. (2.4),

$$G(\vec{r} - \vec{r}') = \int \frac{e^{i\vec{\kappa}(\vec{r} - \vec{r}')}}{\omega^2 - \vec{\kappa}^2 + i0} \frac{d^3\kappa}{(2\pi)^3}, \tag{2.6}$$

where $\vec{k} = \omega\vec{r}/r$. To find the field of radiation, we need to use the asymptotic of (2.6) on large distances from the region where $\varepsilon_\omega(\vec{r})$ is not equal to unit:

$$G(\vec{r} - \vec{r}')\big|_{r\to\infty} \to -\frac{1}{4\pi} \frac{e^{i\omega r}}{r} e^{-i\vec{k}\vec{r}'}. \tag{2.7}$$

Substituting it into (2.5) we obtain the following expression for the radiation field:

$$\vec{E}_\omega^{(rad)}(\vec{r}) = \left(\vec{E}_\omega - \vec{E}_\omega^{(C)}\right)\big|_{r\to\infty} = -\frac{1}{4\pi}\frac{e^{i\omega r}}{r}\left(\omega^2\vec{I} - \vec{k}(\vec{k}\cdot\vec{I})\right) \tag{2.8}$$

where

$$\vec{I} = \int d^3r\, e^{-i\vec{k}\vec{r}}\left(1 - \varepsilon_\omega(\vec{r})\right)\vec{E}_\omega(\vec{r}). \tag{2.9}$$

Computing the flux of Pointing vector into the solid angle element *do* far from the target (see [6], Eq. (66.9)) we find that the radiated wave intensity is equal to

$$\frac{d\mathrm{E}}{d\omega do} = r^2 \frac{1}{4\pi^2} \left| \vec{E}_\omega^{(rad)}(\vec{r}) \right|^2.$$

Substituting (2.7) into the last formula we obtain the following expression for the spectral-angular density of TR:

$$\frac{d\mathrm{E}}{d\omega do} = \frac{\omega^2}{(8\pi^2)^2} \left| \vec{k} \times \vec{I} \right|^2, \tag{2.10}$$

where ω and $\vec{k}$ are the frequency and wave vector of the radiated wave.

Consider TR in the range of high frequencies, where the dielectric permittivity of the target is determined by the formula

$$\varepsilon_\omega(\vec{r}) \approx 1 - \frac{\omega_p^2}{\omega^2}, \quad \omega \gg \omega_p, \tag{2.11}$$

where $\omega_p = \sqrt{4\pi e^2 n(\vec{r})/m}$ is the plasma frequency, m and e are the charge and mass of an electron, $n(\vec{r})$ is the electron density in the target. In this case the solution of Eqs. (2.1), (2.2) could be found as an expansion by the small value $(1 - \varepsilon_\omega)$. In the first order of such expansion (that corresponds to Born approximation) the solution of Eqs. (2.1), (2.2) is Coulomb field of the particle (2.3). So the substitution $\vec{E}_\omega(\vec{r}) = \vec{E}_\omega^{(C)}(\vec{r})$ into (2.9) corresponds to Born approximation in TR theory. It is easy to see that characteristic values of the transverse (perpendicular to $\vec{v}$) component of the Coulomb field of relativistic particle exceed the characteristic values of the longitudinal component in γ times. So, neglecting the terms of the order of γ^{-2}, we can hold on in (2.10) only transverse component of the vector $\vec{I}$ ($\vec{I} \approx \vec{I}_\perp$). In this case

$$\frac{d\mathrm{E}}{d\omega do} = \frac{\omega^4}{(8\pi^2)^2} \left| \vec{I}_\perp^{(B)} \right|^2,$$

$$\vec{I}_\perp^{(B)} = \int d^3 r\, e^{-i\vec{k}\vec{r}} (1 - \varepsilon_\omega(\vec{r})) \vec{E}_\omega^{(C)}(\vec{r})_\perp. \tag{2.12}$$

Keeping in mind the following comparison of the results of Born and eikonal approximations in TR theory and their ranges of validity, consider the simplest problem on TR under normal incidence of the particle to the uniform plate of the thickness a. Born approximation leads us to the following result in this case:

$$\vec{I}_{\perp}^{(B)} = -\frac{4\pi e}{v}\frac{\omega_p^2 / \omega^2}{\frac{\omega}{v}-k_z}\frac{\vec{k}_{\perp}}{k_{\perp}^2 + \left(\frac{\omega}{v\gamma}\right)^2}\left\{\exp\left[i\left(\frac{\omega}{v}-k_z\right)a\right]-1\right\}, \qquad (2.13)$$

$$\frac{dE}{d\omega do} = 2\frac{e^2}{\pi^2}\left(\frac{\omega_p^2}{\omega^2}\right)^2\frac{\theta^2}{\left(\theta^2+\gamma^{-2}\right)^2}\left\{1-\cos\left[\left(\theta^2+\gamma^{-2}\right)\frac{\omega a}{2}\right]\right\}. \qquad (2.14)$$

3.   TR ON A FIBER-LIKE TARGET IN BORN APPROXIMATION

Born approximation in TR theory permits to consider the radiation on the targets of rather complicated configuration. Particularly, TR on fiber-like targets and nanotubes was considered [7,8] using Born approximation. Let us briefly remember that results.

Consider transition radiation of a relativistic particle incident on a thin dielectric fiber at small angle $\psi \ll 1$ with its axis. An atomic string in a crystal or a nanotube could be treated as such fiber.

If the particle interacts with large number of atoms within the length of radiation formation (the coherence length)

$$l_{coh} = \frac{2\gamma^2}{\omega}\frac{1}{1+\gamma^2\theta^2+(\gamma\,\omega_p/\omega)^2}, \qquad \theta \ll 1, \qquad (3.1)$$

where θ is the angle between the wave vector of the radiated wave and the particle velocity, the non-uniformity of the electron density along the fiber axis is not essential for the radiation process. In this case one can use the electron density distribution in the fiber averaged along its axis:

$$n(\vec{\rho}\,') = \frac{1}{L}\int dz'\ n(\vec{r}\,'), \qquad (3.2)$$

where L is the length of the fiber, the z' axis is parallel to the fiber axis, $\vec{\rho}' = (x', y')$ are the coordinates in the transverse plane.

If, in addition, the conditions

$$l_{coh} \gg \frac{2R}{\psi}, \qquad \gamma/\omega \gg R, \qquad (3.3)$$

where R is the transverse size of the fiber, are satisfied then the target can be treated as an uniform infinitely thin fiber. The electron density distribution in this case can be written using delta-function

$$n(\vec{r}') = n_e \delta(x') \delta(y'),$$

where n_e is the electron density per unit length of the fiber.

When the electron is incident under small angle ψ to the fiber axis, it is convenient for calculating the spectral-angular density of radiation to transform the system of coordinates (x', y', z') connected with the fiber to the system of coordinates (x, y, z) in which the z axis is parallel to the particle velocity $\vec{v}$. In the new system of coordinates the electron density distribution can be written in the form

$$n(\vec{r}) = n_e \delta(x - z\psi) \delta(y - y_0) \qquad (3.4)$$

taking into account that the electron moves at distance y_0 to the fiber axis (see Fig.1). The coordinate y here is perpendicular to the fiber axis z' and the particle velocity vector $\vec{v}$.

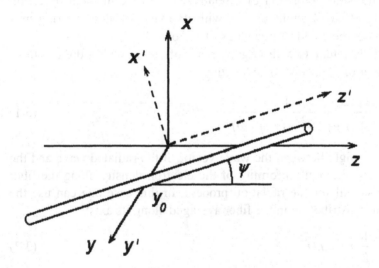

Figure 1. Target position.

Substitution (3.4) into (2.12) gives us the spectral-angular density of TR by the electron incident on the fiber with the given impact parameter y_0. It is convenient to describe the radiation by uniform flux of the particles using the radiation efficiency [6]

$$\frac{dK}{d\omega\,do} = \int dx_0 dy_0 \frac{dE}{d\omega do}, \tag{3.5}$$

where x_0 and y_0 are the coordinates of the incident particles in the plane orthogonal to $\vec{v}$. In the problem under consideration when the beam is incident under small angle to the long fiber the value $dE/d\omega\,do$ do not depend on x_0. In this case the radiation efficiency (3.5) could be written in the form

$$\frac{dK}{d\omega do} = L\psi \int dy_0 \frac{dE}{d\omega do}. \tag{3.6}$$

Here we have used the fact that only the particles with the coordinate x_0 in the frames $\Delta x_0 = L\psi$ participate in the radiation process.

In our case the radiation efficiency could be written in the form

$$\frac{dK}{d\omega do} = \frac{Le^6 n_e^2 \gamma}{m^2 \omega \psi} F(\theta, \varphi), \tag{3.7}$$

where $F(\theta, \varphi)$ is the function that determines the angular distribution of radiation,

$$F(\theta, \varphi) = \frac{1 + 2\left(\gamma\,\theta\cos\varphi - \dfrac{1+\gamma^2\theta^2}{2\gamma\,\psi}\right)^2}{\left[1 + \left(\gamma\,\theta\cos\varphi - \dfrac{1+\gamma^2\theta^2}{2\gamma\,\psi}\right)^2\right]^{3/2}}. \tag{3.8}$$

Here φ is the azimuth angle (the angle between x axis and the projection of wave vector $\vec{k}$ to the plane (x, y)). The surface plot of the function (3.8) for $\psi = 2\cdot10^{-3}$ rad, $\gamma = 2000$ is presented on Fig. 1 (the upper plot). It is easy to see that the angular distribution of radiation intensity possesses the axial symmetry relatively to the fiber axis ($\theta = \psi$, $\varphi = 0$), that can be demonstrated analytically from (3.8). Near the axis of symmetry the intensity of radiation is rather high. For large angles of incidence, $\psi \geq 10\gamma^{-1}$, the angular distribution of intensity takes the shape of a narrow double ring of the radius ψ. Note in connection with this, that the radiation on the fiber can be interpreted as the radiation produced by the perturbation created in the fiber by the relativistically compressed Coulomb field of the incident particle. Such perturbation moves along the fiber with the velocity exceeding the velocity of light. This situation leads to a radiation analogous to Cherenkov one.

The account of the finite value of the fiber radius R leads to the suppression of the radiation under large values of the angle θ. The middle and lower plots on Fig. 2 demonstrate the function F calculated for the case of the fiber with Gaussian distribution of the electron density in the plane orthogonal to the fiber axis,

$$n(\vec{r}) = \frac{n_e}{2\pi R^2} \exp\left[-\frac{(x - z\psi)^2 + (y - y_0)^2}{2R^2}\right], \tag{3.9}$$

for two different values of the mean square radius R.

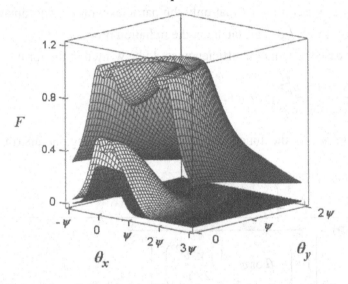

Figure 2. Surface plots of the function $F(\theta, \varphi)$ that determine the angular distribution of transition radiation of relativistic particle on thin fiber-like target for the cases of infinitely thin fiber (Eq. (3.8); upper surface) and the fiber with Gaussian distribution of the electron density (3.9) ($R\omega/\gamma = 0.1\psi\gamma$, middle surface, $R\omega/\gamma = 0.2\psi\gamma$, lower surface); $\psi = 2\cdot 10^{-3}$ rad, $\gamma = 2000$ ($\theta_x = \theta\cos\varphi, \theta_y = \theta\sin\varphi$).

Note that the shape of the angular distribution of the radiation on a fiber-like target depends on the details of the electron density distribution in the fiber. The surface plots of the function F calculated for the cases of Gaussian fiber (3.9), nanotube (in the limitation case of infinitely thin walls of the tube) and uniform cylindrical fiber with the same characteristic radius R are presented on Fig. 3. For the further references, we present here the analytical expression for the function F in the case of uniform cylindrical fiber:

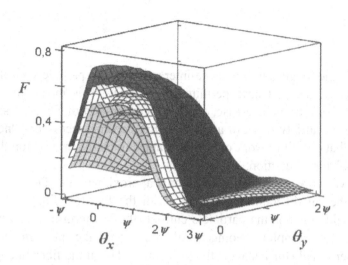

Figure 3. Surface plots of the function $F(\theta,\varphi)$ for the cases of uniform cylindrical fiber (upper surface), fiber with Gaussian distribution of the electron density (lower surface), and nanotube (middle surface) with the same characteristic radius, $R\omega/\gamma = 0.2\psi\gamma$; $\psi = 10^{-3}$ rad, $\gamma = 2000$.

$$F(\theta,\varphi) = \frac{2}{\pi} \int_{-\infty}^{\infty} dq \left(\frac{2J_1\left(\frac{R\omega}{\gamma}\sqrt{q^2 + \left(\frac{1+\gamma^2\theta^2}{2\psi\gamma}\right)^2} \right)}{\frac{R\omega}{\gamma}\sqrt{q^2 + \left(\frac{1+\gamma^2\theta^2}{2\psi\gamma}\right)^2}} \right)^2 \Phi(q,\theta,\varphi), \quad (3.10)$$

where

$$\Phi(q,\theta,\varphi) = \frac{(\gamma\theta\sin\varphi - q)^2 + \left(\gamma\theta\cos\varphi - \frac{1+\gamma^2\theta^2}{2\gamma\psi}\right)^2}{\left[1 + (\gamma\theta\sin\varphi - q)^2 + \left(\gamma\theta\cos\varphi - \frac{1+\gamma^2\theta^2}{2\gamma\psi}\right)^2\right]^2}.$$

4.   EIKONAL APPROXIMATION IN TR THEORY.

As we could see in follows, Born approximation in TR theory is valid under the condition

$$\omega \frac{\omega_p^2}{\omega^2} l \ll 1, \tag{4.1}$$

where l is the length on which the interaction of the particle with the non-uniformity of the dielectric permittivity happens. This inequality violates under l increase or ω decrease, and the consideration of TR process out of the range of validity of Born approximation become necessary. One of the methods that permit to work out of the frames of Born perturbation theory is the eikonal approximation.

However, direct application of eikonal approximation to TR theory leads to a number of difficulties connected with the fact that Eq. (2.1) for the electric field contains the particle's current (the last term in (2.1)) with the fact that for complex geometry of the target the problem becomes multidimensional (for instance, the problem of TR on the fiber-like target is two-dimensional problem).

The attempts to overcome these difficulties based on the construction of Green function for the equation (2.1) have been made [2,9]. The quasi classical Green function for (2.1) was built [2] in the case of one-dimensional problem on TR of relativistic particles in the medium with non-uniform dielectric permittivity. However, that method substantially uses the one-dimensional character of the problem.

The method of construction of Green function for the equation (2.1) with the particle's current that is valid under the number of conditions, which are analogous to the conditions of applicability of the eikonal approximation in quantum mechanics, was proposed [9]. Although that method permits to work out of the frames of Born perturbation theory, its application to particular problems is rather complete and awkward. Only the energy losses of the particle crossing the plate, and in random medium, were calculated on the basis of that method.

In the present article the TR theory based on the method of equivalent photons[10] is developed. In the frames of this theory the transition radiation process is considered as the process of scattering of the particle's electromagnetic field on non-uniformities of the dielectric permittivity of the medium. The field of the particle in this case is presented as the packet of free electromagnetic waves. Such approach permits to consider the equation that determine the evolution of the wave packet in non-uniform medium

$$\left(\Delta + \omega^2 \right) \vec{E}_\omega = \nabla (\nabla \vec{E}_\omega) + \omega^2 \left(1 - \varepsilon_\omega \right) \vec{E}_\omega \tag{4.2}$$

instead of the equation (2.1) with the particle's current. The initial state of the wave packet (before entering the medium) is the expansion of the particle's eigenfield into the set of free electromagnetic waves that permits to build the solution of Eq. (4.2) for multidimensional problem in a simple way. So, let us find the solution of Eq. (4.2) in a form

$$\vec{E}_\omega(\vec{r}) = e^{i\omega z}\vec{\Phi}(\vec{r}). \tag{4.3}$$

Let the function $\vec{\Phi}(\vec{r})$ changes itself in space slow enough to neglect its second derivations in (4.2). In this approximation the equation for $\vec{\Phi}(\vec{r})$ takes the form

$$2i\frac{\partial\vec{\Phi}}{\partial z} = -\omega\Phi_z\vec{e}_z + i\nabla\Phi_z + i\vec{e}_z\,div\,\vec{\Phi} + \omega(1-\varepsilon_\omega)\vec{\Phi}. \tag{4.4}$$

Separating longitudinal and transverse components in (4.4) we obtain

$$\frac{\partial\Phi_x}{\partial x} + \frac{\partial\Phi_y}{\partial y} = -i\omega\,\varepsilon_\omega\Phi_z, \tag{4.5}$$

$$2i\frac{\partial\vec{\Phi}_\perp}{\partial z} = \omega(1-\varepsilon_\omega)\vec{\Phi}_\perp + i\nabla_\perp\Phi_z. \tag{4.6}$$

Substitution Φ_z from (4.5) into (4.6) gives us

$$2i\frac{\partial\vec{\Phi}_\perp}{\partial z} = \omega(1-\varepsilon_\omega)\vec{\Phi}_\perp - \nabla_\perp\frac{1}{\omega\,\varepsilon_\omega}\left(\frac{\partial\Phi_x}{\partial x} + \frac{\partial\Phi_y}{\partial y}\right). \tag{4.7}$$

The second term in the right side of (4.7) could be neglected with the same precision as neglecting of the second derivations of the function $\vec{\Phi}(\vec{r})$ in (4.4) under condition

$$\rho_{eff} \gg \frac{1}{\omega_p}, \tag{4.8}$$

where ρ_{eff} is the characteristic distance in transverse direction on which the value $|\vec{\Phi}_\perp|$ changes itself substantially (it is assumed also that $\varepsilon_\omega(\vec{r})$ changes itself in transverse direction slow enough to neglect the derivation $\nabla_\perp(1/\varepsilon_\omega)$). Under this condition Eq. (4.7) takes the form

$$\frac{\partial\vec{\Phi}_\perp}{\partial z} = -i\frac{\omega}{2}(1-\varepsilon_\omega)\vec{\Phi}_\perp. \tag{4.9}$$

Substituting the solution of the last equation into (4.3) we obtain

$$\vec{E}_{\omega}(\vec{r})_{\perp} = \vec{E}_{\omega}^{(0)}(\vec{r})_{\perp} \exp\left\{-i\frac{\omega}{2}\int_{-\infty}^{z}(1 - \varepsilon_{\omega}(\vec{r}))\,dz\right\}, \tag{4.10}$$

where $\vec{E}_{\omega}^{(0)}(\vec{r})_{\perp}$ is the field of the incident wave packet

$$\vec{E}_{\omega}^{(0)}(\vec{r})_{\perp} = \frac{2e\omega}{v^2\gamma}e^{i\omega z}\frac{\vec{\rho}}{\rho}K_1\left(\frac{\omega\rho}{v\gamma}\right)$$

$$= \frac{ie}{\pi v}e^{i\omega z}\int\frac{\vec{\kappa}}{\omega^2 - \dfrac{\omega^2}{v^2} - \vec{\kappa}^2}e^{i\vec{\kappa}\vec{\rho}}d^2\kappa. \tag{4.11}$$

It is necessary for fulfillment of the condition (4.8) that the characteristic transverse distances on which the function $\varepsilon_{\omega}(\vec{r})$ changes substantially would be not less than the value ρ_{eff}. In other words, our solution is valid only for the target with fuzzy boundaries.

Substituting (4.10) into (2.9) we obtain

$$\vec{I}_{\perp} = \frac{2e\omega}{v^2\gamma}\int d^3r\, e^{i(\omega - k_z)z}e^{-i\vec{k}_{\perp}\vec{\rho}}(1 - \varepsilon_{\omega}(\vec{r}))\frac{\vec{\rho}}{\rho}K_1\left(\frac{\omega\rho}{v\gamma}\right)$$

$$\exp\left\{-i\frac{\omega}{2}\int_{-\infty}^{z}(1 - \varepsilon_{\omega}(\vec{r}))\,dz\right\}. \tag{4.12}$$

The characteristic values ρ_{eff} making the main contribution into the integral (4.12) have the order

$$\rho_{eff} \sim \min(\gamma/\omega, 1/k_{\perp}),$$

where $k_{\perp} \approx \omega\theta$ and θ is the angle on which the radiation is observed ($\theta \ll 1$). So, according to (4.8), Eq. (4.12) is valid in the range of frequencies ω and radiation angles determined by the following inequalities:

$$1 \gg \frac{\omega_p}{\omega} \gg \gamma^{-1}, \qquad \frac{\omega_p}{\omega} \gg \theta. \tag{4.13}$$

The order of value of the argument of the first exponent in (4.12) could be estimated as $\omega\theta^2 l/2$, where l is the target thickness along the direction of the particle velocity. So under the condition

$$\frac{\omega \theta^2 l}{2} \ll 1 \tag{4.14}$$

the first exponent in (4.12) could be replaced by unit. Then after integrating over z we get the following expression for $\vec{I}_\perp$:

$$\vec{I}_\perp^{(eik)} = i \frac{4e}{v^2 \gamma} \int d^2\rho \, e^{-i\vec{k}_\perp \vec{\rho}} \, \frac{\vec{\rho}}{\rho} K_1\left(\frac{\omega \rho}{v\gamma}\right)$$

$$\left\{ \exp\left[-i\frac{\omega}{2} \int_{-\infty}^{\infty} (1 - \varepsilon_\omega(\vec{r})) \, dz \right] - 1 \right\}. \tag{4.15}$$

Substituting it into (2.10) we obtain the spectral-angular density of the transition radiation in the eikonal approximation

$$\frac{dE}{d\omega do} \approx \frac{\omega^4}{(8\pi^2)^2} \left| \vec{I}_\perp^{(eik)} \right|^2. \tag{4.16}$$

This formula is valid in the range of frequencies and radiation angles determined by (4.13) and (4.14).

The argument of the last exponent in (4.12) could be estimated as

$$\frac{\omega}{2} \int_{-\infty}^{\infty} (1 - \varepsilon_\omega(\vec{r})) \, dz \sim \omega \frac{\omega_p^2}{\omega^2} l .$$

If this value is small comparing to unit (that corresponds to (4.1)) then the expansion over the parameter $\omega_p^2 l / \omega$ could be made in (4.15). In the first order of that expansion the value $\vec{I}_\perp^{(eik)}$ coincides with the corresponding result of Born approximation $\vec{I}_\perp^{(B)}$.

The inequality (4.1) violates with l increase and ω decrease. Eq. (4.16) permits to describe TR also out of the range of validity of Born approximation. Indeed, the inequality $\omega_p^2 l / \omega \geq 1$ do not contradict with the conditions (4.13) and (4.14), which determine the conditions of validity of the formula (4.15). Note that (4.13) and (4.14) always could be fulfilled under the particle energy large enough in the range of characteristic transition radiation angles $\theta \sim \gamma^{-1}$.

Eq. (4.15) could be used for consideration of TR on the target of complex configuration, such as dielectric fiber.

5.   TR IN A THIN LAYER OF SUBSTANCE

Consider TR under normal incidence of the ultrarelativistic particle onto thin uniform plate with the thickness a as the simplest example of application of eikonal approximation. In this simplest one-dimensional case the dielectric properties of the target do not depend on ρ, so the condition (4.8) is fulfilled automatically. Computation using (4.15) leads to

$$\vec{I}_{\perp}^{(eik)} = \frac{8\pi e}{v\omega} \frac{\vec{k}_{\perp}}{k_{\perp}^{2} + \left(\dfrac{\omega}{v\gamma}\right)^{2}} \left\{ \exp\left[-i\frac{\omega}{2}\frac{\omega_{p}^{2}}{\omega^{2}}a \right] -1 \right\}, \tag{5.1}$$

and for the spectral-angular density of the radiation on small angles we obtain

$$\frac{dE}{d\omega do} = 2\frac{e^{2}}{\pi^{2}} \frac{\theta^{2}}{\left(\theta^{2} + \gamma^{-2}\right)^{2}} \left\{ 1 - \cos\left[\frac{\omega_{p}^{2}}{\omega^{2}}\frac{\omega a}{2} \right] \right\}. \tag{5.2}$$

Compare the results obtained in Born (2.13), (2.14) and eikonal (5.1), (5.2) approximations with the precise formula for the spectral-angular density of TR on the thin plate. The last one in the range of small radiation angles has the form[1]

$$\frac{dE}{d\omega do} = 2\frac{e^{2}}{\pi^{2}} \left(\frac{\omega_{p}^{2}}{\omega^{2}} \right)^{2} \frac{\theta^{2}}{\left(\theta^{2} + \gamma^{-2}\right)^{2}\left(\theta^{2} + \gamma^{-2} + \dfrac{\omega_{p}^{2}}{\omega^{2}} \right)^{2}}$$

$$\times \left\{ 1 - \cos\left[\left(\theta^{2} + \gamma^{-2} + \frac{\omega_{p}^{2}}{\omega^{2}} \right)\frac{\omega a}{2} \right] \right\}. \tag{5.3}$$

One could obtain this result by substitution into (2.9) the precise expression for the electric field inside the plate, which could be found via "sewing" of the solutions of Eqs. (2.1), (2.2) on the boundaries of the plate. It happens that one could satisfy the boundary conditions only adding to the solutions of Eqs. (2.1), (2.2) with the particle's charge and current the solutions of free equations that correspond to the radiation field. The procedure described leads to the following formula for the total electric field in the medium:

$$\vec{E}_\omega(\vec{r})_\perp = \frac{ie}{\pi v}\int d^2\kappa\, e^{i\vec{\kappa}\vec{\rho}}\,\vec{\kappa}\times \tag{5.4}$$

$$\times\left\{\frac{e^{i\frac{\omega}{v}z}}{\varepsilon_\omega\left(\omega^2\varepsilon_\omega - \frac{\omega^2}{v^2} - \vec{\kappa}^2\right)} + \frac{e^{iz\sqrt{\varepsilon_\omega\omega^2-\vec{\kappa}^2}}}{\omega^2 - \frac{\omega^2}{v^2} - \vec{\kappa}^2} - \frac{e^{iz\sqrt{\varepsilon_\omega\omega^2-\vec{\kappa}^2}}}{\varepsilon_\omega\left(\omega^2\varepsilon_\omega - \frac{\omega^2}{v^2} - \vec{\kappa}^2\right)}\right\}.$$

Substituting this formula into (2.9) we obtain

$$\vec{I}_\perp = \frac{4\pi ie}{v}(1-\varepsilon_\omega)\vec{k}_\perp\int_0^L dz\, e^{-ik_z z}\times \tag{5.5}$$

$$\times\left\{\frac{e^{i\frac{\omega}{v}z}}{\varepsilon_\omega\left(\omega^2\varepsilon_\omega - \frac{\omega^2}{v^2} - \vec{k}_\perp^2\right)} + \frac{e^{iz\sqrt{\varepsilon_\omega\omega^2-\vec{k}_\perp^2}}}{\omega^2 - \frac{\omega^2}{v^2} - \vec{k}_\perp^2} - \frac{e^{iz\sqrt{\varepsilon_\omega\omega^2-\vec{k}_\perp^2}}}{\varepsilon_\omega\left(\omega^2\varepsilon_\omega - \frac{\omega^2}{v^2} - \vec{k}_\perp^2\right)}\right\}.$$

For the dielectric permittivity in the form (2.11) and for small radiation angles Eq. (5.5) takes the form

$$\vec{I}_\perp = \frac{4\pi ie}{v}\frac{\omega_p^2}{\omega^2}\vec{k}_\perp\int_0^L dz\, e^{-ik_z z}\times \tag{5.6}$$

$$\left\{-\frac{e^{i\frac{\omega}{v}z}}{\left(\omega^2\gamma^{-2}+\omega_p^2+\omega^2\theta^2\right)} - \frac{e^{iz\omega\left(1-\frac{\omega_p^2}{2\omega^2}-\frac{\theta^2}{2}\right)}}{\omega^2\gamma^{-2}+\omega^2\theta^2} + \frac{e^{iz\omega\left(1-\frac{\omega_p^2}{2\omega^2}-\frac{\theta^2}{2}\right)}}{\left(\omega^2\gamma^{-2}+\omega_p^2+\omega^2\theta^2\right)}\right\}.$$

Substituting the last result into (2.10) we obtain (5.3).
Under conditions

$$\gamma^2\frac{\omega_p^2}{\omega^2}\ll 1, \tag{5.7}$$

$$\omega\frac{\omega_p^2}{\omega^2}a\ll 1 \tag{5.8}$$

the precise result for the TR intensity (5.3) transforms into (2.14) that corresponds to Born approximation.

Under conditions

$$\left(\theta^2, \gamma^{-2}\right) k < \frac{\omega_p^2}{\omega^2} << 1 \qquad (5.9)$$

the precise result (5.3) transforms into the formula (5.2) corresponding to eikonal approximation.

6.   TR ON THE FIBER-LIKE TARGET IN EIKONAL APPROXIMATION

Consider now TR arising under incidence of fast charged particles on dielectric fiber-like target under small angle ψ to the fiber axis. Let the fiber has the cylindrical shape with radius R and uniform distribution of the electron density, as an example. The effective target thickness along the particle's motion direction in this case is $l \sim 2R/\psi$, so the condition (4.14) takes the form

$$\frac{\omega \theta^2 R}{\psi} << 1. \qquad (6.1)$$

The formula (4.15) in this case gives us

$$\vec{I}_\perp^{(eik)} = \frac{4\pi i e}{v\omega} \int_{-\infty}^{\infty} dy\, e^{-ik_y y} e^{-|y|\sqrt{k_x^2 + (\omega/v\gamma)^2}} \left(\frac{-ik_x \vec{e}_x}{\sqrt{k_x^2 + (\omega/v\gamma)^2}} + \vec{e}_y \operatorname{sgn} y \right) \times$$

$$\times \left\{ \exp\left[-i\frac{\omega}{2}\frac{\omega_p^2}{\omega^2}\frac{2}{\psi}\sqrt{R^2 - (y - y_0)^2} \right] - 1 \right\}, \qquad (6.2)$$

where the z axis is parallel to the particle velocity $\vec{v}$, the fiber axis is parallel to the plane (x, z), y_0 is the impact parameter of the incident particle in relation to the fiber axis. Consider some limitation cases of the formula obtained.

At first, let us find the conditions under which the results of Born and eikonal approximations coincide to each other. As it was mentioned at the end of part 3, the necessary condition of such coincidence is the smallness of the effective target thickness along the particle's motion direction $l \sim 2R/\psi$, that gives the possibility to make an expansion of the last exponent in (6.2) over the parameter

$$\omega \frac{\omega_p^2}{\omega^2} \frac{2R}{\psi} \ll 1. \tag{6.3}$$

In other words, it is necessary to outspread the condition of validity of Born approximation (5.8) to the result obtained in the eikonal approximation.

On the other hand, it is necessary to outspread the condition of smallness of radiation angles (6.1) to the corresponding Born result

$$\vec{I}_\perp^{(B)} = \frac{4\pi e}{v} \frac{\omega_p^2}{\omega^2} \int\limits_{y_0-R}^{y_0+R} dy \, e^{-ik_y y} \, e^{-|y|\sqrt{\frac{\omega^2}{\gamma^2}+\left(k_x-\frac{1}{\psi}\left(\frac{\omega}{v}-k_z\right)\right)^2}} \times$$

$$\times \left(-i\vec{e}_x \frac{k_x - \frac{1}{\psi}\left(\frac{\omega}{v}-k_z\right)}{\sqrt{\frac{\omega^2}{\gamma^2}+\left(k_x-\frac{1}{\psi}\left(\frac{\omega}{v}-k_z\right)\right)^2}} + \vec{e}_y \, \mathrm{sgn}\, y \right) \times \tag{6.4}$$

$$\times \frac{\sin\left(\frac{1}{\psi}\left(\frac{\omega}{v}-k_z\right)\sqrt{R^2-(y-y_0)^2}\right)}{\frac{\omega}{v}-k_z}.$$

It is easy to see that in the first order of the expansion over the small parameters (6.1) and (6.3), under additional condition

$$\psi\gamma \gg 1, \tag{6.5}$$

the expressions for the value $\vec{I}_\perp$ that describe the properties of TR on the fiber in Born and eikonal approximation will coincide:

$$\vec{I}_\perp^{(eik)} \approx \vec{I}_\perp^{(B)} \approx \frac{4\pi e}{v} \frac{\omega_p^2}{\omega^2} \int\limits_{y_0-R}^{y_0+R} dy \, e^{-ik_y y} \, e^{-|y|\sqrt{k_x^2+(\omega/\gamma)^2}} \times \tag{6.6}$$

$$\times \left(\frac{-ik_x \vec{e}_x}{\sqrt{k_x^2 + (\omega/\gamma)^2}} + \vec{e}_y \operatorname{sgn} y \right) \frac{1}{\psi} \sqrt{R^2 - (y - y_0)^2} \,.$$

So, when the conditions (6.1), (6.3) and (6.5) are satisfied, the formulae that describe TR on the fiber in Born approximation are justified not only in the range of frequencies determined by (5.7), but also in more soft range of the radiation spectrum, where the eikonal approximation is applicable (see the condition (5.9)).

Substitution of the value $\vec{I}_\perp$ into (2.10) gives us the spectral-angular density of TR by the electron incident on the fiber with the given impact parameter y_0. After integration over impact parameters we obtain the radiation efficiency in the form (3.7). The surface plot of the function F in the case when the conditions (6.1), (6.3) and (6.5) are satisfied is presented on Fig. 4. The distinct between the results of calculations using the "precise" formula (3.10) and approximated (6.6) is not more than 10 %.

Consider now the limitation case of thick fiber, when its radius R is large comparing to the characteristic transverse size of the Coulomb field of the incident particle $\sim \gamma/\omega$, where the Fourier components of the Coulomb field with the frequency ω are concentrated:

$$R \gg \gamma/\omega \,. \tag{6.7}$$

Note that the characteristic values of y making the main contribution into the integral (6.2) are of the order of value γ/ω whereas the characteristic values of y_0 have the order of value R. In the limitation case of thick fiber (6.7) one could neglect the dependence of the last exponent in (6.2) on y. In this case

$$\vec{I}_\perp^{(eik)} \approx i \frac{4\pi e}{v\omega} \left\{ \exp\left[-i \frac{\omega}{2} \frac{\omega_p^2}{\omega^2} \frac{2}{\psi} \sqrt{R^2 - y_0^2} \right] - 1 \right\} \times$$

$$\times \int_{-\infty}^{\infty} dy \, e^{-ik_y y} e^{-|y|\sqrt{k_x^2 + (\omega/\gamma)^2}} \left(\frac{-ik_x \vec{e}_x}{\sqrt{k_x^2 + (\omega/\gamma)^2}} + \vec{e}_y \operatorname{sgn} y \right) = \tag{6.8}$$

$$= \frac{8\pi e}{v\omega} \frac{\vec{k}_\perp}{k_\perp^2 + (\omega/\gamma)^2} \left\{ \exp\left[-i \frac{\omega}{2} \frac{\omega_p^2}{\omega^2} \frac{2}{\psi} \sqrt{R^2 - y_0^2} \right] - 1 \right\}.$$

This result coincides to the formula (5.1) for TR under normal incidence of the particle onto the plate of the thickness $a = 2\sqrt{R^2 - y_0^2}\,/\psi$ on which the particle effectively interacts with the fiber under the given value of the impact parameter y_0. The analogous result could be obtained also for fibers with inhomogeneous distribution of the electron density in the plane perpendicular to the fiber's axis.

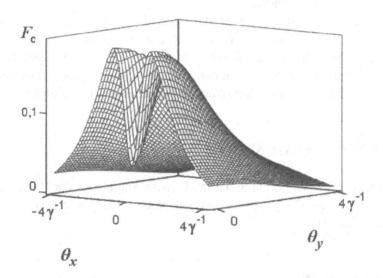

Figure 4. Surface plot of the function $F(\theta,\varphi)$ for the uniform cylindrical fiber in the case when the conditions (6.1), (6.3), (6.5) are satisfied: $\gamma = 2000$, $\psi = 0.1$ rad, $R\omega/\gamma = 0.015\psi\gamma$, $\gamma\,\omega_p/\omega \leq 3$ (for example: $R = 10^{-5}$ cm, $\hbar\omega_p = 20$ eV, $\hbar\omega = 13$ keV).

So, in the limitation case $R \gg \gamma/\omega$ TR on the fiber is equivalent to that on the uniform plate with the thickness and electron density determined by the local effective thickness and average density of the fiber under the given value of y_0.

7.   CONCLUSION

The process of transition radiation of relativistic particles in the medium with non-uniform dielectric permittivity is considered. The approach to description of this process based on the equivalent photons method and eikonal approximation is proposed. The general formulae for TR spectral-angular density for the arbitrary form of the dielectric function of the medium are obtained. That formulae permit to consider TR in the range of small radiation angles in the case when the dielectric permittivity depends on more than one coordinate. It is demonstrated that application of eikonal approximation for that problem makes possible the work out of the frames of Born approximation, which uses the expansion of the radiation fields by the small deviation of the dielectric permittivity from unit. The conditions of validity of Born and eikonal approximations in the problem under consideration are obtained.

TR under the incidence of the particles on the fiber-like target was considered as an example of usage of our formulae in multidimensional problems. Spectral-angular densities calculated for this case using Born and eikonal approximations are compared. The conditions under which the work out of the frames of Born approximation is necessary are obtained.

ACKNOWLEDGEMENTS

This work is supported in part by Russian Foundation for Basic Research (project 03-02-16263) and the internal grant of Belgorod State University (project ВКГ 027-04).

REFERENCES

1. V. L. Ginzburg and V. N. Tsytovich, Transition Radiation and Transition Scattering (Bristol, New York: A. Hilger, 1990).
2. M. L. Ter-Mikaelyan, High-Energy Electromagnetic Processes in Condensed Media (Wiley, New York, 1972).
3. G. M. Garibian, C. Yang, X-ray Transition Radiation (Publ. Acad. of Science of Armenia, Yerevan, 1983 (in Russian)).
4. N. F. Shul'ga and S. N. Dobrovol'skii, JETP 90 (2000) 579.
5. A. I. Akhiezer and N. F. Shul'ga, High-Energy Electrodynamics in Matter (Gordon and Breach, Amsterdam, 1996).
6. L.D. Landau, E.M. Lifshitz, The Classical Theory of Fields (Pergamon, 1972).
7. N. F. Shul'ga and V. V. Syshchenko, Nucl. Instr. and Methods B, 201 (2003) 78.
8. N.F. Shul'ga and V.V. Syshchenko, Phys. Lett. A 313 (2003) 307.
9. A.I. Alikhanyan and V.A. Chechin, Proc. P. N. Lebedev Phys. Inst., 140 (1982) 146 (in Russian).
10. J.D. Jackson, Classical Electrodynamics (John Wiley & Sons, New York, 1998).

IMAGE OF OPTICAL DIFFRACTION RADIATION (ODR) SOURCE AND SPATIAL RESOLUTION OF ODR BEAM PROFILE MONITOR

A.P. POTYLITSYN

Tomsk Polytechnic University, Lenina Ave. 30, 634050 Tomsk, Russian Federation

Abstract: The approach to obtain the image shape of an optical diffraction radiation (ODR) source focused by lens on a detector with taking into account the "pre-wave zone" effect has been developed. The characteristic size of ODR slit image doesn't depend on the Lorentz-factor and is defined by slit width, wavelength lens aperture and is depended on a beam size.

Keywords: Optical diffraction radiation, Pre-wave zone; Beam profile monitor.

1. The authors of [1] have developed an optical transition radiation (OTR) beam profile monitor to measure a transverse size of an electron beam with the resolution $\sigma \sim 2\,\text{mcm}$. The theoretical works [2-4] based on the wave optics have shown that the size of OTR source image (OTR generated by single dimensionless particle) is determined by the amplification of the optical system M, the angular aperture of focusing lens θ_0 and OTR wavelength λ:

$$\sigma \sim \frac{\lambda}{\theta_0} M \qquad (1)$$

For real optical systems ($M = 1$, $\theta_{0\,\text{max}} \sim 0.2$) the minimal resolution σ_{min} runs into the value of a few wavelengths. The similar spatial resolution

H. Wiedemann (ed.), Advanced Radiation Sources and Applications, 149–163.

is sufficient for the diagnostics of the beams with the transverse sizes of the order of 5÷10 mcm.

But as it has been mentioned in[1] the interaction of the beam with small transverse sizes with target may lead to a changing of optical characteristics of OTR target after the passing of a few bunches with the population $\sim 10^{10}\,e^-$ / bunch. That is why the methods of noninvasive diagnostics, where a beam doesn't interact with a target directly are developing rather intensively in recent years [5]. The methods based on the use of optical diffraction radiation (ODR) are the latest ones [6]. The spatial resolution of ODR beam profile monitor at the KEK-ATF extracted electron beam with the Lorentz-factor $\gamma = 2500$ has been investigated in the cited work. The higher the energy of the beam particle is, the bigger the "natural" size of the luminescent area on the target, which is determined by the parameter $\gamma\lambda$. Thus, the distance from the target a, where the similar OTR source with the divergence $\sim \gamma^{-1}$ can be observed as point wise, increases proportionally to the square of the Lorentz-factor [7]:

$$a \gg L_0 = \gamma^2\lambda. \tag{2}$$

In other words, all characteristics of optical devices (including image size) set at the distances less than L_0, (that is, they are situated in pre-wave zone following the terminology [7]) should be calculated with regard to the finite sizes of the luminescent spot.

For example, in case of SLAC FFTB beam with $\gamma = 60000$, the estimate (2) gives $L_0 \sim 1.8$ km for wavelength $\lambda = 0.5$ mcm. So, it is clear, that for a reasonable distance between a target and an optical system ($a \sim 2$ m) the characteristic relationship will be:

$$R = \frac{a}{L_0} = \frac{a}{\gamma^2\lambda} \sim 10^{-3} \ll 1,$$

that is, the monitor is situated in the extremely pre-wave zone.

2. Let us consider the image of OTR source with regard to the "pre-wave zone" effect. For simplification of calculations let us consider backward transition radiation (BTR) for perpendicular particle passing through the target. In real conditions the inclination angle of a target ψ relative to the particle trajectory differs from 90^0, but, as it has been shown in [8,9], the angular characteristics of BTR of ultrarelativistic particles in a wave zone (backward diffraction radiation, BDR, as well) are determined relative to the mirror reflection direction and don't depend on the slope of the target ψ, if the relationship $\psi \gg \gamma^{-1}$ is fulfilled. Following [3], let us write the expression for TR field components generated by ultrarelativistic particle in an infinite perfect target L with the focus distance f on the detector D (the distance between the target and the lens stands a, between the lens and the detector stands b, the standard relationship $\dfrac{1}{a} + \dfrac{1}{b} = \dfrac{1}{f}$ is fulfilled, see Fig.1)

$$E^D_{x,y}(\vec{r}_T) = const \int d\vec{R}_T \int d\vec{R}_L \begin{Bmatrix} \cos\varphi_T \\ \sin\varphi_T \end{Bmatrix} \times \frac{K_1(kR_T/\beta\gamma)}{\beta\gamma} \times$$

$$\times \exp\left[i\frac{k}{2a}R_T^2 \right] \exp\left[i\frac{k}{2a}R_L^2 \right] \exp\left[-i\frac{k}{a}R_L R_T \cos(\varphi_T - \varphi_L) \right] \times \qquad (3)$$

$$\times \exp\left[-i\frac{k}{2f}R_L^2 \right] \exp\left[i\frac{k}{2b}(\vec{R}_L - \vec{R}_D)^2 \right].$$

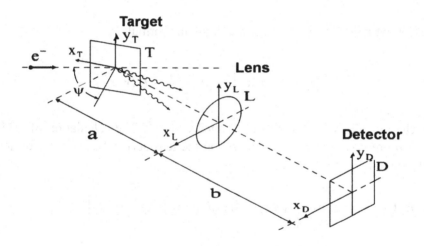

Figure 1. Optical system scheme of OTR beam size monitor.

In (3) the coordinates are indicated by the indices T, L, D in the system on target, lens and detector surfaces subsequently (see Fig.1), K_1 – modified Bessel function, $k = 2\pi/\lambda$ – wave number.

Using the known Bessel's function representation and introducing dimensionless variables

$$r_T = \frac{2\pi R_T}{\gamma\lambda}, \quad r_L = \frac{\gamma}{a}R_L, \quad r_D = \frac{2\pi R_D}{\gamma\lambda} \qquad (4)$$

it is possible to get the expression for the radial component of the transition radiation field for the infinite boundary target:

$$E_D(r_D, R) = const \int r_T dr_T \int r_L dr_L K_1(r_T) J_1(r_T r_L) \times$$

$$\times \exp\left[i\frac{r_T^2}{4\pi R}\right] \times J_1\left(r_L \frac{r_D}{M}\right). \tag{5}$$

Hereafter, we shall make calculations for $M = 1$ (that is, $b = a = 2f$). In a wave zone, where

$$R \gg 1,$$

the exponent can be replaced by unit. In this case the integration over r_T is carried out analytically:

$$E_\infty(r_L) = \int_0^\infty r_T dr_T K_1(r_T) J_1(r_T r_L) = \frac{r_L}{1 + r_L^2}, \tag{6}$$

and for the field (5) we shall get the following formula

$$E_D(r_D) = const \int_0^{r_m} r_L dr_L \frac{r_L}{1 + r_L^2} J_1(r_L r_D),$$

which coincides with the result obtained earlier[3]. The calculation of OTR field in pre-wave zone (for example, on the lens surface) can be performed only numerically:

$$E_L(r_L, R) = const \int_0^\infty r_T d r_T K_1(r_T) J_1(r_T r_L) \exp\left[i\frac{r_T^2}{4\pi R}\right]. \tag{7}$$

For the calculation simplification we shall use approximation

$$r_T K_1(r_T) = (1 + 0.57 r_T - 0.04 r_T^2) e^{-r_T}, \tag{8}$$

which will give an error in some percent in the interval $0 \le r_T \le 5$ (see Fig. 2a).

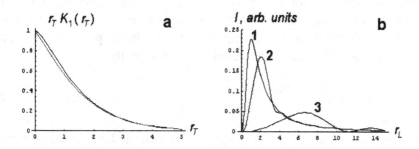

Figure 2. a) Approximation of the dependence $r_T K_1(r_T)$ - upper curve, by the formula (8) – lower curve; b) Radial distribution of OTR intensity on the lens surface for different parameter R values (curve **1** – R =10/2π, **2** – R =1/2π, **3** – R =0.1/2π).

When calculating the integral r_T the integration is carried out in the limits for simplicity $0 \leq r_T \leq 5$.

The intensity dependences $I = |E_L(r_L, R)|^2$ are shown in Fig. 2b for different values R, which are in a good agreement with V. Verzilov's calculations (see Fig. 3 in [7]).

The OTR intensity on the lens $|E_L(r_L, R)|^2$ achieves maximum value at such radius $r_L = r_0$, where the maximal "overlapping" of oscillating subintegral functions $J_1(r_T r_L)$ and $Re[\exp(i r_T^2 / 4\pi R)]$ occurs. As far as the main contribution to the integral (5) will give the range $r_T \leq 1$, this requirement comes to the requirement of the coincidence of the first nulls of the both functions. Thus, we have two relationships

$$r_T r_0 = 3.832,$$

$$\frac{r_T}{4\pi R} = \frac{\pi}{2},$$

(9)

from which we can find the connection between r_0 and R:

$$r_0 = \frac{3.832}{\sqrt{2\pi}\sqrt{R}} = \frac{0.863}{\sqrt{R}}.$$

(10)

Returning to the size variables, we can get:

$$R_L = 0.863\sqrt{a\lambda}.$$

(11)

The approach developed by R.A. Bosch [10,11] gives the closed result for the peak position

$$R_L \approx \sqrt{a\lambda}.$$

The estimate (10) for $R = 1/2\pi$; $0.1/2\pi)$ gives the values $r_0 = 2.16$ and 6.84, which agree reasonable with the exact values $r_0 = 2.05$ and 6.65 (see Fig. 2b). It follows from the first equation (9) that in the pre-wave zone the target range gives the main contribution in OTR intensity

$$R_T \le \sqrt{\frac{a\lambda}{2}} \tag{12}$$

that is, the estimates (11), (12) don't depend on the Lorentz-factor in the case, which is considered.

To obtain the image field (5) on the detector, it is necessary to calculate a double integral. It should be noticed that the "inner" integral in (5) is taken analytically:

$$G(r_T, r_D, r_m) = \int_0^{r_m} r_L d r_L \, J_1(r_T r_L) J_1(r_L r_D) =$$

$$= \frac{r_m}{r_D^2 - r_T^2} \{ r_T J_0(r_m r_T) J_1(r_m r_D) - r_D J_0(r_m r_D) J_1(r_m r_T) \}. \tag{13}$$

Here, the angular lens aperture is defined as r_m (in the angle units γ^{-1}). Thus, the field on the detector is calculated through the single integral:

$$E_D(r_D, R) = \int r_T d r_T \, K_1(r_T) G(r_T, r_D, r_m) \exp\left[i \frac{r_T^2}{4\pi R} \right] =$$

$$\int_0^5 d r_T \left(1 + 0{,}57 r_T - 0{,}04 r_T^2 \right) \exp\left[-r_T + i \frac{r_T^2}{4\pi R} \right] G(r_T, r_D, r_m) \tag{14}$$

In Fig. 3a normalized shapes of OTR images on the detector $I = |E_D(r_L, R)|^2$ are shown for $r_m = 50$, 100 in wave zone ($R = 1$). For the Lorentz-factor $\gamma = 1000$ the indicated lens aperture values correspond to the angles $\theta_m = 0.05$ and $\theta_m = 0.1$, for which the quantity $|E_D|^2$ has been calculated in [3].

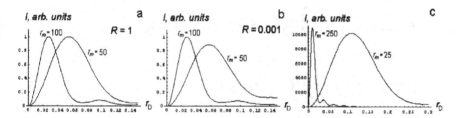

Figure 3. a) Normalized shape of OTR source image on the detector plane for lens aperture r_m =100 (left curve) and r_m = 50 (right curve) in wave zone (R = 1); b) OTR source image in «pre-wave zone» (R = 0.001) for the same conditions and with the same normalizing factors; c) OTR source image at the fixed lens diameter for different distances a between the target and the lens (R = 0.01, r_m = 250 - right curve; R = 0.1, r_m = 25 - left).

The results obtained by different methods coincide with a good accuracy. Thus, for example, for aperture θ_m = 0.1 rad the maximum in the radial distribution corresponds to the radius ρ_L^{max} = 4.4λ (see [3]). As it follows from Fig. 3a (left curve) the nondimensional variable value r_D= 0.0275 corresponds to the maximum for r_m = 100, that for γ = 1000 corresponds to the radius R^{max} = 1000/2π × 0.0275 λ = 4.38 λ. At aperture decreasing twice, the radius r_D^{max} increases in two times.

The images in a pre-wave zone (R = 0.001) calculated by formula (14) are shown in Fig. 3b. It can be noticed that the normalized distribution shape on the detector depends on the parameter R very slightly if the lens aperture θ_m remains constant.

In a real case, at the change of the distance a between the lens and the target for the same parameters of the task, not only the parameter R changes, but lens aperture r_m at its fixed diameter changes also.

OTR spot images for R = 0.1, r_m =25 and R = 0.01, r_m= 250 are shown in Fig. 3c) that corresponds to the fixed lens diameter at the distance change in 10 times. The right curve is multiplied by 100 times for the sake of convenience. As it follows from the figure, the spot size (monitor resolution) is decreased in 10 times at the lens approach to the target (the angular aperture is increased in the same number of times).

3. The approach developed in the previous paragraph gives the opportunity to obtain the "image" of a round hole with the radius ρ_0, in the center of which the charged particle flies and generates the optical diffraction radiation (ODR) (see [12]). ODR field on the detector is calculated by the formula, which is analogous to (14):

$$E_D^{DR}(r_D,R,\rho_0) = \int\limits_{\rho_0}^{5} dr_T \left(1 + 0.57r_T - 0.04r_T^2\right) \times$$

$$\times exp\left[-r_T + i\frac{r_T^2}{4\pi R}\right] G(r_T,r_D,r_m)$$

(15)

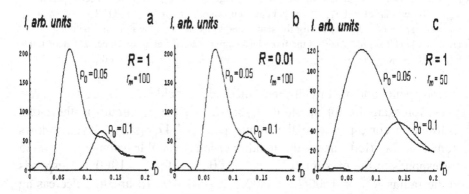

Figure 4. a) The image of the round hole in wave zone ($\rho_0 = 0.05$ – left curve; $\rho_0 = 0.1$ - right) for aperture $r_m =100$, $R = 1$; b) The image of the round hole in "pre-wave zone" ($\rho_0 = 0.05$ – left curve; $\rho_0 = 0.1$ - right) for aperture $r_m = 100$ and $R = 0.01$; c) The hole image ($\rho_0 =0.05$, $\rho_0 = 0.1$) in wave zone ($R=1$) with a poor resolution ($r_m = 50$).

Using field (15) the similar "images" of the ODR source (round hole) for $\rho_0 = 0.05$ and $\rho_0 = 0.1$ in a wave zone ($R =1$) were obtained (see Fig. 4a). If the spatial resolution of OTR (in other words, the image size of ODR from the hole with the infinite small radius) is determined by the quantity λ/θ_0 (or in our variables, quantity $\Delta r_D \sim 0.0275$ for $r_m = 100$), it should be expected, that at $\rho_0 < \Delta r_D$ the ODR source image will coincide practically with OTR one, but for $\rho_0 > \Delta r_D$ one may obtain the hole image on the detector. As it follows from the figure 4a), the intensity of radiation on the detector at $r_D < \rho_0$ is practically lacking and achieves the maximum at $r_D^m \approx r_T + \Delta r_D^{DR}$, with the maximum width, which is determined by the quantity $\Delta r_D^{DR} \approx 0.04$ in both cases.

The analogous dependences for the pre-wave zone are shown in Fig. 4b). It can be noted that for ODR the pre-wave zone effect reveals slightly in the image plane. But the lens aperture, as in OTR case, influences the width of maximal near hole edge image (see Fig. 4c). At aperture decreasing two times (up to $r_m = 50$) the width of the distributions increases more than two times in comparison with $r_m = 100$ (see Fig. 4a,b). In this case the hole image with $\rho_0 = 0.05$ is too smoothed.

Using the formulas (13), (15) we can get ODR characteristics in a pre-wave zone for the disk of finite sizes having substituted the upper limit in integral (15) for the outer disk radius ρ_{max}.

In wave zone the use of the formula (5) is the more simple method. We can write instead of the formula (6)

$$E^{DR}(r_L) = \int_{\rho_0}^{\rho_{max}} r_T dr_T \, K_1(r_T) J_1(r_T r_L) =$$

$$= -\frac{r_T}{1+r_L^2}[K_0(r_T)J_1(r_T r_L) + r_L K_1(r_T)J_0(r_T r_L)] \Big|_{\rho_0}^{\rho_{max}}. \tag{16}$$

For the hole radius $\rho_0 \ll 1$ with a good accuracy $\rho_0 K_0(\rho_0) \approx 0$, $\rho_0 K_1(\rho_0) \approx 1$, and, consequently, at $\rho_{max} \to \infty$

$$E_\infty^{DR}(r_L) \approx \frac{r_L}{1+r_L^2} J_0(\rho_0 r_L). \tag{17}$$

The formula (17) is in a good agreement with the result obtained earlier[13].

But at the violation of axial symmetry of the task (for example, for a particle flying with a finite offset relative the center of the hole), the formulas obtained above don't work.

4. ODR beam size monitor [6] is based on the measuring of the degree of the deformation of ODR angular distribution from the slit in a wave zone generated by electron beam with finite transverse size, in comparison with the distribution from the infinite narrow beam.

To estimate whether it is possible to obtain the information about transverse sizes of the beam passing through the slit using the ODR slit image at the focusing on the detector, is of interest. In this case it is necessary to use Cartesian coordinates in the initial formulas.

As before, we shall introduce Cartesian coordinates on the target, lens and detector using T, L, D indices. Dimensionless variables will be introduced analogously (4):

$$\begin{Bmatrix} x_T \\ y_T \end{Bmatrix} = \frac{2\pi}{\gamma\lambda} \begin{Bmatrix} X_T \\ Y_T \end{Bmatrix}, \begin{Bmatrix} x_L \\ y_L \end{Bmatrix} = \frac{y}{a} \begin{Bmatrix} X_L \\ Y_L \end{Bmatrix}, \begin{Bmatrix} x_D \\ y_D \end{Bmatrix} = \frac{2\pi}{\gamma\lambda} \begin{Bmatrix} X_D \\ Y_D \end{Bmatrix}. \tag{18}$$

Thus, instead of (3) we shall have:

$$\left.\begin{matrix} E_x^D(x_D, y_D) \\ E_y^D(x_D, y_D) \end{matrix}\right\} = const \int dx_T dy_T \int dx_L dy_L \times$$

$$\times \begin{Bmatrix} x_T \\ y_T \end{Bmatrix} \frac{K_1\left(\sqrt{x_T^2 + y_T^2}\right)}{\sqrt{x_T^2 + y_T^2}} \exp\left[i(x_T x_L + y_T x_L)\right] \times \tag{19}$$

$$\times \exp\left[-i\left(x_L \frac{x_D}{M} + y_L \frac{y_D}{M}\right)\right].$$

In (19), as before, $R = a/\gamma^2 \lambda$; M – magnification. For the rectangular lens:

$$-x_m \le x_L \le x_m, -y_m \le y_L \le y_m \tag{20}$$

this integral is calculated very simply. In this case the "inner" double integral is taken analytically again:

$$\int_{-x_m}^{x_m} dx_L \int_{-y_m}^{y_m} dy_L \exp\left[-ix_L\left(x_T + \frac{x_D}{M}\right)\right] \exp\left[-iy_D\left(y_T + \frac{y_D}{M}\right)\right] =$$

$$= 4 \frac{\sin\left[x_m\left(x_T + \frac{x_D}{M}\right)\right]}{x_T + \frac{x_D}{M}} \frac{\sin\left[y_m\left(y_T + \frac{y_D}{M}\right)\right]}{y_T + \frac{y_D}{M}} = \tag{21}$$

$$= G_x(x_T, x_D, x_m) G(y_T, y_D, y_m).$$

Thereafter we shall consider the case $M = 1$ again. Thus the expression (19) is reduced to the double integral over the target surface:

$$\left.\begin{matrix} E_x^D(x_D, y_D) \\ E_y^D(x_D, y_D) \end{matrix}\right\} = const \int dx_T dy_T \begin{Bmatrix} x_T \\ y_T \end{Bmatrix} \frac{K_1\left(\sqrt{x_T^2 + y_T^2}\right)}{\sqrt{x_T^2 + y_T^2}} \times$$

$$\times \exp\left[i\frac{x_T^2 + y_T^2}{4\pi R}\right] G_x(x_T, x_D, x_m) \cdot G_y(y_T, y_D, y_m). \tag{22}$$

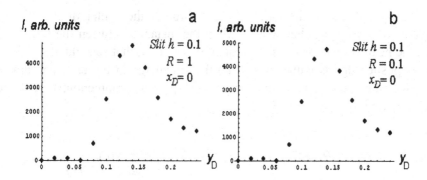

Figure 5. a) The ODR slit image for slit width $h = 0.1$ in wave zone for aperture $x_m = y_m = 50$ ($y_D = 0, R = 1$); b) The same image for pre-wave zone ($R = 1$).

5. At particle flight in the center of the slit of width $2h$, directed along the axis x, ODR field is calculated by formula (22) at the integration over y_T in the limits $\{-5, -h\}$, $\{h, 5\}$. In Fig. 5a) the slit image (y_d - distribution) in wave zone ($R = 1$) for the aperture $x_m = y_m = 50$ is shown. The slit half-width $h = 0.1$ has been chosen rather small.

In Fig. 5b) the analogous distribution, but in the pre-wave zone ($R = 0.1$) is shown. It can be noted that both distributions coincide with the accuracy less than a few percent in a full analogy with a round hole. The maximum width achieves the quantity $\Delta y_D \approx 0.07$ in both cases. The distributions along the slit edge on the detector (the x_D - distribution) calculated for the maximum at $y_D = 0.14$ are given in Fig. 6 a, b) for $R = 1$ and 0.1 accordingly. In comparison with the intensity distribution "across" the slit, the distributions "along" the slit are too blurred.

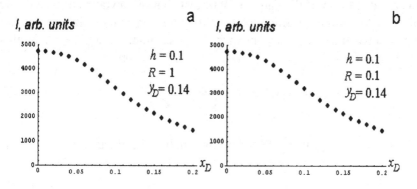

Figure 6. a) The intensity distribution of the slit image in wave zone along the slit edge for $y_D = 0.14$; b) The same intensity distribution for pre-wave zone ($R = 0.1$).

Fig. 5, 6 illustrate the particle flight through the center of the slit. At asymmetric case (in other words, when the distance between the trajectory and the nearest edge of the slit is equal to $h - d$, in that case the distance to the opposite edge is equal to $h + d$) the slit image in the used coordinate system (the axis z coincides with the particle momentum) will be asymmetric too.

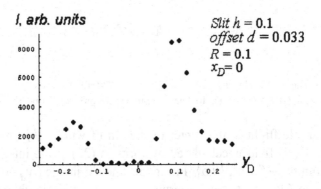

Figure 7. The image of the "narrow" slit ($h = 0.1$) for particle flight with offset $d = 0.033$, (R = 0.1).

In Fig. 7 the slit image for asymmetric flight ($h = 0.1$; $d = 0.033$) is shown. Different distances to the slit edges (0.067 and 0.133) are fixed on the image clearly.

If the beam flying through the slit has a finite size the shape of the slit image will be distorted. It should be expected that the distortion degree of a perfect distribution "carries" the information about beam size in the same way as the distortion of the angular ODR distribution in wave zone gives the opportunity to get the information about the transverse beam size. It is rather simple to investigate this dependence at approximation $x_m \to \infty$. In this case the function $G_x(x_T, x_D, x_m)$ turns into δ-function:

$$G_{x_L}(x_T, x_D, \infty) = \delta\left(x_T + \frac{x_D}{M}\right)$$ (23)

which takes one integration. As before, we shall make calculations for $M = 1$. Thus,

$$\begin{Bmatrix} E^D_{x_L}(x_D,y_D) \\ E^D_{y_L}(x_D,y_2) \end{Bmatrix} = const \int dy_T \begin{Bmatrix} -x_D \\ y_T \end{Bmatrix} \frac{K_1\left(\sqrt{x_D^2+y_T^2}\right)}{\sqrt{x_D^2+y_T^2}} \times$$

$$\times exp\left[i\frac{x_D^2+y_D^2}{4\pi R}\right] G_{y_L}\left(y_D,y_T,y_m\right).$$

$$(24)$$

For the sake of simplicity we shall consider the slit image for a "rectangular" beam, described by the distribution (y_b is a coordinate of a beam particle along Y_T - axes).

$$F(y_b) = \begin{cases} \dfrac{1}{2d} \ , & -d \leq y_b \leq d \\ 0 & y_b < -d, \ \ y_b > d \end{cases}$$

In this case the averaging on the beam size may be performed in the simplest way.

Let's consider the beam size effect on the ODR slit image for the SLAC FFTB beam with transverse size ~ 10 mcm. In this extremely relativistic case for the wavelength $\lambda = 0.5$ mcm the pre-wave zone parameter R is equal to 0.001 and slit half width $h = 0.003$ looks as desirable (in usual units slit width $2h$ is equal to 28 mcm).

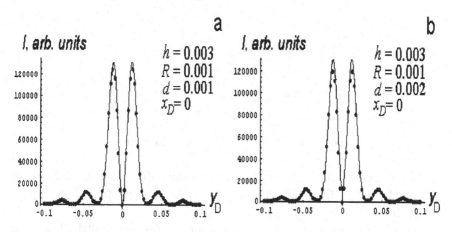

Figure 8. a) The narrow slit image in the extremely pre-wave zone ($R = 0.001$; $h = 0.003$) for the finite size beam – $0.001 \leq y_b \leq 0.001$ (line). Points – dimensionless beam; b). The same for the beam with sizes $-0.002 < y_b < 0.002$.

In Fig. 8a the calculated slit image with the beam size $d = 0.001$ (line) in comparison with the perfect image (points) is shown. Results presented on

Fig. 8b obtained for two times wider beam ($d = 0.002$). Calculations were performed for lens aperture $y_m = 200$.

It should be noticed that the maximum positions don't depend on the beam size, but the radiation intensity ratio between the maximal and minimal values depends on the slit "filling" by the beam. At the increasing of the beam size the particle beam contribution with impact-parameter less than h rises faster for DR from the nearest edge, than the intensity of the radiation from the farthest is decreased. As it follows from Fig. 8a, 8b the ratio

$$\eta = \frac{I_{min}}{I_{max}},$$

where $I_{min(max)}$ – the intensity of the radiation in the minimum (maximum) is determined by the beam size d and increases with beam size growth (see Fig. 9). As it follows from the figure, this dependence may be measured experimentally, and it can give the possibility to get the information about beam size.

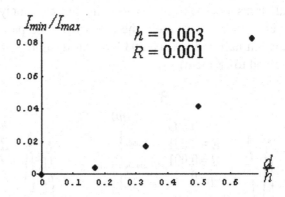

Figure 9. The dependence of η ratio on the beam size.

6. The considered geometry for $\gamma = 60\,000$ and $\lambda = 0.5$ mcm corresponds to the slit width 28 mcm and the beam size $2d = 9$ mcm (Fig. 8a) and $2d = 18$ mcm (Fig. 8b).

Extremely pre-wave zone has been considered for $R = 0.001$, that is for the distance between the target and the lens $L = 1.8$ m. To achieve the magnification ~ 10 (to obtain the slit image on the detector about 280 mcm of size that can be easily measured by CCD with pixel size ~ 20 mcm) it is necessary to use the short focus lens placed at a short distance.

In conclusion, we can note the following:

i) the image shape of ODR source on an image plane of an optical system is defined by lens aperture and practically independent on a distance between target and lens (in other words one may neglect the pre wave zone effect for large lens aperture).

ii) in the considered scheme of the ODR beam size monitor the slit width h is determined by the chosen wave length λ and the lens aperture θ_0, but doesn't depend on the Lorentz-factor: $h \geq \lambda / \theta_0$;

iii) the sensitivity of a proposed method is increasing with growth of the slit filling by the beam (with growth of the ratio d / h);

iv) the additional focusing of the image along the slit edge increases the image brightness (it gives the possibility to decrease the requirements on CCD sensitivity) at insignificant loss of information.

References

1. M. Ross, S. Anderson, J. Frisch, K. Jobe, D. McCormick, B. McKee, J. Nelson, T. Smith, H. Hayano, T. Naito, N. Terunuma, A very high resolution optical transition radiation beam profile monitor, SLAC-PUB-9280, July 2002.
2. V.A. Lebedev. Diffraction-limited resolution of the optical transition radiation monitor. *Nucl. Instrum. and Meth. A* **372**, 344-348 (1996).
3. M. Castellano, V. Verzilov. Spatial resolution in optical transition radiation beam diagnostics. *Phys. Rev. ST-AB* **1**, 062801 (1998).
4. X.Artru, R. Chehab, K. Honkavaara, A. Variola. *Nucl. Instrum. and Meth. B* **145**, 160-168 (1998).
5. M. Ross, Review of diagnostics for next generation linear accelerator. SLAC-PUB-8826, May 2001.
6. P.V. Karataev, Investigation of optical diffraction radiation for non-invasive low-emittance beam size diagnostics, in PhD Thesis, Tokyo Metropolitan University, 2004.
7. V.A. Verzilov. Transition radiation in the pre-wave zone. *Phys. Letters A* **273**, 135-140 (2003).
8. A..P. Potylitsyn. Transition radiation and diffraction radiation. Similarities and differences. *Nucl. Instrum. and Meth. B* **145**, 169-179 (1998).
9. N. Potylitsyna-Kube, X. Artru. Diffraction radiation from ultrarelativistic particles passing through a slit. Determination of the electron beam divergence. *Nucl. Instrum. and Meth. B* **201**, 172-183 (2003).
10. R.A. Bosch. Focusing of infrared edge and synchrotron radiation. *Nucl. Instrum. and Meth. A* **431**, 320-333 (1999).
11. R.A. Bosch. *Phys. Rev.* ST-AB, **5**, 020701 (2002).
12. B.M. Bolotovskii, E.A. Galst'yan. Diffraction and diffraction radiation. *Usp. Fiz. Nauk.* **43** (8), 809 (2000).
13. A.P. Potylitsyn. Scattering of coherent diffraction radiation by a short electron bunch. *Nucl. Instrum. and Meth. A* **455**, 213-216 (2000).

SPONTANEOUS AND STIMULATED PHOTON EMISSION IN CRYSTALINE UNDULATORS

A. V. Korol[1]*, A. V. Solov'yov[2,4†], W. Greiner[3,4]

[1]*Department of Physics, St.Petersburg State Maritime Technical University, Leninskii prospect 101, St. Petersburg 198262, Russia*

[2]*A.F.Ioffe Physical-Technical Institute of the Academy of Sciences of Russia, Polytechnich-eskaya 26, St. Petersburg 194021, Russia*

[3]*Institut fur Theoretische Physik der Johann Wolfgang Goethe-Universitat, 60054 Frankfurt am Main, Germany*

[4]*Frankfurt Institute for Advanced Studies der Johann Goethe-Universitat, 60054 Frankfurt am Main, Germany*

Abstract   The electromagnetic radiation generated by ultra-relativistic positrons channeling in a crystalline undulator is discussed. The crystalline undulator is a crystal whose planes are bent periodically with the amplitude much larger than the inter-planar spacing. Various conditions and criteria to be fulfilled for the crystalline undulator operation are established. Different methods of the crystal bending are described. We present the results of numeric calculations of spectral distributions of the spontaneous radiation emitted in the crystalline undulator and discuss the possibility to create the stimulated emission in such a system in analogy with the free electron laser. A careful literature survey covering the formulation of all essential ideas in this field is given. Our investigation shows that the proposed mechanism provides an efficient source for high energy photons, which is worth to study experimentally.

1.   Introduction

We discuss a new mechanism of generation of high energy photons by means of a planar channeling of ultra-relativistic positrons through a periodically bent crystal. The *feasibility* of this scheme was explicitly demonstrated in [1, 2]. In these papers as well as in our subsequent publications [3]-[13] the idea of this new type of radiation, all essential conditions and limitations which

*E-mail: korol@rpro.ioffe.rssi.ru
†E-mail: solovyov@fias.uni-frankfurt.de

H. Wiedemann (ed.), Advanced Radiation Sources and Applications, 165–189.
© 2006 *Springer. Printed in the Netherlands.*

must be fulfilled to make possible the observation of the effect and a crystalline undulator operation were formulated in a complete and adequate form for the first time. A number of corresponding novel numerical results were presented to illustrate the developed theory, including, in particular, the calculation of the spectral and angular characteristics of the new type of radiation.

The aim of this paper is to review briefly the results obtained so far in this newly arisen field as well as to carry out a historical survey of the development of all principal ideas and related phenomena. The larger and much more detailed version of this review, currently being available as an electronic preprint [14], will be soon published in International Journal of Modern Physics. The necessity of such a review is motivated by the fact that the importance of the ideas suggested and discussed in [1]-[13] has been also realized by other authors resulting in a significant increase of the number of publications in the field within the last 3-5 years [15]-[26] but, unfortunately, often without proper citation [19]-[26]. In this paper and in [14] we review all publications known to us which are relevant to the subject of our research.

The main phenomenon to be discussed in this review is the radiation formed in a crystalline undulator. The term 'crystalline undulator' stands for a system which consists of two essential parts: (a) a crystal, whose crystallographic planes are bent periodically, and (b) a bunch of ultra-relativistic positively charged particles undergoing planar channeling in the crystal. In such a system there appears, in addition to a well-known channeling radiation, the radiation of an undulator type which is due to the periodic motion of channeling particles which follow the bending of the crystallographic planes. The intensity and the characteristic frequencies of this undulator radiation can be easily varied by changing the energy of beam particles and the parameters of crystal bending.

The mechanism of the photon emission by means of a crystalline undulator is illustrated by figure 1. The short comments presented below aim to focus on the principal features of the scheme. At the moment we do not elaborate all the important details but do this instead in section 2.

The (yz)-plane in the figure is a cross section of an initially linear crystal, and the z-axis represents the cross section of a midplane of two neighbouring non-deformed crystallographic planes (not drawn in the figure) spaced by the interplanar distance d. Two sets of black circles denote the nuclei which belong to the periodically bent neighbouring planes which form a periodically bent channel. The amplitude of the bending, a, is defined as a maximum displacement of the deformed midplane (thick solid line) from the z-axis. The quantity λ stands for a spatial period of the bending. In principle, it is possible to consider various shapes, $y(z)$, of the periodically bent midplane. In this review we will mainly discuss the harmonic form of this function, $y(z) = a\sin(2\pi z/\lambda)$. For further referencing let us stress here that we will mainly consider the case when the quantities d, a and λ satisfy the strong

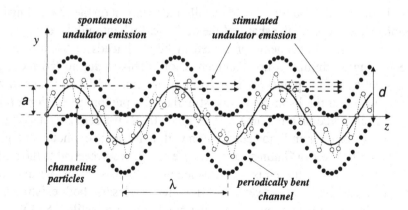

Figure 1. Schematic representation of spontaneous and stimulated radiation in a periodically bent crystal. The y- and z-scales are incompatible!

double inequality: $d \ll a \ll \lambda$. Typically $d \sim 10^{-8}$ cm, $a \sim 10 \dots 10^2\, d$, and $a \sim 10^{-5} \dots 10^{-4}\, \lambda$.

The open circles in figure 1 denote the channeled ultra-relativistic particles. Provided certain conditions are met, the particles, injected into the crystal, will undergo channeling in the periodically bent channel. Thus, the trajectory of a particle (represented schematically by the dashed line) contains two elements. Firstly, there are oscillations inside the channel due to the action of the interplanar potential, - the channeling oscillations. This mode is characterized by a frequency Ω_{ch} dependent on the projectile type, energy, and the parameters of the interplanar potential. Secondly, there are oscillations because of the periodicity of the distorted midplane, - the undulator oscillations, whose frequency is $\omega_0 \approx 2\pi c/\lambda$ (c is the velocity of light which approximately is the velocity of an ultra-relativistic particle).

The spontaneous emission of photons which appears in this system is associated with both of these oscillations. The typical frequency of the emission due to the channeling oscillations is $\omega_{\mathrm{ch}} \approx 2\gamma^2\Omega_{\mathrm{ch}}$ where γ is the relativistic Lorenz factor $\gamma = \varepsilon/mc^2$. The undulator oscillations give rise to photons with frequency

$$\omega \approx 4\gamma^2\omega_0/(2+p^2), \tag{1}$$

where the quantity p, a so-called undulator parameter, is:

$$p = 2\pi\gamma\frac{a}{\lambda}. \tag{2}$$

If the strong inequality $\omega_0 \ll \Omega_{\mathrm{ch}}$ is met, then the frequencies of the channeling radiation and the undulator radiation are also well separated, $\omega \ll \omega_{\mathrm{ch}}$. In this case the characteristics of the undulator radiation (the intensity and spectral-angular distribution) are practically independent on the channeling oscillations but depend on the shape of the periodically bent midplane.

For $\omega_0 \ll \Omega_{ch}$ the scheme presented in figure 1 leads to the possibility of generating a stimulated undulator emission. This is due to the fact that photons emitted at the points of the maximum curvature of the midplane travel almost parallel to the beam and thus, stimulate the photon generation in the vicinity of all successive maxima and minima. In [2] we demonstrated that it is feasible to consider emission stimulation within the range of photon energies $\hbar\omega = 10 \dots 10^4$ keV (a Gamma-laser). These energies correspond to the range $10^{-8} \dots 10^{-4}$ μm of the emission wavelength, which is far below than the operating wavelengths in conventional free-electron lasers (both existing and proposed) based on the action of a magnetic field on a projectile [28–30].

However, there are essential features which distinguish a seemingly simple scheme presented in figure 1 from a conventional undulator. In the latter the beam of particles and the photon flux move in vacuum whereas in the proposed scheme they propagate through a crystalline media. The interaction of both beams with the crystal constituents makes the problem much more complicated from theoretical, experimental and technical viewpoints. Taking into consideration a number of side effects which accompany the beams dynamics, it is not at all evident *a priori* that the effect will not be smeared out. Therefore, to prove that the crystalline undulator as well as the radiation formed in it are both feasible it is necessary to analyze the influence, in most cases destructive, of various related phenomena. Only on the basis of such an analysis one can formulate the conditions which must be met, and define the ranges of parameters (which include the bunch energy, the types of projectiles, the amplitude and the period of bendings, the crystal length, the photon energy) within which all the criteria are fulfilled. In full this accurate analysis was carried out very recently and the feasibility of the crystalline undulator and the Gamma-laser based on it were demonstrated in an adequate form for the first time in [1, 2] and in [3]-[14].

From the viewpoint of this compulsory programme, which had to be done in order to draw a conclusion that the scheme in figure 1 can be transformed from the stage of a purely academic idea up to an observable effect and an operating device, we have reviewed critically in [14] some of the recent publications as well as the much earlier ones [27, 31–37]. A brief summary of a part of this critical review is presented in section 2.

2.   Feasibility of a Crystalline Undulator

In this paper we do not pretend to cover the whole range of problems concerning the channeling effect, the channeling and the undulator radiation, and refer to a much more detailed discussion of all these issues in our recent review

[14]. This section is devoted solely to a brief description of the the conditions which must be fulfilled in order to treat a crystalline undulator as a feasible scheme for devising, on its basis, new sources of electromagnetic radiation. These conditions are:

$$
\left\{
\begin{array}{ll}
C = (2\pi)^2 \dfrac{\varepsilon}{qU'_{\max}} \dfrac{a}{\lambda^2} \ll 1 & \text{stable channeling} \\[2mm]
d \ll a \ll \lambda & \text{large-amplitude regime} \\[2mm]
N = \dfrac{L}{\lambda} \gg 1 & \text{large number of undulator periods} \\[2mm]
L < \min\left[L_d(C), L_a(\omega)\right] & \text{account for the dechanneling} \\
& \quad \text{and photon attenuation} \\[2mm]
\dfrac{\Delta\varepsilon}{\varepsilon} \ll 1 & \text{low radiative losses}
\end{array}
\right. \tag{3}
$$

As a supplement to this system one must account for the formulae (1) and (2), which define the undulator parameter and the frequencies of the undulator radiation.

Provided all conditions (3) are met for a positively charged particle channeling through a periodically bent crystal then

- within the length L the particle experiences stable planar channeling between two adjacent crystallographic planes,

- the characteristic frequencies of the undulator radiation and the ordinary channeling radiation are well separated,

- the intensity of the undulator radiation is essentially higher than that of the ordinary channeling radiation,

- the emission spectrum is stable towards the radiative losses.

For each type of the projectile and its energy, for a given crystal and crystallographic plane the analysis of the system (3) is to be carried out in order to establish the ranges of a, λ and ω within which the operation of the crystalline undulator is possible.

Most of these important conditions were realized and carefully investigated for the first time in [1]-[11], where the realistic numerical calculations of the characteristics of the radiation formed in crystalline undulator were performed as well. We consider the set of analytical and numerical results obtained by us in the cited papers as a proof of the statement that the scheme illustrated in figure 1 can be transformed from a purely academic idea to an observable effect and an operating device.

For positron channeling, in particular, we found the optimal regime in which the spontaneous undulator radiation is most stable and intensive, and demonstrated that this regime is realistic [2, 4, 5]. This regime is characterized by the following ranges of the parameters: $\gamma = (1\ldots10) \times 10^3$, $a/d = 10\ldots50$, $C = 0.01\ldots0.2$, which are common for all the crystals which we have investigated. These ranges ensure that the energy of the first harmonic ω_1 (see (1)) lies within the interval $50\ldots150$ keV and the length of the undulator can be taken equal to the dechanneling length $L_d(C)$.

The importance of exactly this regime of operation of the crystalline undulator was later realized by other authors. In particular, in recent publications by Bellucci et al.[23–25], where the first practical realization of the crystalline undulator was reported, the parameters chosen for a Si crystal were as follows: $\varepsilon = 0.5\ldots0.8$ GeV for a positron (i.e. $\gamma = (1\ldots1.6) \times 10^3$), $a = 20\ldots150$ Å(i.e. $a/d = 10\ldots80$), $L = L_{\rm d}$. These are exactly the values for which we predicted the strong undulator effect. However, in these papers, where the authors mention all the conditions (3) and stress their importance, there is no proper reference to our works. Instead, our paper [2], labeled as Ref. [10] in [23], was cited as follows: 'With a strong world-wide attention to novel sources of radiation, there has been broad theoretical interest [4-12] in compact crystalline undulators...' (page 034801-1 in the cited paper). This was the only referencing to the paper [2], in which we clearly formulated, for the first time, the conditions (3) and carried out a detailed analysis aimed to prove why this regime is most realistic. None of it was done in the papers [27, 31–33, 37] (labeled in [23] as Refs. [4],[6],[7],[8] and [9], correspondingly). Moreover, we state that one will fail to construct a crystalline undulator based on the estimates presented in [27, 31–33, 37]. In what follows we carry out critical analysis of the statements and the estimates made in the cited papers.

Historically, the paper by Kaplin et al.[27] was the first where the idea of a crystalline undulator based on the action of the transverse acoustic wave was presented. However, a number of ambiguous or erroneous statements makes it impossible to accept the thesis that the concept of a crystalline undulator was correctly described in this two-page paper. To be precise in our critique below we use the exact citations taken from the English edition of [27]. In the citations the italicizing is done by us.

Our first remark concerns the type of projectile which the authors propose to use in the undulator. The first paragraph of the paper reads:
'Radiation by relativistic *electrons and positrons*, which occurs during channeling in single crystals, has been observed experimentally and is being extensively studied at the present time[1-4].'

This is the *only* place in the text where the term 'positron' is used. In the rest of the paper the projectile is called either 'a particle', or a 'relativistic electron' as in the one before last paragraph of the paper (p. 651). Thus, it is absolutely unclear to the reader, which particle is to be used; for a positron it is possible to construct an undulator, however if an electron is considered, then the rest of the paper does not make any sense.

The concept of a periodically bent crystal and its parameters is formulated as follows (p. 650, right column): 'Still higher intensity can be achieved by using instead of a uniformly curved crystal one deformed in such a way that the radiation from different portions of the particle trajectory adds coherently. This can be accomplished by giving a crystalline plate a wavelike shape in such a way that the sagitta A satisfies the relation $4A\gamma/\lambda_0 < 1$ in relation to the quarter period λ_0 of the bending. *For large values of the dechanneling depth L_0 this will provide a high radiated power from the crystalline undulator (wiggler).* For rather thin crystalline plates with a simple bend one can produce $\lambda_0 \sim 4$ mm ... *We recall that the channeling depth in centimeters is approximately $L_0 = E$ (GeV), as follows from experiments.*'

Note, that no citation is made when referring to the experiments which result in 'L_0 (cm) = E (GeV)'. For a positron (see [5] and [13]) this relation overestimates the dechanneling length by more than an order of magnitude, for an electron it is even farther from the reality. Therefore, the idea to construct a positron-based undulator with the period $\lambda = 4\lambda_0 = 1.6$ cm is absolutely unrealistic.

The parameters of the undulator based on the action of the acoustic wave are presented in the left column on p. 651: 'To obtain radiation in the optical region in a transparent crystal or to generate very hard γ rays, it has been proposed to use ultrasonic vibrations to deform the crystal lattice... For example, one can obtain γ rays with the energy up to $\omega = 0.14 - 14$ MeV for $\varepsilon = 1$ GeV and $\lambda_0 = 10 - 0.1$ μm.'

Note, that none of the following characteristics, - the type of the projectile, the crystal, the acoustic wave amplitude (in our notations 'sagitta A' is called 'amplitude a'), are specified. *Assuming* that the positron channeling is implied, let us analyze the above-mentioned values from the viewpoint of the condition for a stable channeling (the first equation in (3)). The parameter C can be written in the form: $C \approx 40/\lambda^2 \, (\varepsilon \, d/q U'_{max}) \, (a/d)$ with λ in μm, ε in GeV, d in Å, and $q U'_{max}$ in GeV/cm. Let us estimate the ratio a/d for the range $\lambda = 4\lambda_0 = 0.4 - 40$ μm and for (110) planes in Si and W, for which $d_{Si} = 1.92$ Å, $d_W = 2.24$ A, $(q U'_{max})_{Si} = 6.9$ Gev/cm, $(q U'_{max})_W = 57$ Gev/cm [38]. For $\varepsilon = 1$ GeV and the lowest λ-value one gets $C \approx 250(\varepsilon \, d/q U'_{max}) \, (a/d)$, which means that, for both crystals, to satisfy the condition $C \ll 1$ it is necessary to consider $a \ll d$. Thus, this is a low-amplitude regime, for which the intensity of the undulator radiation is negligibly small. The upper limit of λ is more realistic to ensure the condition $C \ll 1$ for the amplitudes $a \gg d$. However, this analysis is not performed by the authors.

Our final remark concerns the statement (the last paragraph in the left column on p. 651): 'A lattice can be deformed elastically up to $A = 1000$ Å...' This is true, but when referring to the crystalline undulator with the amplitude $a = 10^{-5}$ cm one has to supply the reader (and a potential experimentalist) with the estimates of the corresponding values of λ and N. Let us carry out these estimates (note, this was *not* done in the paper). The channeling condition (see (3)) can be written as follows: $\lambda = \lambda_{min}/\sqrt{C} > \lambda_{min}$, where $\lambda_{\min}$ is the absolute minimum of λ (for given a, ε and a crystal) which corresponds to $C = 1$ (i.e. to the case when the dechanneling length $L_d(C)$ effectively equals to zero, see [2, 4, 5, 14]. It is equal to $\lambda_{\min} = 2\pi\sqrt{a}\,(\varepsilon/qU'_{\max})^{1/2}$. For a 1 GeV positron channeling in Si and W crystals along the (110) plane, which is bent periodically with $a = 10^{-5}$ cm, the values of $\lambda_{\min}$ are: 7.6×10^{-3} cm for Si and 7.5×10^{-3} cm for W. These values already exceed the upper limit of 40 μm mentioned by Kaplin *et al.*. Choosing the length of the crystal to be equal to the dechanneling length and using equation

$$L_d(C) = (1 - C)^2 L_d(0) \tag{4}$$

where $L_d(0)$ is the dechanneling length in a straight crystal, to estimate $L_d(C)$ one estimates the number of undulator periods $N = L_d(C)/\lambda = C^{1/2}(1 - C)^2 L_d(0)/\lambda_{\min}$.

For a positively charged projectile a good estimate for $L_d(0)$ is [5, 39]:

$$L_d(0) = \gamma\,\frac{256}{9\pi^2}\,\frac{M}{Z}\,\frac{a_{\mathrm{TF}}}{r_0}\,\frac{d}{\Lambda} \tag{5}$$

where $r_0 = 2.8 \times 10^{-13}$ cm is the electron classical radius, Z, M are the charge and the mass of a projectile measured in units of elementary charge and electron mass, and a_{TF} is the Thomas-Fermi atomic radius. The quantity Λ stands for a 'Coulomb logarithm' which characterizes the ionization losses of an ultra-relativistic particle in amorphous media (see e.g. [40, 41]):

$$\Lambda = \begin{cases} \ln\dfrac{2\varepsilon}{I} - 1 & \text{for a heavy projectile} \\[2ex] \ln\dfrac{\sqrt{2}\varepsilon}{\gamma^{1/2}I} - 23/24 & \text{for } e^+ \end{cases} \tag{6}$$

with I being the mean ionization potential of an atom in the crystal.

The largest value of N is achieved when $C = 0.2$, giving $C^{1/2}(1 - C)^2 \approx 0.29$. Hence, $N \le N_{\max} = 0.29 L_d(0)/\lambda_{\min}$. Using formulae (5) and (6) one calculates the dechanneling lengths in straight crystals: $L_d(0) = 6.8 \times 10^{-2}$ cm $L_d(0) = 3.9 \times 10^{-3}$ cm for W. Finally, one derives that the 'undulator' suggested in the cited paper contains $N \le 2.6$ periods in the case of Si, and $N \le 1.5$ for a tungsten crystal.

Thus, because of the inconsistent and ambiguous character of the paper [27] we cannot agree that the feasibility of a crystalline undulator was demonstrated in this paper in a manner, sufficient to stimulate the experimental study of the phenomenon.

None of the essential conditions, summarized in (3), were analyzed in [27]. For the first time such an analysis was carried out in [1, 2] and developed further in our subsequent publications. In this connection we express disagreement with utterly negligent and unbalanced style of citation adopted by Avakian *et al.*in [19] and other publications [20–22] by this group, and by Bellucci *et al.*[23–25].

Much of our criticism expressed above in connection with [27] refers also to the paper by Baryshevsky *et al.*[31]. Our main point is: the concept of the crystalline undulator based on the action of an acoustic wave was not convincingly presented. From the text of the paper it is not at all clear what channeling regime, axial or planar, should be used. The only reference to the regime is made in last part of the paper (on p. 63), which is devoted to the quantum description of the spectral distribution of the undulator radiation. This part starts with the sentence: 'Let us consider, for example, planar channeling'. The question on whether the axial channeling is also suitable for a crystalline undulator is left unanswered by the authors. Neither it is clearly stated what type of projectile is considered. Indeed, in all parts of the paper, where the formalism is presented, the projectile is called as a 'particle'. The reference to a positron is made in the introductory paragraph, where the effect of channeling radiation is mentioned, and on p. 62, where the numerical estimates of the intensity of the undulator radiation are presented. The limitations due to the dechanneling effect are not discussed. As a consequence, the regime, for which the estimates are made, can hardly be called an undulator one. Indeed, on p. 62 the ratio of the undulator to the channeling radiation intensities is estimated for a 1 GeV positron channeled in Si (*presumably*, the planar channeling is implied). The amplitude of the acoustic wave (labeled $r_{0\perp}^s$) is chosen to be equal to 10^{-5} cm. The period λ is not explicitly given by the authors. However, they indicate the frequency of the acoustic wave, $f = 10^7$ s^{-1}. Hence, the reader can deduce that $\lambda = v/f = 4.65 \times 10^{-2}$ cm, if taking the value $v = 4.65 \times 10^5$ cm/s for the sound velocity in Si [42]. The values of ε, a and λ, together with the maximal gradient of the interplanar field $(qU'_{max})_{Si} = 6.9$ Gev/cm [38], allows one to calculate $C = 2.65 \times 10^{-2}$ (see (3)), and, consequently, to estimate the dechanneling length $L_d(C) = 6.47 \times 10^{-2}$ cm. As a result, we find that the number of the undulator periods in the suggested system is $N = 1.4$, which is not at all $N \gg 1$ as it is implied by the authors (this is explicitly accented by them in the remark in the line just below their Eq. (2) on p. 62). Another point of criticism is that the classical formalism, used to derive the Eq. (2), is applicable only for the dipole case, i.e. when the undulator parameter is small, $p^2 \ll 1$. However, the estimates which are made refer to a strongly non-dipolar regime: $p^2 = (2\pi\gamma a/\lambda)^2 = 7.3$. As a consequence, the estimate of the energy of the largest emitted harmonic, carried out by the authors on p. 62, is totally wrong. Exactly in their regime the harmonics with a low number will never emerge from the crystal due to photon attenuation.

Papers [32, 33, 37] considered *only* the case of small amplitudes, $a \ll d$, when discussing the channeling phenomenon in periodically bent crystalline structures. As a result, in [32, 37] the attention was paid not to the undulator radiation (the intensity of which is negligibly small in the low-amplitude regime, see section 3), but to the influence of the periodicity of the channel bending on the spectrum of the channeling radiation. Similar studies were carried out in [34–36, 43, 44]. These effects are irrelevant from the viewpoint of the crystalline undulator problem discussed here. Another issue, which we want to point out, is that the authors of [32, 37] did not distinguish between the cases of an electron and a positron channeling. The limitations due to the dechanneling effect were not discussed. In [33] the idea of using a superlattice (or a crystal bent by means of a low-amplitude acoustic wave) as an undulator for a free electron laser was explored. The main focus was made on the regime when the undulator radiation is strongly coupled with the ordinary channeling radiation. This regime is different from the subject of the present discussion. The essential role of the large-amplitude regime of the crystalline undulator was not demonstrated in these papers.

3.   Crystalline Undulator Radiation

To illustrate the crystalline undulator radiation phenomenon, let us consider the spectra of spontaneous radiation emitted during the passage of positrons through periodically bent crystals. The results presented below clearly demonstrate the validity of the statements made in [1–5] and summarized in section 2 above, that the properties of the undulator radiation can be investigated separately from the ordinary channeling radiation.

The calculated spectra of the radiation emitted in the forward direction (with respect to the z-axis, see figure 1) in the case of $\varepsilon = 0.5$ GeV planar channeling in Si along (110) crystallographic planes and for the photon energies from 45 keV to 1.5 MeV are presented in figures 2 [4]. The ratio a/d was varied within the interval $a/d = 0 \ldots 10$ (the interplanar spacing is 1.92 Å). The case $a/d = 0$ corresponds to the straight channel. The period λ used for these calculations equals to 2.33×10^{-3} cm. The number of undulator periods and crystal length were fixed at $N = 15$ and $L = N\lambda = 3.5 \times 10^{-2}$ cm. These data are in accordance with the values allowed by (3).

The spectra correspond to the total radiation, which accounts for the two mechanisms, undulator and channeling. They were calculated using the quasi-classical method [38, 45]. Briefly, to evaluate the spectral distribution the following procedure was adopted (for more details see [4, 8, 11]). First, for each a/d value the spectrum was calculated for individual trajectories of the parti-

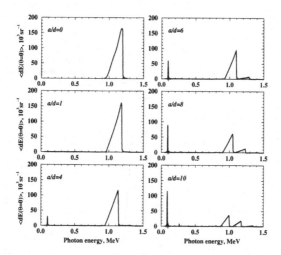

Figure 2. Spectral distribution of the total radiation emitted in the forward direction ($\vartheta = 0°$) for the $\varepsilon = 0.5$ GeV ($\gamma \approx 10^3$) positron channeling in Si along the (110) crystallographic planes calculated at different a/d ratios. Other parameters are given in the text. The crystal length is $L = 3.5 \times 10^{-2}$ cm.

cles. These were obtained by solving the relativistic equations of motion with both the interplanar and the centrifugal potentials taken into account. We considered two frequently used [46] analytic forms for the continuum interplanar potential, the harmonic and the Moliere potentials calculated at the temperature $T = 150$ K to account for the thermal vibrations of the lattice atoms. The resulting radiation spectra were obtained by averaging over all trajectories. Figures 2 correspond to the spectra obtained by using the Moliere approximation for interplanar potential.

The first graph in figure 2 corresponds to the case of zero amplitude of the bending (the ratio $a/d = 0$) and, hence, presents the spectral dependence of the ordinary channeling radiation only. The asymmetric shape of the calculated channeling radiation peak, which is due to the strong anharmonic character of the Moliere potential, bears close resemblance with the experimentally measured spectra [47]. The spectrum starts at $\hbar\omega \approx 960$ keV, reaches its maximum value at 1190 keV, and steeply cuts off at 1200 keV. This peak corresponds to the radiation in the first harmonic of the ordinary channeling radiation (see e.g. [48]), and there is almost no radiation in higher harmonics.

Increasing the a/d ratio leads to the modifications in the radiation spectrum. These are: (i) the lowering of the channeling radiation peak, (ii) the gradual increase of the intensity of undulator radiation due to the crystal bending.

The decrease in the intensity of the channeling radiation is related to the fact that the increase of a leads to lowering of the allowed maximum value of the channeling oscillations amplitude a_c (this is measured with respect to the

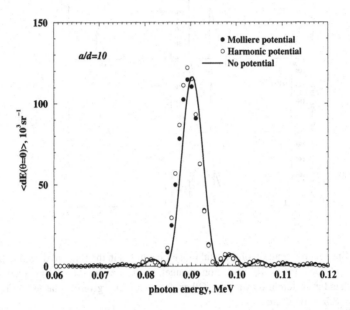

Figure 3. Comparison of different approximations for the interplanar potentials used to calculate the total radiative spectrum in vicinity of the first harmonic of the undulator radiation. The ratio $a/d = 10$, other parameters as in figure 2.

centerline of the bent channel) [3, 39]. Hence, the more the channel is bent, the lower the allowed values of a_c are, and, consequently, the less intensive the channeling radiation is, being proportional to a_c^2 [38].

The undulator radiation related to the motion of the particle along the centerline of the periodically bent channel is absent in the case of the straight channel (the graph $a/d = 0$), and is almost invisible for comparatively small amplitudes (see the graph for $a/d = 1$). Its intensity, which is proportional to $(a/d)^2$, gradually increases with the amplitude a. For large a values ($a/d \sim 10$) the intensity of the first harmonic of the undulator radiation becomes larger than that of the channeling radiation. The undulator peak is located at much lower energies, $\hbar\omega^{(1)} \approx 90$ keV, and has the width $\hbar\Delta\omega \approx 6$ keV which is almost 40 times less than the width of the peak of the channeling radiation.

It is important to note that the position of sharp undulator radiation peaks, their narrow widths, and the radiated intensity are, practically, insensitive to the choice of the approximation used to describe the interplanar potential. In addition, provided the first two conditions from (3) are fulfilled, these peaks are well separated from the peaks of the channeling radiation. Therefore, if one is only interested in the spectral distribution of the undulator radiation, one may disregard the channeling oscillations and to assume that the projectile moves along the centerline of the bent channel [1, 2]. This statement is illus-

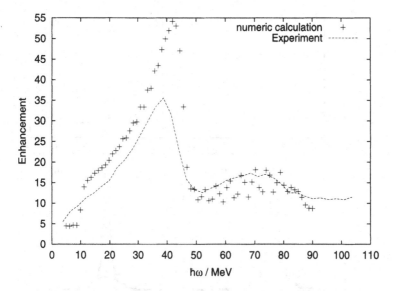

Figure 4. Comparison of the experimentally measured spectrum [49, 47] and the results of the calculation [8, 11] for 6.7 GeV positrons in Si(110).

trated by 3 [4] where we compare the results of different calculations of the radiative spectrum in vicinity of the first harmonic of the undulator radiation in the case $a/d = 10$. All parameters are the same as in figure 2. The filled and open circles represent the results of evaluation of the total spectrum of radiation accompanied by numerical solution of the equations of motion for the projectile within the Moliere (filled circles) and the harmonic (open circles) approximations for the interplanar potential. The solid line corresponds to the undulator radiation only. For the calculation of the latter it was assumed that the trajectory of a positron, $y(z) = a \sin(2\pi\lambda/z)$, coincides with the centerline of the bent channel. It is clearly seen that the more sophisticated treatment has almost no effect on the profile of the peak obtained by means of simple formulae describing purely undulator radiation [1, 2]. Moreover, the minor changes in the position and the height of the peak can be easily accounted for by introducing the effective undulator parameter and (in the case of the harmonic approximation) the effective undulator amplitude [3].

To check the numerical method, which was developed in [4] for the calculation of the total emission spectrum of ultra-relativistic positrons in a crystalline undulator, we calculated the spectrum of the channeling radiation for 6.7 GeV positrons in Si(110) integrated over the emission angles. Figure 4 shows the experimental data [47, 49] and the results of our calculations [8, 11] normalized to the experimental data at the right wing of the spectrum. The height of the first harmonic is overestimated in our calculations. The calculations per-

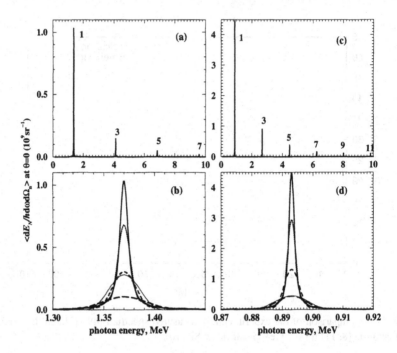

Figure 5. Spectral distribution (in 10^9 sr^{-1}) of the undulator radiation at $\vartheta = 0$ for 5 GeV positron channeling along periodically bent (110) planes in Si ((a) and (b)) and W ((c) and (d)) crystals. The a/d ratio is equal to 10. Other parameters used are presented in table 1. The upper figures (a) and (c) reproduce $\langle dE_N/\hbar\, d\omega\, d\Omega_n \rangle$ in the wide ranges of ω and correspond to $N = 4\,N_d$. The numbers enumerate the harmonics (in the case of the forward emission the radiation occurs only in odd harmonics). The profiles of the first harmonic peak (figures (b) and (d)) are plotted for $N = 4\,N_d$ (solid lines), $N = 2\,N_d$ (dotted lines), $N = N_d$ (dashed lines), $N = N_d/2$ (long-dashed lines).

Table 1. The values of the parameter C, undulator period λ, the dechanneling length $L_d(C)$, the number of undulator periods $N_d = L_d(C)/\lambda$ within $L_d(C)$, the undulator parameter p, and the fundamental harmonic energy used for the calculation of the spectra presented in figures 5.

Crystal	C	λ	$L_d(C)$	N_d	p	$\hbar\omega_1$
		μm	cm			MeV
Si	0.15	63.0	0.321	51	1.87	1.37
W	0.05	42.2	0.637	151	3.26	0.89

formed in [49] gave a similar result. This disagreement arises likely due to the neglect of multiple collisions which were accounted for neither in [8, 11] nor in [49]. However, the shape and the location of the first harmonic of the channeling radiation are described quite well.

The intensity and the profile of the peaks of the undulator radiation are defined, to a great extent, by the magnitude of the dechanneling length. In [5] a more sophisticated, than in [1, 2], theoretical and numerical analysis of this influence was presented. In particular, we solved the following problems: (a) a simple analytic expression was evaluated for the spectral-angular distribution of the undulator radiation which contains, as a parameter, the dechanneling length L_d, (b) the simulation procedure of the dechanneling process of a positron in periodically bent crystals was presented, (c) the dechanneling lengths were calculated for 5 GeV positrons channeling in Si, Ge and W crystals along the periodically bent crystallographic planes, (d) the spectral-angular and spectral distributions of the undulator radiation formed in crystalline undulator were calculated in a broad range of the photon energies and for various a, λ and C.

To illustrate the results obtained in [5], in figures 5(a)-(d) we present the spectral distribution of the undulator radiation emitted along the undulator axis, $\hbar^{-1}\langle dE_N/d\omega\, d\Omega_\mathbf{n}\rangle_{\vartheta=0^\circ}$, for 5 GeV positron channeling along (110) planes in Si and W crystals. The spectra correspond to the ratio a/d, where $d = 1.92$ Å for Si and $d = 2.45$ Å for W. The values of other parameters, used in the calculations, are given in table 1. The values of the dechanneling lengths, $L_d(C)$, were obtained in [5] by means of the simulation procedure of the dechanneling process of a positron in periodically bent crystals.

The upper figures, 5(a) and (c), illustrate the spectral distributions in Si and W over a wide range of emitted photon energy, and corresponds to the crystal length, L, exceeding the dechanneling length by a factor of 4: $L = 4L_d(C)$. Each peak corresponds to emission in the odd harmonics, the energies of which follow from the relation $\omega_k = k\,\omega_1$, $k = 1, 3, \ldots$. The difference in the magnitudes of the undulator parameters for Si and W (see table 1) explains a number of the harmonics visible in the spectra. It is seen that all harmonics are well separated: the distance $2\hbar\omega_1$ between two neighbouring peaks is 2.74 MeV for Si and 1.78 MeV in the case of W, whilst the width of each peak $\hbar\Delta\omega$ is ≈ 8.7 keV for Si and ≈ 2.5 keV for W.

Figures 5(b) and (d) exhibit, in more detail, the structure of the first harmonic peaks. For the sake of comparison we plotted the curves corresponding to different values of the undulator periods. It is seen that for $N > N_d$ the intensity of the peaks is no longer proportional to N^2, as it is in the case of the ideal undulator without the dechanneling of the particles [50]. For both Si and W crystals, the intensities of the radiation calculated at $N \longrightarrow \infty$ exceed those at $N = 4N_d$ (the thick full curves in the figures) only by several per cent. Thus, the full curves correspond to almost saturated intensities which are the maximal ones for the crystals used, projectile energies and the parameters of the crystalline undulator. For a more detailed discussion see paper [5].

4.   Stimulated Emission from a Crystalline Undulator

As demonstrated in [1, 2], the scheme illustrated by figure 1 allows one to consider a possibility to generate stimulated emission of high energy photons by means of a bunch of ultra-relativistic positrons moving in a periodically bent channel. The photons, emitted in the forward direction ($\vartheta = 0$) at the points of the maximum curvature of the bent channel, travel parallel to the beam and, thus, stimulate the photon generation in the vicinity of all successive maxima and minima. This mechanism of the radiation stimulation is similar to that known for a free-electron laser (see, e.g. [51]), in which the periodicity of a trajectory of an ultra-relativistic projectile is achieved by applying a spatially periodic magnetic field. Also from the theory of FEL it is known [52], that the stimulation occurs at the frequencies of the harmonics of the spontaneous emission, $\omega_k = k\omega_1$, $k = 1, 2, \ldots$. The frequency of fundamental harmonic, ω_1, is defined in (1). In [1, 2] and, also, in a more recent paper [13] it was shown, that it is possible to separate the stimulated photon emission in the crystalline undulator from the ordinary channeling radiation in the regime of large bending amplitudes $a \gg d$. This scheme of the stimulated photon emission allows to generate high energy photons up to MeV region and, thus, we call it a gamma laser. As a further step in developing the ideas proposed in these papers, a study, carried out in [19], was devoted to the investigation of the influence of the beam energy spread on the characteristics of the stimulated emission in crystalline undulators.

For low amplitudes, $a < d$, the idea of using a periodically bent crystal as an undulator for a free electron laser was explored in [33]. In this regime the intensity of the undulator radiation is small compared with the channeling radiation. However, it is possible to match the undulator frequency to that of the channeling motion. This results in a resonant coupling of the emissions via the two mechanisms, which leads to the enhancement of the gain factor.

Let us review the results obtained in [1, 2, 13]. To do this we first outline the derivation of the general expression for the gain factor in an undulator, and, after accounting for the conditions (3), estimate gain for the crystalline undulator. For the sake of simplicity we consider the stimulated emission for the fundamental harmonic only, and, also, consider the emission in the forward direction. In the formulae below, we use the notation ω instead of ω_1 for the fundamental harmonic frequency.

The gain factor $g(\omega)$ defines the increase in the total number $\mathcal{N}$ of the emitted photons at a frequency ω due to stimulated emission by the particles of the beam: $d\mathcal{N} = g(\omega)\mathcal{N}\,dz$. The general expression for the quantity $g(\omega)$ is

$$g(\omega) = n\,[\sigma_e(\varepsilon, \varepsilon - \hbar\omega) - \sigma_a(\varepsilon, \varepsilon + \hbar\omega)]\,, \qquad (7)$$

where $\sigma_e(\varepsilon, \varepsilon - \hbar\omega)$ and $\sigma_a(\varepsilon, \varepsilon + \hbar\omega)$ are the cross sections of, correspondingly, the spontaneous emission and absorption of the photon by a particle of the beam, n stands for the volume density (measured in cm^{-3}) of the beam particles. By using the known relations between the cross sections $\sigma_{e,a}$ and the spectral-angular intensity $dE/d\omega\, d\Omega$ of the emitted radiation [40], one derives the following expression for the gain:

$$g = -(2\pi)^3 \frac{c^2}{\omega^2}\, n\, \frac{d}{d\varepsilon}\left[\frac{dE}{d\omega\, d\Omega}\right]_{\vartheta=0} \Delta\omega\, \Delta\Omega. \tag{8}$$

Here $\Delta\omega$ is the width of the first harmonic peak, and $\Delta\Omega$ is the effective cone (with respect to the undulator axis) into which the emission of the ω-photon occurs. Note that expression (8) is derived under the assumption that the photon energy is small compared to the energy of the particle, $\hbar\omega \ll \varepsilon$.

The total increase in the number of photons over the length L is

$$\mathcal{N} = \mathcal{N}_0\, e^{G(\omega)L}, \tag{9}$$

where $G(\omega) = g(\omega)L$ is the total gain on the scale L. The expression for $G(\omega)$ follows from (8) (the details of derivation can be found in [2, 38]):

$$G(\omega) = n\,(2\pi)^3 r_0\, \frac{Z^2}{M}\, \frac{L^3}{\gamma^3 \lambda} \cdot \begin{cases} 1 & \text{if } p^2 > 1 \\ p^2 & \text{if } p^2 < 1 \end{cases}. \tag{10}$$

Note the strong inverse dependence on γ and M which is due to the radiative recoil, and the proportionality of the gain to L^3 and to the squared charge of the projectile Z^2.

The main difference, of a principal character, between a conventional FEL and a FEL-type device based on a crystalline undulator is that in the former the bunch of particles and the photon flux both travel in vacuum whereas in the latter they propagate in a crystalline medium. Consequently, in a conventional FEL one can, in principle, increase infinitely the length of the undulator L. This will result in the increase of the total gain and the number of undulator periods N. The limitations on the magnitude of L in this case are mainly of a technological nature.

The situation is different for a crystalline undulator, where the dechanneling effect and the photon attenuation lead to the decrease of n and of the photon flux density with the penetration length and, therefore, result in the limitation of the allowed L-values. The reasonable estimate of L is given by the condition $L < \min\left[L_d(C), L_a(\omega)\right]$. In turn, this condition, together with the estimate

$$L_a(\omega) \approx \begin{cases} \infty & \text{for } \hbar\omega \ll I_0 \leq 10\text{eV} \\ \ll 10^{-2}\ \text{cm} & \text{for } \hbar\omega = 10^{-2}\ldots 10\text{eV} \\ = 0.01\ldots 10\ \text{cm} & \text{for } \hbar\omega > 10\text{eV} \end{cases} \tag{11}$$

defines the ranges of photon energies for which the operation of a crystalline undulator is realistic. These ranges are:

- High-energy photons: $\hbar\omega > 10$ keV when $L_a > 0.01$ cm;

- Low-energy photons: $\hbar\omega < I_0 \leq 10$ eV.

Here I_0 stands for a (first) ionization potential of the crystal atom.

In the regime of high-energy photons (the gamma laser regime) the stimulation of the emission must occur during a single pass of the bunch of the particles through the crystal. Indeed, for such photon energies there are no mirrors, and, therefore, the photon flux must develop simultaneously with the bunch propagation. In the theory of FEL this principle is called 'Self-Amplified Spontaneous Emission' (SASE) [51, 53] and is usually referred as the FEL operation in the high gain regime, which implies that $G(\omega) > 1$ to ensure that the exponential factor in (9) is large. In this case the quantity $\mathcal{N}_0$ denotes the number of photons which appear due to the spontaneous emission at the entrance part of the undulator.

For $\hbar\omega < I_0 \leq 10$ eV there is no principal necessity to go beyond 1 for the magnitude of $G(\omega)$ during a single pass. Indeed, for such photons there is a possibility to use mirrors to reflect the photons. Therefore, the emitted photons, after leaving the undulator can be returned back to the entrance point to be used for further stimulation of the emission by the incoming projectiles.

Below we present the results of numerical calculations of the parameters of the undulator (the first harmonic energy and the number of periods) and of the volume density n needed to achieve $G(\omega) = 1$. The calculations were performed for relativistic positrons, muons, protons and heavy ions and took into account all the conditions summarized in (3). The results presented correspond to the cases of the lowest values of n needed to ensure $G(\omega) = 1$ and which, simultaneously, produce the largest available values of N.

High-energy photons: the gamma-laser regime

A detailed analysis of the conditions (3) demonstrates, that to optimize the parameters of the stimulated emission in the range $\hbar\omega > 10$ keV in the case of a positron channeling one should consider the following ranges of parameters: $\gamma = (1 \ldots 5) \times 10^3$, $a/d = 10 \ldots 20$, $C = 0.1 \ldots 0.3$, which are common for all the crystals which we have investigated. For these ranges the energy of the first harmonic lies within the interval $50 \ldots 150$ keV, and the length of the undulator can be taken equal to the dechanneling length because of the inequality $L_d(C) < L_a(\omega)$, valid for such ω.

The results of calculations are presented in figure 6, where the dependences of the first harmonic energy, $\hbar\omega$, the number of undulator periods, N, and the ratio $G(\omega)/n$ versus C are presented for various crystals. The data correspond to the ratio $a/d = 20$ except for the case of Si for which $a/d = 10$.

For each crystal the curves $\hbar\omega$ and $G(\omega)/n$ were truncated at those C values for which N becomes less than 10 (see the graph in the middle). It is seen

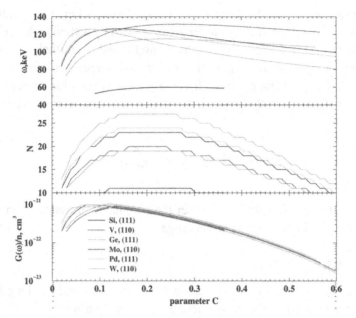

Figure 6. First harmonic energy, ω, number of undulator periods, N, and the ratio $G(\omega)/n$ in cm^3 versus C for 0.5 GeV positron channeling in various channels as indicated.

from the bottom graph, that $G(\omega)/n$ is a rapidly varying function of C. For all the channels this function attains its maximum value $\approx 10^{-21}$ cm^3 at $C \approx$ 0.1. The maximum value of $G(\omega)/n$ defines the magnitude of the volume density of a positron bunch needed to achieve total gain $G(\omega) = 1$. It follows from the graph that to achieve the emission stimulation within the range $\hbar\omega =$ 50...150 keV it is necessary to reach the value $n = 10^{21}$ cm^{-3} for a positron bunch of the energy of several GeV.

At first glance the idea of using a crystalline undulator based on the channeling of heavy, positively charged particles looks very attractive. Indeed, as it is seen from (5), the dechanneling length for a heavy particle is $M/Z \gg 1$ times larger than for a positron with the same value of γ. This factor, being cubed in (10), could lead to a noticeable increase of the total gain.

However, the allowed undulator period increases with M: $\lambda > \lambda_{\min} \propto \sqrt{M}$. In turn, this results in a decrease in the first harmonic energy $\omega \approx 4\pi c\gamma^2/\lambda \propto 1/\sqrt{M}$ (for a heavy projectile $p^2 \ll 1$ and the term p^2 can be disregarded when calculating ω, see (1)). For realistic values of relativistic factor, $\gamma \leq 10^3$, this leads to the following restriction on the photon energy:

$$\hbar\omega \leq \hbar\omega_{\max} \approx \begin{cases} 50 \text{ keV} & \text{for } \mu^+ \\ 10 \text{ keV} & \text{for } p \\ < 1 \text{ keV} & \text{for a heavy ion} \end{cases}$$

For a proton and an ion the range of $\hbar\omega$ is exactly that where the attenuation is very strong. Therefore, the crystal length is defined by a small value of $L_a(\omega)$. In the case of μ^+ the upper limit of $\hbar\omega$ is higher but, nevertheless, it leads to a condition $L_a(\omega) \ll L_d(C)$, so that the crystal length also must be chosen as $L = L_a(\omega)$. Although in such conditions it is possible to construct an undulator with sufficiently large N, the total gain factor becomes very small:

$$G(\omega) \sim n \cdot \left\{ \begin{array}{ll} 10^{-22} & \text{for } \mu^+ \\ 10^{-26} & \text{for } p \end{array} \right.$$

Therefore, it is not realistic to consider the stimulated emission from a heavy projectile in the high-energy photon range.

Low-energy photons

For $\hbar\omega \leq I_0$ the photon attenuation becomes small and the length of the undulator is defined by the dechanneling length of a particle.

To illustrate the regime of low-energy stimulated emission during the positron channeling, in figure 7 we present the dependences of N and $G(\omega)/n$ on the relativistic factor. The data correspond to a fixed ratio $a/d = 5$ and to a fixed energy of the first harmonic, $\hbar\omega = 5$ eV, which is lower than the atomic ionization potentials for all crystals indicated in the figures. The undulator length was chosen to be equal to the dechanneling length, which is the minimum from $L_d(C)$ and $L_a(\omega)$. Two graphs in figure 7 demonstrate that although the number of the undulator periods is, approximately, independent on the type of the crystal, the magnitude of the total gain is quite sensitive to the choice of the channel. The highest values of $G(\omega)/n$ (and, correspondingly, the lowest densities n needed for $G(\omega) = 1$) can be achieved for heavy crystals.

Analysis of (10) together with (3) shows that, in the case of a heavy projectile, to obtain the largest possible values of $G(\omega)$ during a single pass through a crystal the following regime can be considered: (a) moderate values of the relativistic factor, $\gamma \sim 10 \ldots 100$; (b) $C = 0.25$ which turns out to be the optimal value; (c) $Z \gg 1$, i.e. the best choice is to use a bunch of heavy ions.

In figure 8 we present the dependences of N and $G(\omega)/n$ on γ within the range specified above. All curves, which were obtained for different a/d ratios, refer to the case of U^{+92} channeling in W along the (110) crystallographic planes. The energy of the emitted photon is fixed at 5 eV.

It is seen from the graphs that for all the crystals the most optimal range of relativistic factor is $\gamma \sim 10 \ldots 30$ where both the number of the undulator periods and the magnitude of $G(\omega)/n$ noticeably exceed the corresponding values in the case of a positron channeling, see figure 7.

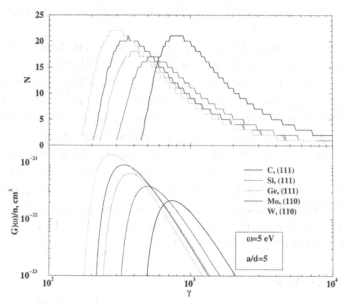

Figure 7. Values of N and $G(\omega)/n$ versus γ for a positron-based crystalline undulators in a low-ω region calculated for various channels as indicated.

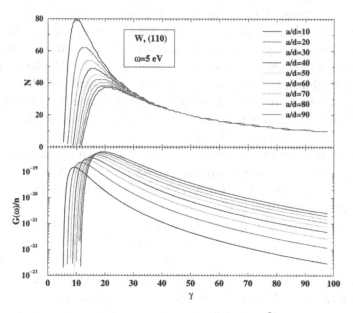

Figure 8. Number of periods, N, and values of $G(\omega)/n$ in cm^3 versus γ for an ion-based crystalline undulator in a low-ω region calculated at different a/d values as indicated. The data are presented for bare U ion channeling along the (110) planes in a tungsten crystal.

A similar analysis, carried out for the case of a proton channeling, demonstrates that for the same value of γ the magnitudes of $G(\omega)/n$ are several times higher than those for a heavy ion.

The results presented in this section show, that the stimulated emission in the range $\hbar\omega < I_0 \sim 10$ eV can be discussed for all types of positively-charged projectiles. To achieve the value $G(\omega) = 1$ within the SASE mode it is necessary to consider the densities from $n \approx 5 \times 10^{18}$ cm^{-3} for heavy ion beams up to $n \approx 5 \times 10^{21}$ cm^{-3} for a positron beam. However, this large numbers can be reduced by orders of magnitudes if one considers the multi-pass mode of the FEL. Indeed, there exist the mirrors which allow reflections of low energy photons. Thus, the emitted photons can be returned back to the entrance point and used further to stimulate the emission generated by the particle of the long bunch. The number of passes, equal approximately to L_b/L (here L_b is the bunch length), can be very large (up to 10^4). Therefore, volume density can be reduced by the factor $L/L_b \ll 1$.

5. Conclusions

In this paper we have discussed the feasibility of the crystalline undulator and the gamma laser based on it. We have presented the detailed review covering the development of all essential aspects of these important ideas.

Firstly, we note that it is entirely realistic to use a crystalline undulator for generating *spontaneous* radiation in a wide range of photon energies. The parameters of such an undulator, being subject to the restrictions mentioned above, can easily be tuned by varying the shape function, the energy and the type of a projectile and by choosing different channels. The large range of energies available in modern colliders for various charged particles, both light and heavy, together with the wide range of frequencies and bending amplitudes in crystals allow one to generate the crystalline undulator radiation with the energies from eV up to the MeV region.

Secondly, it is feasible to obtain *stimulated* emission by means of a crystalline undulator. For a single-pass laser (SASE mode) high volume densities $n \geq 10^{20}$ cm^{-3} of a positron bunch are needed to obtain the stimulated emission in the range $\hbar\omega > 10$ keV.

Stimulated emission in the low-ω range ($\hbar\omega < I_0 \sim 10$ eV) can be discussed for all types of positively-charged projectiles. In this case the large values of the beam densities required for the lasing effect can be reduced by orders of magnitudes if one considers the multi-pass mode of the FEL.

The crystalline undulators discussed in this paper can serve as a new efficient source for the coherent high energy photon emission. As we have pointed out, the present technology is nearly sufficient to achieve the necessary conditions to construct not only crystalline undulator, but also the stimulated photon emission source. The parameters of the crystalline undulator and the gamma laser based on it differ substantially from what is possible to achieve with the undulators constructed on magnetic fields.

This review clearly demonstrates that experimental effots are needed for to verify the numerous theoretical predictions. Such efforts will certainly make this field of endeavour even more facinating than it is already and will possibly lead to the practical development of a new type of tunable and monochromatic radiation sources.

Finally, we mention that not all theoretical issues for the described system have been solved so far. Thus, the analysis of dynamics of a high-density positon beam channeling through a periodically bent crystal in the presence of an induced high intensity photon flux is to be performed in a greater detail. This and many more other interesting theoretical problems are still open for future investigation.

References

[1] Korol A.V., Solov'yov A.V. and Greiner W. (1998). *J.Phys.G: Nucl.Part.Phys.* 24:L45-53; (1997). Institute for Theoretical Physics, Frankfurt am Main University, Preprint-UFTP467.

[2] Korol A.V., Solov'yov A.V. and Greiner W. (1999). *Int. J. Mod. Phys. E* 8:49-100.

[3] Korol A.V., Solov'yov A.V. and Greiner W. (2000). *Int. J. Mod. Phys. E* 9:77-105.

[4] Krause W., Korol A.V., Solov'yov A.V. and Greiner W. (2000). *J.Phys.G: Nucl.Part.Phys.* 26:L87-L95.

[5] Korol A.V., Solov'yov A.V. and Greiner W. (2001). *J.Phys.G: Nucl.Part.Phys.* 27:95-125.

[6] Greiner W., Korol A.V. and Solov'yov A. V. (2000). Proceedings of Int. Meeting on *Frontiers of Physics* (Kuala Lumpur, Malaysia, 26-29 October 1998), Editors S.P.Chia and D.A.Bradley, (Singapore: World Scientific) p. 60-92.

[7] Greiner W., Korol A.V. and Solov'yov A. V. (2001). 16th Int. Conf. *Application of Accelerators in Research and Industry* (Denton,Texas, USA, 1-5 November (2000). AIP Conference Proceedings, editors J.L. Duggan and I.L. Morgan, 576:17-20.

[8] Krause W., Korol A.V., Solov'yov A.V. and Greiner W. (2001). *Nucl.Instrum.Methods* A 475:441-444.

[9] Krause W., Korol A.V., Solov'yov A.V. and Greiner W. (2001). Proceedings of Int. Conf. *Fundamental and Applied Aspects in Modern Physics* (Luderitz, Namibia, November 13-17, 2000). Editors S.H. Connell and R. Tegen (Singapore: World Scientific) p. 115-122.

[10] Krause W., Korol A.V., Solov'yov A.V. and Greiner W. (2001). in 'Electron-Photon Interaction in Dense Media' NATO Science Series II. Mathematics, Physics and Chemistry vol. 49, ed. H. Wiedemann (Kluwer Acad. Publishers) p. 263-276.

[11] Korol A.V., Krause W., Solov'yov A.V., Greiner W. (2002). *Nucl.Instrum.Methods* A 483:455-460.

[12] Korol A.V., Solov'yov A.V. and Greiner W. (2003). Proc. Symp. *Channeling - Bent Crystals - Radiation Processes* (Frankfurt am Main, Germany, June 2003), eds. W. Greiner, A.V. Solov'yov, S. Misicu (Hungary: EP Systema Bt., Debrecen) pp. 135-154.

[13] Korol A.V., Solov'yov A.V. and Greiner W. (2003). Proc. Symp. *Channeling - Bent Crystals - Radiation Processes* (Frankfurt am Main, Germany, June 2003), eds. W. Greiner, A.V. Solov'yov, S. Misicu (Hungary: EP Systema Bt., Debrecen) pp. 155-169.

[14] Korol A.V., Solov'yov A.V. and Greiner W. (2004). *Int. J. Mod. Phys. E* 13:867-916 http://xxx.lanl.gov/ [arXiv: physics/0403082 (2004)].

[15] Avakian R.O., Gevorgian L.A., Ispirian K.A. and Ispirian R.K. (1998). *Pis'ma Zh. Eksp. Teor. Fiz.* 68:437-441 (*JETP Lett.* 68:467-471 (1998)).

[16] Zhang Q-R (1999). *Int. J. Mod. Phys. E* 8:493.

[17] Mikkelsen U. and Uggerhfij E. (2000). *Nucl.Instrum.Methods* B 160:435-439.

[18] Aganiants A.O., Akopov N.Z., Apyan A.B. *et al.*(2000). *Nucl.Instrum.Methods* B 171:577-583.

[19] Avakian R.O., Gevorgian L.A., Ispirian K.A., Ispirian R.K. (2001). *Nucl.Instrum.Methods* B 173:112-120.

[20] Avakian R.O., Avetyan K.T., Ispirian K.A., Melikyan E.G. (2001). in 'Electron-Photon Interaction in Dense Media' NATO Science Series II. Mathematics, Physics and Chemistry vol. 49, ed. H. Wiedemann (Kluwer Acad. Publishers) p. 277.

[21] Avakian R.O., Avetyan K.T., Ispirian K.A., Melikyan E.G. (2002). *Nucl.Instrum.Methods* A 492:11-13.

[22] Avakian R.O., Avetyan K.T., Ispirian K.A., Melikyan E.G. (2003). *Nucl.Instrum.Methods* A 508:496-499.

[23] Bellucci S., Bini S., Biryukov V.M., Chesnokov Yu.A. *et al.*(2003). *Phys.Rev.Lett.* 90:034801.

[24] Bellucci S., Bini S., Giannini S. *et al.*(2003). *Crystal undulator as a novel compact source of radiation* Electronic preprint arXiv:physics/0306152 (http://xxx.lanl.gov/).

[25] Bellucci S. (2003). Proc. Symp. *Channeling - Bent Crystals - Radiation Processes* (Frankfurt am Main, Germany, June 2003), eds. W. Greiner, A.V. Solov'yov, S. Misicu (Hungary: EP Systema Bt., Debrecen) pp. 171-178.

[26] Korhmazyan N.A., Korhmazyan N.N. and Babadjanyan N.E. (2004). *Zh. Tekh. Fiz.* 74:123-125.

[27] Kaplin V.V., Plotnikov S.V. and Vorob'ev S.V. (1980). *Zh. Tekh. Fiz.* 50:1079-1081 (*Sov. Phys. – Tech. Phys.* 25:650-651 (1980)).

[28] *Conceptual Design of a 500 GeV e^+e^- Linear Collider with Integrated X-ray Laser Facility*, (1997), eds. R. Brinkmann et al, DESY 1997-048, ECFA 1997-182 (Hamburg, DESY).

[29] Rullhusen P., Artru X. and Dhez P. (1998). *Novel Radiation Sources using Relativistic Electrons* (Singapore: World Scientific).

[30] Colson W. B. (2001). *Nucl.Instrum.Methods* A 475:397-400

(see also http://www.physics.nps.navy.mil/table1.pdf).

[31] Baryshevsky V.G., Dubovskaya I.Ya. and Grubich A.O. (1980). *Phys.Lett.* A 77:61-64.

[32] Ikezi H., Lin-liu Y.R. and Ohkawa T. (1984). *Phys.Rev.* B 30:1567-1569.

[33] Bogacz S.A. and Ketterson J.B. (1986). *J.Appl.Phys.* 60:177-188.

[34] Mkrtchyan A.R., Gasparyan R.A. and Gabrielyan R.G. (1986). *Phys.Lett.* A 115:410-412.

Mkrtchyan A.R., Gasparyan R.A. and Gabrielyan R.G. (1987). *Zh. Eksp. Teor. Fiz.* 93: 432-436 (*Sov. Phys. – JETP* 66:248-250 (1987)).

[35] Amatuni A.T. and Elbakyan S.S. (1988). *Zh. Eksp. Teor. Fiz.* 94:297-301 (*Sov. Phys. – JETP* 67:1903-1905 (1988)).

[36] Mkrtchyan A.R., Gasparyan R.H., Gabrielyan R.G. and Mkrtchyan A.G. (1988). *Phys.Lett.* A 126:528-530.

[37] Dedkov V.G. (1994). *Phys.Status Solidi* (b) 184:535-542.

[38] Baier V.N., Katkov V.M. and Strakhovenko V.M., (1998). *High Energy Electromagnetic Processes in Oriented Single Crystals* (Singapore: World Scientific).

[39] Biruykov V.M., Chesnokov Yu.A. and Kotov V.I. (1996). *Crystal Channeling and its Application at High-Energy Accelerators* (Berlin: Springer).

[40] Berestetskii V.B., Lifshitz E.M. and Pitaevskii L.P. (1982). *Quantum Electrodynamics* (Oxford: Pergamon).

[41] Sternheimer R.M. (1966). *Phys.Rev.* 145:247.
Komarov F.F. (1979). *Phys. Stat. Sol.* (b) 96:555.

[42] Mason W.P. (1972). Acoustic properties of Solids, *American Institute of Physics Handbook*, 3rd edn (New York: McGraw-Hill).

[43] Grigoryan L.Sh., Mkrtchyan A.R., Mkrtchyan A.H. *et al.*(2001). *Nucl.Instrum.Methods* B 173:132-141.

[44] Grigoryan L.Sh., Mkrtchyan A.R., Khachatryan H.F. *et al.*(2003). *Nucl.Instrum.Methods* B 212:51-55.

[45] Baier V.N. and Katkov V.M. (1967). *Zh. Eksp. Teor. Fiz.* 53:1478 (*Sov. Phys. – JETP* 26:854 (1968)).
Baier V.N. and Katkov V.M. (1968). *Zh. Eksp. Teor. Fiz.* 55:1542 (*Sov. Phys. – JETP* 28:807 (1969)).

[46] Gemmell D.S. (1974). *Rev.Mod.Phys.* 46:129.

[47] Uggerhfij E. (1993). *Rad. Eff. Def. Solids* 25:3-21.

[48] Kumakhov M.A. and Komarov F.F. (1989). *Radiation From Charged Particles in Solids* (New York: AIP).

[49] Bak J.F., Ellison J.A., Marsh B. *et al.*(1985). *Nucl.Phys.* B 254:491-527.
Bak J.F., Ellison J.A., Marsh B. *et al.*(1988). *Nucl.Phys.* B 302:525-558.

[50] Bazylev V.A. and Zhevago N.K. (1990). *Usp. Fiz. Nauk* 160:47 (*Sov. Phys. – Uspekhi* 33: 1021).
(1987). *Radiation of Fast Charged Particles in Matter and External Fields* (Moscow: Nauka).

[51] Saldin E.L., Schneidmiller E.A. and Yurkov M.V. (1995). *Phys. Rep.* 260:187.

[52] Madey J.M.J (1971). *J.Appl.Phys.* 42:1906.

[53] Bonifacio R., Pellegrini C., and Narducci L.M. (1984). *Opt. Commun.* 50:373.

ACCELERATOR TESTS OF CRYSTAL UNDULATORS

V.M. Biryukov[@1], A.G. Afonin[1], V.T. Baranov[1], S. Baricordi[3], S. Bellucci[2], G.I. Britvich[1], V.N. Chepegin[1], Yu.A. Chesnokov[1], C.Balasubramanian[2], G. Giannini[2], V. Guidi[3], Yu.M. Ivanov[4], V.I. Kotov[1], A. Kushnirenko[1], V.A. Maisheev[1], C. Malagu[3], G. Martinelli[3], E. Milan[3], A.A. Petrunin[4], V.A. Pikalov[1], V.V. Skorobogatov[4], M. Stefancich[3], V.I. Terekhov[1], F. Tombolini[2], U.I. Uggerhoj[5]

[1]*Institute for High Energy Physics, 142281 Protvino, Russia;* [2]*INFN - Laboratori Nazionali di Frascati, P.O. Box 13, 00044 Frascati, Italy;* [3]*Department of Physics and INFN, Via Paradiso 12, I-44100 Ferrara, Italy;* [4] *St. Petersburg Institute for Nuclear Physics, Russia;* [5]*Aarhus University, Denmark;*

Abstract: A series of Silicon crystal undulator samples were produced based on the approach presented in PRL 90 (2003) 034801, with the periods of undulation from 0.1 mm to 1 mm, and the number of periods on the order of 10. The samples were characterized by X-rays, revealing the sine-like shape of the crystal lattice in the bulk. Next step in the characterization has been the channeling tests done with 70 GeV protons, where good channeling properties of the undulated Silicon lattice have been observed. The photon radiation tests of crystal undulators with high energy positrons are in progress on several locations: IHEP Protvino, LNF Frascati, and CERN SPS. The progress in the experimental activities and the predictions from detailed simulations are reported.

Keywords: crystal channeling; undulator; radiation.

1. INTRODUCTION

Many areas of modern science and engineering require intense sources of high-energy X-ray and gamma beams. Efficient sources of such radiation are electron or positron beams traversing an undulated magnetic field in accelerators. However, the usual electromagnetic undulators have a

---

[@] Corresponding author, biryukov@mx.ihep.su

191

H. Wiedemann (ed.), Advanced Radiation Sources and Applications, 191–200.

relatively large period of magnetic structure, which limits the energy of generated radiation. The energy of a photon emitted in an undulator is in proportion to the square of the particle Lorentz factor γ and in inverse proportion to the undulator period L: $\hbar\omega \approx 2\pi\hbar\gamma^2 c/L$. Typically, at the modern accelerators the period of undulator in the synchrotron light sources is a few centimeters [1]. With a strong worldwide attention to novel sources of radiation, there has been broad interest [2-11] to compact undulators based on channeling in crystals, where a strong periodic electric field exists by nature. The crystalline undulators with periodically deformed crystallographic planes offer electromagnetic fields in the order of 1000 Tesla and could provide a period L in sub-millimeter range. This way, a hundred-fold gain in the energy of emitted photons would be reached, as compared to a usual undulator.

The use of bent crystals for channeling extraction of beams from accelerators has been under development at several laboratories [12-17]. A collaboration of researchers working at the 70-GeV accelerator of IHEP has achieved a substantial progress in the efficiency of crystal-assisted beam deflection: extraction efficiency larger than 85% is obtained at intensity as high as 10^{12} protons [17]. Since 1999, the use of bent crystals to extract beams for the high energy physics program became regular in all accelerator runs of U-70. About one half of all extracted beams with intensities up to 10^{12} at IHEP are now obtained with channeling crystals. The efficient practical use of crystal channeling elements for steering of particle beams at modern accelerators has been another major motivation for research towards a crystal channeling undulator.

2.   THE REALIZED TECHNIQUE OF CU

The first idea on how to make an undulated crystal lattice was ultrasound, suggested in 1979 by Kaplin et al. [4]. Later on, in 1984, Ikezi et al. [6] proposed a $Si_{1-x}Ge_x/Si$ graded composition lattice with periodical modulation of *Ge* content x. More recently, Avakian et al. [11] suggested to apply a periodic surface strain on crystal wafer. These ideas have been theoretically developed [4-11] but they are still pending realization. Recently [2,3], we demonstrated experimentally by means of X-rays that microscratches on the crystal surface make sufficient stresses for creation of a crystalline undulator (CU) by making a series of scratches with a period of sub-millimeter range. This method is based on an interesting observation in our earlier 70-GeV proton channeling experiments ref. [12], p.120, where it was found that accidental micro-scratches on a crystal surface cause a significant deflection of high-energy particles near the scratch (see photo, Fig. 1).

On this photo, a fragment of the end face of crystal and an image of the beam deflected by this crystal are presented. One can see that the image of the deflected beam strongly reflects the character of the surface scratches. This effect is explained by the fact that the protons near a scratch are channeled by deformed crystal planes. Reconstruction analysis of the angles of deflected particles shows that deformation of the crystallographic planes penetrates to substantial depths, down to a few hundred microns as depicted in Fig. 2(a). Therefore, this effect could be profitably used for creation of a CU by making a periodic series of micro grooves on the crystal surface as shown on Fig. 2(b).

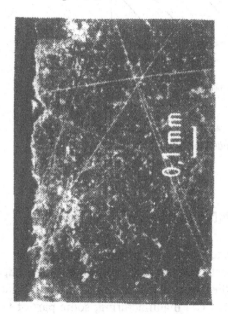 

Figure 1. The microphotography of crystal end face with scratches (left) and image of the 70 GeV proton beam deflected by this crystal shown at a distance of 1 m (right).

Presently a series of undulators (Silicon (110) oriented wafers) was manufactured with the following parameters: the length along the beam from 1 to 5 mm, thickness across the beam 0.3 to 0.5 mm, 10 periods of oscillation with the step from 0.1 to 0.5 mm, and the amplitude on the order of 20 - 150 Å. The scratching was done in IHEP and Ferrara.

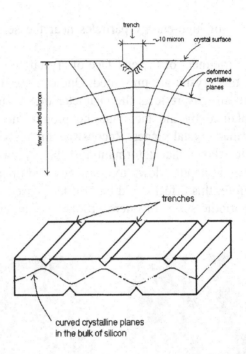

Figure 2. Left: The angular distortion of crystal planes near a surface scratch (groove). Right: The scheme of proposed crystalline undulator.

3.   HIGH-ENERGY CHANNELING TESTS OF CU

Firstly, the undulators were tested and characterized with X-rays as described in refs. [2,3]. The X-ray tests showed that a sinusoidal-like shape of crystalline planes goes through the bulk of the crystal with appropriate amplitudes. In the proposed method of CU manufacturing, some part of the crystal lattice is disrupted by scratches and is not suitable for channeling, therefore we tested our crystal undulators directly for channeling of 70 GeV protons. The tests were made with our usual technique, carefully described in ref. [17]. Four CU wafers of the thickness of 0.3-0.5 mm, with about 10 periods, were bent by means of the devices described in ref. [17] by the angle of about 1 mrad and installed in a circulating accelerated beam as shown in Fig. 3.

This angle of bending is sufficient to separate the circulating and the deflected (by the crystal) beams in space. The beam deflection effect due to channeling was measured by secondary emission detector, located in the vacuum chamber of the accelerator near to circulating beam.

On Fig. 4(a) the profile of the beam deflected by one of the tested crystals is shown as a function of the orientation of this crystal. The crystal was 5 mm along the beam and 0.3 mm across and had 10 periods of sinusoidal deformations with a step of 0.5 mm and the amplitude of about 100 Å. The top left profile corresponds to the disoriented crystal when only a scattering tail of the beam is seen. On the following picture (top right) the crystal orientation is optimal. In this case the efficiency of extraction (the ratio of the fully deflected channeled beam to the total losses of the circulating beam) is equal to 31 %. In the same conditions, a usual bent crystal (without scratches) of a similar size has extracted about 45 % of the

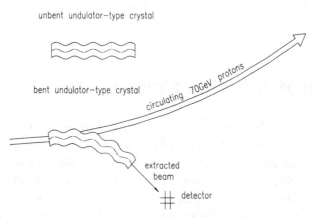

Figure 3. The scheme of extraction of a circulating high-energy proton beam with a bent undulator-type crystal.

beam. Thus the measurements showed that all CU crystals deflect protons with good efficiency and at least 70 % of the crystal cross-section is available for channeling despite of the distortions caused by scratches. Actually, part of the overall reduction of 30% should be due to additional centrifugal forces from undulation in crystal lattice, therefore the crystal cross-section available for channeling is significantly larger than 70% of the total.

The presence of periodic deformations with large amplitude in the crystal lattice, confirmed earlier by X-ray tests, and the experimentally established transparence of the lattice for channeled high-energy particles allows one to start a direct experiment on photon production from positron beam in a crystalline undulator.

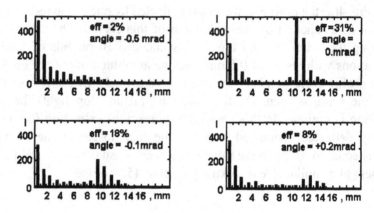

Figure 4. Deflected beam profile and the efficiency of extraction versus the crystal orientation angle.

4.   PHOTON EMISSION EXPERIMENT

The present collaboration has several appropriate sites for an experiment on generation of photons in a crystalline undulator. One site is Laboratori Nazionali di Frascati with the positron beam energy 500-800 MeV. Another site is IHEP where one can arrange positron beams with the energy higher than 2 GeV. The new opportunity has occurred recently at CERN SPS positron beam line H4 under the proposal [18] at the energy of about 10 GeV and higher. The progress in LNF is described in ref. [3].

Fig. 5 shows the equipment layout for realization of experiment on the secondary beam line 22 in IHEP Protvino. Crystalline undulator is placed into vacuum box, inside which a remotely controlled goniometer is positioned. The goniometer provides a horizontal translation of the crystal for its exposure into the beam within the limits of about 100 mm, with a step size of 0.05 mm. Also, it provides an alignment of the crystal within ±30 mrad with a step size of about 0.010 mrad.

A cleaning magnet is positioned right after the vacuum box. The vacuum system is ended by a 10 m long tube, 200 mm in diameter, which has a 0.1-0.2 mm thick Mylar window at the end. Small finger scintillator counters, one installed before the crystal and another one 12 m downstream, select beam particles in a narrow angular interval of about ⌒0.1 mrad and form a trigger signal for spectrometer of photons. As a detector of photons, a crystal of NaI (Tl) of ∅1 cm ×10 cm is used. Part of the above described equipment is seen on photos of Figs. 6 and 7.

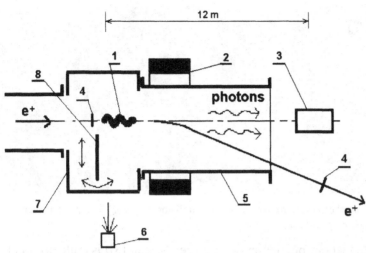

Figure 5. Layout of photon emission experiment at beam line 22: 1 – crystal undulator, 2 – cleaning magnet, 3 – photon detector, 4 – scintillator counters, 5 – vacuum pipe with thin exit window, 6 – laser for crystal alignment, 7 – vacuum box with flanges and electric plugs, 8 – goniometer for crystal rotation and transmission.

Figure 6. The vacuum box with CU and cleaning magnet of the setup at IHEP.

Figure 7. The photon spectrometer (1) and scintillator trigger (2) mounted on mechanical drivers in the setup at IHEP.

Notice that our prediction (Fig. 8, see also ref. [3]) is quite different from that of Solov'ev et.al. (e.g. Fig. 2 in ref. [9]) who predict the spectrum to be essentially zero everywhere but a very narrow undulator peak and a peak of channeling radiation, in vast contrast with our results.

The commissioning of the set-up with 3 GeV positron beam has shown that the main systems work normally and background conditions on photon detector are appropriate. The first results with necessary statistics are expected in the next run of the IHEP accelerator (planned near the end of 2004). The expected spectra of the photons generated in CU are shown in

Fig. 8. Calculations were carried out for the (110) plane of a silicon single crystal with realistic sizes and take into account the following factors: channeling radiation and dechanneling process, radiation of the above-barrier positrons, absorption of gamma-quanta in the undulator bulk [3]. The calculated number of photons in the range 100–600 KeV is 0.15 per one positron channeled through the crystalline undulator. The spectral density of undulator radiation is five times greater than that of usual channeling radiation.

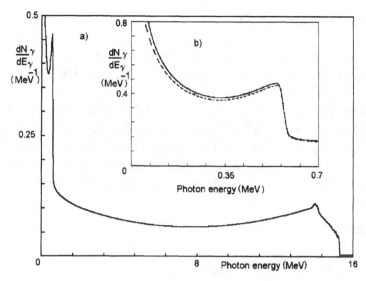

Figure 8. The expected photon spectrum for 3 GeV positrons in the range 0–16 MeV (a) and 0–0.7 MeV (b). The dashed curve is the photon spectrum, where absorption process of photons in the body of undulator is taken into account. The curves are normalized on one positron passing through the undulator within channeling angle.

5. CONCLUSION

Real crystal undulator devices have been created. The tests of these devices with X-rays and high-energy channeled protons were performed, and the experimental conditions necessary for observation of photon emission from positron beam in crystalline undulator were prepared. Creation of a new source of radiation with high spectral density, superior to the usual channeling radiation is expected on the basis of crystal undulator. Crystalline undulator would allow to generate photons with the energy on the order of 1 MeV at the synchrotron light sources where one has at the moment only 10 KeV, and for this reason crystal undulators have interesting prospects for application.

ACKNOWLEDGEMENTS

This work was partially supported by INFN - Gruppo V, as NANO experiment, by the Italian Research Ministry MIUR, National Interest

Program, under grant COFIN 2002022534, by INTAS-CERN grants 132-2000 and 03-52-6155 and by RFBR Grant No. 01-02-16229. The invitation of the talk and support from the Organizers of the workshop for the corresponding author (V.M.B.) is gratefully acknowledged. Armenian hospitality has greatly contributed to the success of the workshop.

References

1. Beam Line, v. **32**, no. 1 (2002)
2. S. Bellucci et al., Phys. Rev. Lett. **90**, 034801 (2003)
3. S. Bellucci et al., Phys. Rev. ST AB **7**, 023501 (2004)
4. V.V. Kaplin, S.V. Plotnikov, and S.A. Vorobiev, Zh. Tekh. Fiz. **50**, 1079-1081 (1980).
5. V.G. Baryshevsky, I.Ya. Dubovskaya, and A.O. Grubich, Phys. Lett., **77A**, 61-64 (1980)
6. H. Ikezi, Y.R. Lin-Liu, and T. Ohkawa, Phys. Rev., **B30**, 1567-1568 (1984).
7. S.A. Bogacz and J.B. Ketterson, J. Appl. Phys. **60**, 177-188 (1986). S.A. Bogacz, Particle Accelerators, **42** (1993) 181.
8. G.B. Dedkov, Phys.Stat.Sol. (b) **184**, 535-542 (1994).
9. W. Krause, A.V. Korol, A.V. Solov'ev, W. Greiner. Arxiv:physics/0109048, Proc. of NATO ARW "Electron-photon interaction in dense media" (Nor Amberd, 2001), see also these Proceedings.
10. U. Mikkelsen and E. Uggerhoj, Nucl. Instr. and Meth., **B160**, 435- 439 (2000).
11. R.O. Avakian, K.T. Avetyan, K.A. Ispirian and E.G. Melikyan. Nucl. Instr. and Meth., **A492**, 11-13 (2002).
12. V.M. Biryukov, Yu.A. Chesnokov, and V.I. Kotov, Crystal Channeling and Its Application at High-Energy Accelerators (Springer: Berlin, 1997). See also http://crystalbeam.narod.ru/
13. H. Akbari et al. Phys. Lett. B **313** (1993) 491-497
14. R.A. Carrigan et al., Phys. Rev. ST Accel. Beams **1**, 022801 (1998)
15. A.G. Afonin et al. JETP Lett. **74**, 55-58 (2001)
16. A.G. Afonin et al. Instrum. Exp. Tech. **45**(4), 476 (2002)
17. A.G. Afonin et al. Phys. Rev. Lett. **87**, 094802 (2001)
18. U.I. Uggerhoj, invited talk at the "Relativistic Channeling and Related Coherent Phenomena" workshop. March 23-26 (2004) INFN, Frascati, to appear in NIM B

SURFACE EFFECTS ON A DIAMOND TARGET AT NON-LINEAR EMISSION OF SPONTANEOUS RADIATION FROM 4.3 GEV ELECTRONS

A. Aganyants

Yerevan Physics Institute, Alikhanian Brothers Str. 2, Yerevan 375036, Armenia

1.   Introduction

In recent papers [1-3] radiation with non-linear dependence on electron beam intensity has been reported. It was expressed an opinion that observed phenomenon is connected with atomic excitations of crystalline medium (by the same electrons) and their correlation in non-equilibrium conditions. As known, atomic excitations in a crystal is a collective process that is defined by its Debye temperature. What else can promote this process? Apparently, it is the crystal boundary. Each atom in the depth of the crystal is under the influence of all its neighbors and the resulting forces have not primary direction. However, near the surface there are unbalanced collective forces, especially on the edge of a crystal because of spatial asymmetry. Relativistic electrons entering an oriented crystal will interact with tightly connected atomic fields and must radiate especially strong on the surface of a plate.

2.   Experimental Conditions

The experiment for verification of this statement was performed by means of the internal 4.3 GeV electron beam of Yerevan synchrotron. Electron beam had a divergence of 0.05 mrad before single crystal-diamond target installed

H. Wiedemann (ed.), Advanced Radiation Sources and Applications, 201–204.

inside the synchrotron vacuum chamber close to the electron equilibrium orbit. An orbit excitation magnet placed near the target provided slow dumping of the electron beam on the crystal. The produced radiation entered the experimental hall, where it was detected by pair spectrometer or other more simple scintillation counters. In this experiment moveable diamond radiator with transversal size of 2 mm could be displaced together with electron beam in transversal direction relative to aperture of collimator 3.3×3.3 mm^2, which was placed on the distance of 9.4m from radiator. So collimator took gamma beam only from defined part of diamond from its centre up to an edge. Monitoring of the total energy of the radiation was carried out by means of Wilson quantameter.

3.   Measurements and Results

3.1.   Integral measurements with simple detectors.

In this case the energy of photons was not measured. First detector is situated under the angle 6° relative to gamma beam direction and the same second one did under the angle 19° Both counters detected any particle from thin aluminum target. Table 1 presents the counting rates of the detectors versus coordinate of diamond displacement **r** relative to collimator aperture. The angle between crystal planes (110) and the electron beam direction was 0°. It is seen increase of the detector counting rate when collimator "sees" only an edge of the crystal.

Table 1 Counting rates of detectors normalized to quantameter data depending on the crystal displacement **r** relative to the collimator aperture.

r, mm	Detector 6°	Detector 19°
0	16000	995
0.5	25100	1300
2	38400	1920

3.2. Differential measurements with the pair spectrometer.

Growth of gamma emission on the edge of the target did not observed when the measurements were performed by means of the pair spectrometer. Fig. 1 presents the counting rates of the pair spectrometer versus the coordinate of diamond displacement **r** relative to collimator aperture. The angle between crystal planes (110) and electron beam direction was 0°. The spectrometer detected gamma-quanta with the energy of ≈50 MeV. It is observed contrary picture here. The photon yield drops at the crystal edge.

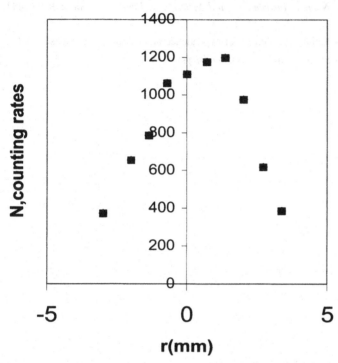

Figure1. Counting rates of the pair spectrometer **N** normalized to quantameter data depending on the crystal displacement **r** relative to the collimator aperture. Statistical errors are less 5%.

Similar results were obtained also when the angle between crystal planes (110) and electron beam direction was 1.3 mrad, i.e. in the range of coherent bremsstrahlung. So these contrary results most probably testify to changing the spectrum and the emission angles of radiation on the crystal edge.

Acknowledgement

V.Karibyan and Yu. Vartanov participated also in measurements presented by Fig 1.

References

1. A Aganyants. Influence of excited crystalline medium on interaction processes of ultrarelativistic electrons. *Laser and Particle Beams*, Vol. 20, Number 2 (2002).

2. A.Aganyants, V.Garibyan,Yu.Vartanov. Intensity-dependent electron beam losses in oriented monocrystals. *Nuclear Instruments and Methods in Physics Research B* 168,493-497(2000).

3. A.Aganyants.New radiation with non-linear dependence on electron beam intensity. *NATO Science Series. II Mathematics, Physics and Chemistry*-Vol.49, p.309-312.

X-RAY TRANSITION AND CHERENKOV RADIATION PRODUCED BY HIGH-ENERGY PARTICLES IN MULTILAYERS AND THEIR APPLICATIONS

M.A. Aginian, L.A. Gevorgian, K.A. Ispirian and I.A. Kerobyan
Yerevan Physics Institute, Brothers Alikhanian 2, 375036, Yerevan, Armenia

Abstract: A new formula for radiation produced in semitransparent stacks of two alternating thin layers with different dielectric constants $\varepsilon_{1,2}$ less as well as greater than 1 is given. When $\varepsilon_{1,2} < 1$ it coincides with the well known formula for the X-ray transition radiation, while in the frequency region of anomalous dispersion where one can have $\varepsilon_{1,2} > 1$ it provides the properties of the so called X-ray Cherenkov radiation produced in multilayers. The obtained numerical results are used for comparison with the results obtained with the help of other formulae as well as for discussing new experiments

Keywords: Transition radiation/X-ray Cherenkov radiation/multilayers.

1. INTRODUCTION

After the pioneering work [1] on TR it becomes clear that the intensity of TR produced when a charged particle passes a single interface between two media with different dielectric constant $\varepsilon_{1,2}$, is low, about 0.01 photon. To increase the intensity of optical (OTR), and X-ray TR (XTR) it is necessary to use radiators containing many layers, i.e. multilayers (see [2-5]).

In the beginning the theory of XTR produced in a stack of layers has been developed [2] using the WKB or quasiclassical approximation according to which the wavelength, λ, is much less than the radiator period, $a + b$, and/or

205

H. Wiedemann(ed.), Advanced Radiation Sources and Applications, 205–216.
© 2006 *Springer. Printed in the Netherlands.*

the difference of the dielectric constants, $\Delta\varepsilon = \varepsilon_1 - \varepsilon_2$, is much less than $\varepsilon_{1,2}$ of the layers with thickness a and b. Soon the coherent summation method (CSM) has been developed [6,7] in which XTR from various interfaces was coherently summed in the assumption that the reflection not only from one interface, but also from all the interfaces is small. If the WKB approximation is valid[2] for periods a + b $\geq 10^{-5}$ Z cm, where Z is the atomic number of the substances of the layers, CSM has no limitation on the period. The rapid development of the theory of XTR was connected with the construction and use of XTR detectors for measuring the Lorentz factors of ultra relativistic particles [8,9] as well as with the possibility of production intense and cheap X-ray beams [10]. Recently it has been experimentally shown that thin multilayers consisting of many pairs of two alternating 0.5 – 5000 nm thick layers of two materials, which found wide application as X-ray mirrors, X-ray monochromators, etc, can be used as effective radiators for production of X-ray quasi monochromatic XTR[11] as well as of parametric x-ray radiation (PXR or XTR produced under Bragg conditions) [12] beams.

The development of the theory of OTR produced in multilayers went more slowly. It is known [13,14] that when the thickness a of the single plate decreases and becomes equal or of the order of the emitted photon wavelength, a $\rightarrow$ λ, the spectral-angular distributions of the TR and Cherenkov radiation (CR) are mixed essentially, and this fact has been observed experimentally in the work [15]. The δ-function form angular distribution of the usual CR becomes wider. As it has been mentioned above to enhance the intensity of the produced TR photons it is necessary to take a stack of many thin plates as radiator. The properties of the radiation produced in such transparent stratified radiators have been theoretically considered for the first time in the work [16]. Solving the Maxwell equations for an infinite radiator it has been derived expressions for the fields and for the radiated intensity. The authors of [16] have shown that when some conditions are fulfilled the intensity of the emitted so-called parametric Cherenkov radiation is enhanced. However the derived expressions are complicated, and no satisfactory experimental results on the optical radiation produced in transparent stacks exist.

Using simple formulae for TR from one interface [17] expressions for the radiation fields arising when a charged particle passes through a stack of N plates have been constructed. Using the results [17] the expression of the intensity of TR produced in stack of N plates with thickness a and vacuum spacing b has been written in [3]. The difference between the results of[17] and other CSM formulae given in [3] is in the fact that in [17] the reflection of the radiation from the surfaces has been taken into account. The analysis carried out in [3,17,18] has shown that the formulae obtained by CSM are in agreement with the complicated expressions obtained accurately in the work [16] for an infinite laminar medium.

Taking into account the possibility of obtaining soft X-ray beams with the help of mutilayers of submicron layers, the CSM of XTR has been developed further in [19] taking into account the absorption at K, L edges as well as the scattering. The authors of [19] have proposed an optimization procedure which allows to choose for the given required photon energy the optimal material couples, their thickness and number of the layers (or the total thickness of the radiator). The complicated quantum mechanical theory of XTR and CR has been developed in some recent works (see [20] and references therein). It has been shown [20] that the difference between the quantum and CSM results often is not negligible. Though it has been claimed in [20] that CSM does not give the possible Cherenkov radiation contribution when the index of refraction becomes greater than unity (see below), nevertheless following the formulae of [19] one can see that the Cherenkov contribution also are taken into account. The disadvantage of [19] is in the fact that as CSM theory the refraction at the interfaces is neglected, while [20] is quantum theory and not available for complete understanding.

On the other hand in the work [21] it has been shown that in some substances X-ray Cherenkov radiation (XCR) can be generated by relativistic charged particles in thin layers of absorbing materials in frequency regions of abnormal dispersion near the K and L edges of absorption where the refractive index becomes greater than unity. This fact has been experimentally confirmed in [22] and some other works (see [23]) and is of interest as a method for production of monochromatic photon beams, especially, in the region of "water window" from 284 eV up to 540 eV.

This work is devoted to the analysis of the properties of XTR and XCR produced by ultrarelativistic particles in multilayers consisting of two solid state layers using a new CSM formula which is the generalization of the formula for stacks in vacuum [3] and takes into account the reflection from the interfaces. Some possible applications are discussed.

2.　THE CSM FORMULAE FOR XTR AND XCR RADIATION PRODUCED IN MULTILAYERS

Let a charged particle passes through a multilayer in vacuum, consisting of N pairs of layers of two materials of thickness a and b with dielectric constants ε_1 and ε_2, perpendicularly to the surfaces of the layers. Then according to the CSM formula given in [24] the energy-angular distribution of the radiated photons produced after the N-th pair is described by

$$\frac{d^2 N_{ph}}{dw d\theta} = \frac{2\beta^2}{137\pi} \frac{1}{w} \frac{\left|\varepsilon_1 - \varepsilon_2\right|^2 u_2^2 \sin^3}{\left|1 - u_1^2\right|^2 \left|1 - u_2^2\right|^2} \left|F_N(w,\theta)\right|^2 \tag{1}$$

where

$$F_N(w,\theta) = B \frac{U_{N-1}(\xi)}{Q_N(\xi)} + \frac{B + D \exp[ib_0(1+u_2)]}{2Q_N(\xi)[\cos(a_0 + b_0) - \xi]} f_N(w,\theta) \tag{2}$$

$$f_N(w,\theta) = U_{N-2}(\xi) - U_{N-1}(\xi)\exp[-i(a_0 + b_0)] + \exp[-iN(a_0 + b_0)] \tag{3}$$

$$\xi = \cos(a_0 u_1)\cos(b_0 u_2) - \frac{1}{2}\left[\frac{u_1 \varepsilon_2}{u_2 \varepsilon_1} + \frac{u_2 \varepsilon_1}{u_1 \varepsilon_2}\right]\sin(a_0 u_1)\sin(b_0 u_2)$$

$$Q_N = Q_1 U_{N-1} - 4U_{N-2}\varepsilon_1\varepsilon_2 u_1 u_2 \exp[i(a_0 u_1 + b_0 u_2)]e$$

$$Q_1 = (\varepsilon_1 u_2 + \varepsilon_2 u_1)^2 - (\varepsilon_1 u_2 - \varepsilon_2 u_1)^2 \exp(2ia_0 u_1)$$

$$B = 2u_1(1+u_2)(1 - \beta^2\varepsilon_1 - u_2)\exp[-ia_0(1-u_1)] - (\varepsilon_1 u_2 + \varepsilon_2 u_1)(1+u_1)(1 - \beta^2\varepsilon_2 - u_1) +$$
$$(\varepsilon_1 u_2 - \varepsilon_2 u_1)(1-u_1)(1 - \beta^2\varepsilon_2 + u_1)\exp(2ia_0 u_1),$$

$$D = 2u_1(1+u_2)(1 - \beta^2\varepsilon_1 - u_2)\exp[ia_0(1+u_1)] -$$
$$(\varepsilon_1 u_2 + \varepsilon_2 u_1)(1+u_1)(1 - \beta^2\varepsilon_2 + u_1)\exp(2ia_0 u_1) +$$
$$(\varepsilon_1 u_2 - \varepsilon_2 u_1)(1-u_1)(1 - \beta^2\varepsilon_2 + u_1)$$

and

$$a_0 = \frac{a\omega}{v} = \frac{2\pi a}{\beta\lambda}, \quad b_0 = \frac{b\omega}{v} = \frac{2\pi b}{\beta\lambda} \tag{4}$$

$$u_{1,2} = \beta(\varepsilon_{1,2} - \sin^2\theta)^{1/2} \tag{5}$$

In the above formulae $w = h\varpi$ is the energy of the emitted photon and $U_N(\xi)$ are second kind Chebishev polynoms of the argument ξ, satisfying the recurrent relations

$$U_N - 2\xi U_{N-1} + U_{N-2} = 0 \tag{6}$$

In expressions (1)-(6) $\varepsilon_{1,2} = \varepsilon'_{1,2} + i\varepsilon''_{1,2}$ are the complex dielectric constants of the layers. For various values of w the values of $\varepsilon'_{1,2}$ are given in tables, while $\varepsilon''_{1,2}$ can be determined with the help of relation $\mu_{1,2} = \varpi\varepsilon''_{1,2}/c$, where $\mu_{1,2}$ are the linear absorption coefficients.

3. XTR

In the pure X-ray region far from the absorption edges and when $\omega \gg \omega_{p1,2}$, where $\omega_{p1,2}$ the plasma frequencies the dielectric constants are equal to $\varepsilon_{1,2} = 1 - \omega_{p1,2}^2/\omega^2$,. For reasonable finite N one can assume that it is small not only the reflection from one interface, but also the total reflection from all the boundaries is small, i.e. N $|\varepsilon_{1,2} - 1| \ll 1$. When $\gamma \gg 1$ assuming $\theta \ll 1$ and expanding various terms in (1)-(6) into series one obtains the following formulae for XTR

$$d^2 N_{RTR}/d\Omega d\omega = F_1 F_2 F_3, \qquad (7)$$

where

$$F_1 = d^2 N/d\Omega d\omega = \alpha\omega\theta^2 (Z_1 - Z_2)^2/16\pi^2 c^2 \qquad (8)$$

is TR intensity from one interface with $Z_{1,2}$ formation zones of XTR in the first and second layers, respectively;

$$F_2 = 1 + \exp(-\mu_1 a) - 2\exp(-\mu_1 - a/2)\cos(2a/Z_1) \qquad (9)$$

is a factor taking into account the absorption and interference of TR from single layer of the first medium,

$$F_3 = \frac{1 + \exp(-2M\sigma) - 2\exp(M\sigma)\cos(2MX)}{1 + \exp(-\sigma) - 2\exp(-\sigma/2)\cos(2X)} \qquad (10)$$

is a factor taking into account the absorption and interference in the entire stack; $\sigma = \mu_1 a + \mu_2 b$ and $X = a/Z_1 + b/Z_2$. The formation zones are given by the formulae

$$Z_{1,2} \approx \frac{4\beta c}{\omega\left[\gamma^{-2} + (\omega_{p1,2}/\omega)^2 + \theta^2\right]}. \qquad (11)$$

The formulae (7)-(11) for XTR are used in experiments with submicron [11] as well as with thicker layers [2-11] and are in good agreement with the experimental data.

As it has been mentioned above the experimental results [11,12] show that the X-ray mirror like stacks of submicron layers can serve as effective radiators for production of quasi monochromatic X-ray beams. Here we want to show additionally that just as OTR they can find application for XTR beam diagnostics (see the review [25] and references therein), for instance, for measurement of the energy of particles in a beam with angular spread less than $1/\gamma$. Since the angle of the multiple scattering in such radiators with total thickness $L_R \approx 1$ μm $\approx 10^{-5} X_0$ (X_0 is the radiation length of the radiator material) $\theta_{ms} \approx 21$ MeV/(mc^2 γ) $(L_R / X_0)^{1/2} < 1/\gamma$, the angular distribution of XTR will not be destroyed. Such energy measurement can be carried out either measuring the angular distribution of TR as in case of OTR beam diagnostics or by a new method measuring such spectral distributions of XTR photons under certain angles as it is shown in Fig. 1 because despite OTR the spectral distribution of XTR strongly depends on θ and γ.

At very high energies when $\gamma > 500$ and the necessary angular collimation is very difficult one cannot use this proposed method. Nevertheless XTR produced in thin multilayers despite its small intensity can be used for the measurement of the energy of particle beams in the way, which is used for particle identification. As the results of our calculations in Fig. 2 show, the yield of XTR energy W and number N of photons integrated over angles and in certain photon energy interval depends strongly on the particle energy γ. Therefore measuring the number of XTR photons produced by a beam one and knowing the number of the beam particle one can measure the energy of the beam particle tough the number of XTR photons per particle is less than one. Moreover, taking into account that in the case of heavy relativistic ions XTR intensity is proportional to Z^2 one can measure the energy of single ions as it is discussed in the work [26].

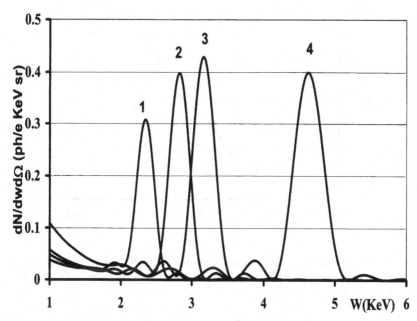

Figure 1. The spectral distribution of XTR produced by electrons with $\gamma = 30, 35, 40$ and 100 (curves 1, 2, 3 and 4) in a multilayer consisting of M = 10 pairs of nickel and carbon layers with thickness a = 175 nm and b = 221 nm, respectively and used in [11].

4.   XCR

As it has been mentioned in the Introduction due to the fact that dielectric constant of some materials can be greater than unity in X ray regions close to absorption edges quasi monochromatic XCR can be produced by particles passing through these materials [4]. One can calculate the intensity of such XCR produced in a thick media taking into account the absorption and transition radiation using the formulae [4]. Having multilayers of such materials one can calculate the intensity of the expected XCR using the CSM formulae of the works [3,6,7] and especially [19] or the results of quantum scattering theory model (STM) [20]. However, the CSM formulae of [3,6,7,19] do not take into account the reflections from the interfaces of the multilayer, which can be significantly large in the case of XCR, while the use of quantum mechanical results of [20] or other works (see the references of [20]) is very complicated. Therefore in this section we shall use the above CSM formulae (1)-(6), which despite the previous ones take into account the reflection.

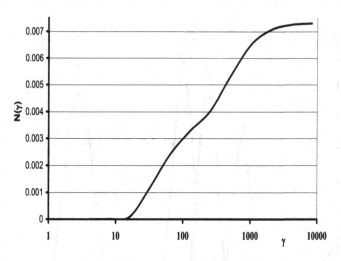

Figure 2. The γ -dependence of XTR photons with energies (2-20) keV coming out from a multilayer consisting of N=300 pairs of tungsten and carbon with thickness a = 0.5 nm and b = 0.7 nm, respectively and used in[12].

The necessary values of $\varepsilon_{1,2}$ for large intervals of photon energies have been calculated using the data on atomic form factors given in [27] according to the recipes described in [4]. Figure 3 shows the angular distributions calculated with the help of new and CSM formulae. As it is seen at angles $\theta = 11.9^0$, 22.0^0, 33.3^0 and 42.0 there are peaks. Comparing with the results of [20], namely with the Fig. 3 of [20] it is necessary to note with an accuracy ~20% that the places and intensities of the last three peaks are in agreement with those in the work [20] calculated using STM and CSM methods. However instead of the high peaks at ~12^0 in Fig. 3 there is only a peak in Fig. 3 STM calculations at ~1.9^0 which they interpret as TR.

Such discrepancy between the results calculated with various methods can be explained, for instance, by the fact that the formulae (1)-(6) take into account the reflection from the interfaces. These and other results need in more detailed consideration.

Nevertheless we made some other calculations too. Figure 4 shows the spectral distributions calculated by the formulae (1)-(6) at the radiation angles 11.9^0 and 22.3^0, which show that XCR with its interference phenomena gives the dominant contribution. CSM results obtained when $\varepsilon < 1$ are not applicable. Figure 5 shows the angular distribution of XCR integrated over the frequencies in the photon energy interval 60-140 eV which can be detected by detectors of type of CCD. After integration over angles one has bout 0.005 and 0.002 of photons emitted around the angles $\theta_1 = 11.8^0$ and $\theta_2 = 22.3^0$.To obtain the real detected angular distribution of

XCR it is necessary to take into account also the detection efficiency of the detectors, which will be done, in our next work.

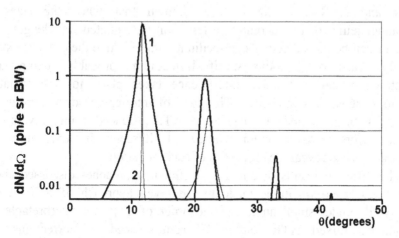

Figure 3. The angular distributions of XCR produced by 15 MeV electrons in SiMo multilayer consisting of N=101 pair with thick nesses $a=b=$ 65 nm at $w=$ 99.8 eV. The results of calculations with the help of the formulae (1) – (6) (curves 1, thick solid) and with the help of formulae (7) – (11) (curves 2, thin solid).

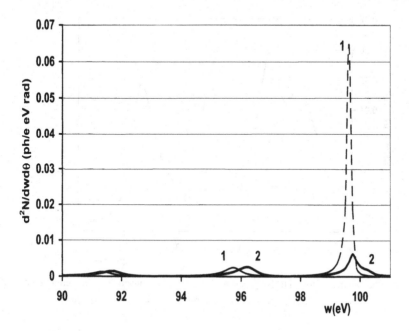

Figure 4. The spectral distribution of the radiation under angles $\theta_1 = 11.88^0$ (curve 1, dashed) and $\theta_2 = 22.3^0$ (curve 2, solid).

5.   DISCUSSION

It is difficult to study analytically the behavior of the radiation, when XTR and XCR are mixed. It is known that when the sizes of inhomogeneities of the medium are less than the radiated wavelengths, the radiator can be considered as one medium with effective dielectric constant. In this case of periodic stratified medium, probably, complicated constructive and destructive interference takes place for the radiation produced at various interfaces. This type of interference differs from that which has been studied earlier [1-5] for XTR produced in multilayers with thicker layers in X-ray region where $\varepsilon < 1$. It seems that only numerical calculations can reveal the properties of such radiation.

The different results obtained by different approaches discussed above are due to approximations made in the given approach. If XTR results observed[11] under small angles of the order of $1/\gamma$ are interpretable the parametric mixed XTR and XCR results need in corrections and experimental verification. This is necessary since according to the theoretical results given in already published works as well as in this work thin multilayer of the X-ray mirror type can serve as efficient radiators for production of soft X-ray beams.

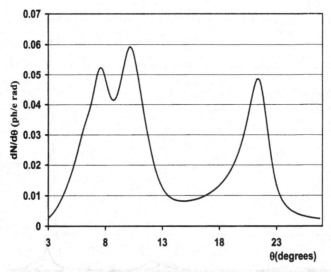

Figure 5. The calculated angular distribution of XCR photon produced in the energy interval (60-140) eV without taking into accounts the parameters of the photon detector.

The proposed applications of XTR produced in thin multilayers for beam diagnostics and relativistic heavy ion detection are supplement for the existing methods. The possible application of XCR for particle detection will be discussed in another publication.

Acknowledgement

This work is supported by the project NFSAT Ph-092-02/CRDF 12046.

References

1. V.L. Ginzburg and I.M. Frank, Radiation of uniformly moving electron passing from one medium to another, J.Phys. USSR, 9 (1945) 353; *Zh. Ekxp.Teor.Fiz.* **16** (1946) 15.
2. M.L. Ter-Mikaelian, High Energy Electromagnetic Processes in Condensed Media (Wiley Interscience, New York, 1972).
3. G.M. Garibian and Yan Shi, Rentgenovskoe Perekhodnoe Izluchenie (Publishing House of Academy of Science of Armenia, 1983, in Russian).
4. V.A. Bazilev and N.K. Zhevago, Izluchenie Bistrikh Chastits vo Veschestve I vo Vneshnikh Polyakh (Moscow, Nauka, 1987, in Russian).
5. P. Rulhusen, X. Artru and P. Dhez, Novel Radiation Sources Using Relativistic Electrons (World Scientific, Singapore, 1998).
6. G.M. Garibian, L.A. Gevorgian and Yan Shi, The calculation of X-ray transition radiation generated in regular and irregular layered media, *Zh. Eksper. Teor. Fiz.* **66**, 552 (1974); *Nucl. Instr. and Meth.* **125**, 133 (1975).
7. X. Artru, G.B. Yodh, G. Menessier, Practical theory of the multilayered transition radiation detector, *Phys. Rev.* **D12**, 1289 (1975).
8. A.I. Alikhanian, F.R. Arutyunian, K.A. Ispirian and M.L. Ter-Mikaeliam, A possibility of detecting of high-energy particles, *Zh. Eksp. Toer. Fiz.* **41**, 2001 (1961).
9. F.R. Arutyunian, K.A. Ispirian and A.G. Oganesian, Coherent radiation of cosmic-ray muons with energies 700-6000 GeV in layered medium, *Yad. Fizika*, **1**, 842 (1965).
10. M.A. Piestrup, J.O. Kephart, H. Park, R.K. Klein, R.H. Pantell, P.J. Ebert, M.J. Moran, B.A. Dahling and B.L. Berman, *Phys. Rev. A* **32**, 917 (1985).
11. K. Yamada, T. Hosokawa and H. Takenaka, Observation of soft X-Rays of single-mode RTR from multilayert target with submicron period, *Phys. Rev. A* **59**, 3673 (1999).
12. V.V. Kaplin, S.R. Uglov, V.N.Zabaev, M.A.Piestrup, M.A. Gary, N.N. Nasonov and M.K. Fuller, Observation of bright monochromatic X-rays generated by relativistic electrons passing through a multilayer mirror, *Appl. Phys. Lett.* **76**, 3647 (2000).
13. I.E. Tamm, *J. Phys. USSR* **1**, 439 (1939).
14. I.M.Frank, Izluchenie Vavilova-Cherenkova, Voprosi Teorii (Nauka, Moscow, 1988).
15. A.P. Kobzev, On the direction properties of the Vavilov-Cherenkov radiation, *Yadernaya Fizika*, **27**, 1256 (1978).
16. Ya.B. Fainberg and N.A. Khizhnyak, Energy losses of charged particles passing through layered dielectric *Zh. Eksp. Teor. Fiz.* **32**, 883 (1957).
17. V.A. Arakelian and G.M. Garibian, XTR produced in stack of plates by CSM, *Izvestia Akad. Nauk Arm. SSR, Fizika*, **4**, 339 (1969).
18. V.A. Arakelian, The energy losses of charged particles moving through stack of plates, *Radiofizika*, **29**, 320 (1986).

19. A.E. Kaplan, C.T. Law and P.L. Shkolnikov, X-ray narrow-line TR source based on low-energy electron beams traversing multilayer, *Phys. Rev. E* **52,** 6795 (1995).
20. B. Lastdrager, A. Tip and J. Verhoeven, Theory of Cerenkov and transition radiation from layered structures, *Phys. Rev. E* **61,** 5767 (2000).
21. V.A. Bazilev, V.I. Glebov, E.I. Denisov, N.K. Zhevago and A.S. Khlebnikov, Cherenkov radiation as a intense X-ray source, *Pisma Zh. Eksp. Toer. Fiz.* **24,** 406 (1976).
22. V.A.Bazilev, V.I. Glebov, E.I. Denisov et al., X-ray Cherenkov radiation. Theory and experiment, *Zh. Eksp. Teor. Fiz.* **81,** 1664 (1981).
23. W. Knults, M.J. van der Wiel, J. Luiten and J. Verhoeven, High-brightness, compact soft X-ray source based on Cherenkov radiation, *Appl. Phys. Lett.*; Proc. of SPIE **393,** 5196 (2004).
24. M.A. Aginian, L.A. Gevorgian, K.A. Ispirian V.G. Khachatryan, UV Radiation produced in a stack of thin plates and UV ring Transition radiation detectors, *Nucl Instr. And Meth. A* **522,** 99 (2004).
25. K.A. Ispirian, New application of none-optical transition radiation for beam diagnostics, *Nucl Instr. and Meth. A* **522,** 5 (2004).
26. N.Z. Akopov, K.A. Ispirian, M.K. Ispirian and V.G. Khachatryan, Ring transition radiation detectors, in: *Proc. of the Intern. Workshop TRD of 3-rd Millennium*, edited by N. Giglietto, P. Spinelli, (Bari, Italy, Frascati, 2001) Physics Series. Vol. XXV, p.59.
27. B.L. Henke, E.M. Gullikson and L.C. Davis, X-ray interactions: photoabsorption, scattering, transmission and reflection at E=50-30000 eV, Z=1-92, *Atom. Data and Nucl. Data Tables* **54,** 181 (1993); see also http://www.cxro.lbl.gov/optical_constants.

X-RAY CHERENKOV RADIATION

K.A. Ispirian

Yerevan Physics Institute, Br. Alikhanian 2, 375036 Yerevan, Armenia

Abstract: After a short review of the published few theoretical works on X-ray Cherenkov radiation (XCR) produced in some materials with positive dielectric constant near the K-, L-, M- absorption edges it is considered the existing and possible experimental results and discussed some applications.

Keywords: Cherenkov radiation/Transition radiation/

1. INTRODUCTION

Intense beams of quasimonochromatic soft X-ray photons are required for many scientific and industrial applications. In particular, for the biological studies with the help of soft X-ray microscope it is necessary photons with energies in the water window region, 280-543 eV. At present such beams are produced at synchrotron radiation facilities with controllable, good parameters and by means of various methods, such as high harmonic generation by irradiation by femtosecond lasers without good reproducibility and with high pedestal. At present [1] transition radiation (TR), channeling radiation (ChR) and parametric X-ray radiation are considered as candidates for a compact source for high brightness cheaper beams which can be available at smaller institutes, factories and hospitals.

Still in 1972 it has been [2] proposed to use the X-ray TR contained in the beams of bremsstrahlung radiation providing higher intensity, than the single pass synchrotron radiation, while the spectra of soft photons [3] calculated with the help of measured real and imaginary part of the dielectric constant have shown enhancements near the absorption edges. In the works [4,5] it has been experimentally shown that the Cherenkov radiation

H. Wiedemann (ed.), Advanced Radiation Sources and Applications, 217–234.

produced by charged particle in noble gases also can serve as an intense source in the region of VUV and soft X-rays photon beams.

In 1976 V.A. Bazylev, N.K. Zhevago and others [6] have shown that for some materials the dielectric constant can be larger than 1, $\varepsilon > 1$, in narrow X-ray regions due to the anomalous resonance dispersion near the absorption edges, and therefore Cherenkov radiation can be produced in such regions. Using the behavior of the susceptibility, $\chi = 1 - \varepsilon$ they have calculated X-ray Cherenkov radiation (XCR) intensity and have predicted a narrow spectral distribution for this radiation. XCR has been observed for the first time in 1981 [7] using a carbon thin foil and 2 GeV electrons. Up to now it has been experimentally studied only in a few works [8-11] using ~10 μ m thick single layers.

Still in [12] it has been shown that the Cherenkov radiation produced in multilayers has parametric character, and its intensity can be enhanced if some conditioned are satisfied. This circumstance has been theoretically and experimentally proven for X-ray transition radiation (see [13-16]). Despite of theoretical efforts (see [17,18] and references therein) interference effects have never been observed for Cherenkov radiation and XCR. As it is well known when the thickness of the layers is small, of the order of the emitted wavelength the Cherenkov and transition radiation are mixed. As the results of recent calculations [18] show when the dielectric constant even of one multilayer material is greater than unity in the angular distribution of X-rays there are peaks under large angles which are characteristic for Cherenkov radiation. One can call such radiation XTR, XCR or following [12] parametric Cherenkov radiation. The results of [17] and similar works have been obtained using the coherent summation method (CSM) without taking into account the reflection of the radiation from the interfaces, while a quantum mechanical approach using scattering theoretical method (STM) has been developed in [18]. In the new unpublished work [19] it has been derived a CSM formula valid for optical as well as for X-ray region taking into account the reflection from the interfaces. Naturally the results of [17], [18] and [19] differ each from other since the adopted methods and approximations are different, and experimental results are necessary to separate the correct theory.

In the work [8] by measuring the angular distributions of XCR at various incidence angles it has been shown that for grazing incidence XCR intensity produced in Si with E_L = 99.82 eV and carbon with E_K = 284.2 eV by 75 MeV electrons can be more than for the perpendicular incidence. With the help of notions introduced in [6, 7] they have given some considerations for the qualitative interpretation of the experimental data. The most important results of [9] is the evidence that XCR photons with energy ~10 eV can be produced with an intensity of ~ 1 10^{-3} photon per electron with as low an energy as 5 MeV. The measured spectral distributions of XCR are in satisfactory agreement with the theoretical ones calculated for the first time for Si by the authors.

These and later works [10, 11] the result of which will be discussed bellow have confirmed that 1) XCR intensity is sufficiently high ~ 10^{-4} – 10^{-3} photon per electron, 2) it is naturally very monochromatic with spectral width of few eV, 3) its production is not connected with crystalline radiators requiring certain orientation in goniometers and 4) it is less fastidious to the parameters of the electron beam. These properties make the use of XTR very perspective for series of applications, especially, in the water window region.

Further after reviewing the physics and theoretical results produced in infinite media without absorption, in more realistic single layers taking into account the absorption and possible development of XCR produced in multilayers we shall consider the experimental results of [10,11] and discuss XCR applications.

2. THEORY OF XCR

2.1 Real and Imaginary Parts of Susceptibility, χ' and χ''

In general the dielectric constant ε and the susceptibility χ of the amorphous substances depend on frequency, are complex and connected with each other by the relation

$$\varepsilon(\omega) = \varepsilon'(\omega) + \varepsilon''(\omega) = 1 + \chi(\omega) = 1 + \chi'(\omega) + \chi''(\omega). \quad (1)$$

Omitting ω one can write

$$-\chi' = 1 - \varepsilon', \quad (2)$$

$$\chi'' = \varepsilon''. \quad (2')$$

For $\omega \gg \omega_P = \sqrt{4\pi N_e e^2 / m}$, where N_e is the electron density, and $w_P(eV) = \hbar\omega_P \approx 19\sqrt{\rho Z / A}$ is the energy corresponding to the plasma frequency ω_P for material with density ρ (g/cm^3) and A and Z are atomic weight and number

$$\varepsilon = 1 - \omega_P^2 / \omega^2 + i\mu_l c / \omega, \tag{3}$$

where μ_l is the linear absorption coefficient Therefore, χ' and χ'', the real and imaginary parts of $\chi = \chi' + i\chi''$, can be determined by the relations

$$\chi' = -\omega_P^2 / \omega^2, \tag{4}$$

$$\chi'' = \mu_l c / \omega. \tag{4'}$$

However near the absorption edges K, L, M...it is reasonable to determine χ' and χ'' using the so-called atomic factors $f = f_1 + if_2$ which are the coefficient multiplying which with the X-ray scattering amplitudes for single electrons one find the amplitude for the given atoms. The values for f_1 and f_2 are tabulated in [20], and the necessary relations are

$$\chi'' = \frac{4\pi N_{at} r_0 c^2}{\omega^2} f_2, \quad \text{or} \quad \chi'' = K \frac{\rho}{A} \frac{f_2}{w^2(eV)}, \tag{5}$$

$$\chi' = \frac{2}{\pi} \oint_0 \frac{x\chi''(x)}{x^2 - w^2} dx. \tag{6}$$

In (5) K = 830.15 eV^{2}cm^3, when w is measured in eV.

Without discussing the difficulties of calculations of χ' and χ'' (see [15]) for illustration on Fig. 1a we show the behavior of χ' and χ'' for C [7]. Fig. 2 shows the calculated [11] dependence of χ' and χ'' for Si and Ti. As it is seen, the values of χ' (and therefore ε) become positive only in a narrow region around the K-edge with maximal values of the order of $\sim 10^{-3}$, which exceeds the values of χ'' by a about a factor of 10^3.

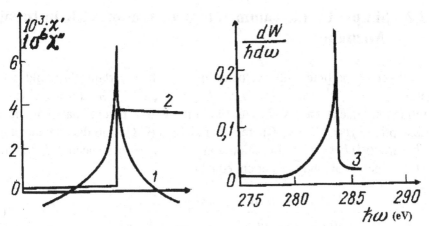

Figure 1. Dependence of χ' and χ'' a) and of XCR intensity b) produced by 1 GeV electrons in carbon with ρ =1.5 g/cm^3 upon photon energy [15].

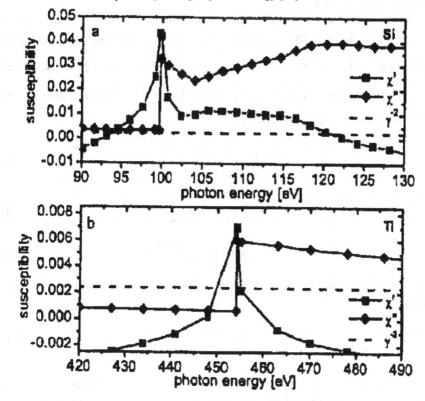

Figure 2. The dependence of χ' and χ'' for Si and Ti [11].

2.2   Simple Frank-Tamm Interpretation of XCR in Infinite Medium

When a particle with velocity greater than certain threshold value $v > v_{thr} = c/\sqrt{\varepsilon'}$ or $\beta = v/c > \beta_{thr} = 1/\sqrt{\varepsilon'} = 1/n$ ($n = \sqrt{\varepsilon'}$ is the refraction index) passes through an infinite transparent material without absorption ($\mu = \chi'' = 0$) Cherenkov radiation (CR) is produced under angle $\theta = \arccos(1/\beta n)$ with formation zone $Z_f \to \infty$ and spectral distribution of the Cherenkov photons per cm equal to.

$$\frac{d^2 N^{CR}}{d l d\omega} = \frac{1}{\hbar\omega}\frac{d^2 W^{CR}}{d l d\omega} = \frac{\alpha}{c}\left(1 - \frac{1}{\beta^2 n^2}\right), \tag{7}$$

where $\alpha = e^2/\hbar c = 1/137$.

From (7) we conclude that in certain X-ray regions where $n^2 = \varepsilon \gg 1$, XCR will be produced [15] by relativistic particles with $\gamma = E/mc^2 = (1 - \beta^2)^{-1/2} \gg 1$, and again $Z_f \to \infty$. For XCR the spectral distribution, $d^2 N/d l d\omega$, threshold, γ_{thr}^{XCR}, radiation angle, and θ^{XCR} are given by the expressions

$$\frac{d^2 N}{d l d\omega} = \frac{\alpha}{c}(\chi' - \gamma^{-2}), \tag{8}$$

$$\gamma_{thr}^{XCR} = 1/\sqrt{\chi'}, \tag{9}$$

$$\theta^{XCR} = \sqrt{\chi' - \gamma^{-2}}, \tag{10}$$

If one takes into account the absorption the formation zone becomes finite [15].

$$Z_f^{XCR} = \frac{\lambda}{\chi' - \gamma^{-2}}. \tag{11}$$

Only XCR produced in the last part of a semi-infinite radiator thickness $(1-5)/\mu$ comes out and the spectral distribution of XCL from a radiator with thickness L is given by

$$\left|\frac{dN}{d\omega}\right|_L = \frac{\alpha}{c}(\chi' - \gamma^{-2})\frac{1 - \exp(-\mu L)}{\mu}. \tag{12}$$

For radiator with $L >> L_{abs} = 1/\mu$ which is usually taken for technical and intensity reasons

$$\left.\frac{dN}{d\omega}\right|_{L>>L_{abs}} = \frac{\alpha}{\omega}\frac{\chi' - \gamma^{-2}}{\chi''}. \tag{13}$$

Therefore XCR intensity is higher if $\chi' - \gamma^{-2} >> \chi''$.

2.3 Joint Ginzburg-Frank-Garibian Interpretation of XCR and XTR in a Layer

Taking a radiator with thickness $L >> L_{abs} = 1/\mu$ and detecting the radiation in vacuum from a particle passing through such a layer or from semi-infinite medium with ε to vacuum with $\varepsilon = 1$ one will have XCR and X-ray transition radiation (XTR), the summary intensity of which is given by Ginzburg-Frank-Garibian formula [14-16]

$$\frac{d^2N}{d\omega d\Omega} =$$

$$\frac{\alpha\beta^2|\varepsilon - 1|^2}{\pi^2\omega}\frac{\sin^2\theta\cos\theta}{\left|\varepsilon\cos\theta + \sqrt{\varepsilon - \sin^2\theta}\right|^2}\frac{\left|1 - \beta^2 - \beta\sqrt{\varepsilon - \sin^2\theta}\right|^2}{\left|(1 - \beta^2\cos^2\theta)(1 - \beta\sqrt{\varepsilon - \sin^2\theta})\right|^2}. \tag{14}$$

For $\gamma >> 1$, $\theta << 1$ and assuming $\mu = \chi'' = 0$ (no absorption) one obtains Garibian's formula for XTR

$$\frac{d^2N}{d\omega d\Omega} = \frac{\alpha\theta^2}{\omega\pi^2}\left|\frac{1}{\gamma^{-2} - \chi' + \theta^2} - \frac{1}{\gamma^{-2} + \theta^2}\right|^2. \tag{15}$$

The expressions (14) and (15) contain XTR and XCR. The expression (14) gives ∞ when $\left(1 - \beta\sqrt{\varepsilon - \sin^2\theta}\right) = 0$, which taking into account the Snell's law, gives the Cherenkov condition $1 - \beta\sqrt{\varepsilon}\cos\theta_{med} = 0$ or $\cos\theta_{med} = 1/\beta n$, where θ_{med} is the angle of radiation in the semi-infinite medium.(15) gives ∞ when $\gamma^{-2} - \chi' + \theta^2 = 0$ which means that XCR has maximum at angles (10). XTR has maximum at $\theta \approx 1/\gamma$.

When $\mu \neq 0$ (there is absorption) without taking into account the multiple scattering the spectral-angular distribution of XCR after a layer with thickness L is given by [15]

$$\frac{d^2N}{d\omega d\Omega} = \frac{\alpha c\theta^2}{\pi^2\omega}|A(\omega,\theta)|, \tag{15'}$$

where

$$A = \left[\frac{1}{\gamma^{-2} - \chi + \theta^2} - \frac{1}{\gamma^{-2} + \theta^2} \right] \left\{ 1 - \exp\left[-\frac{i\omega L}{2}\left(\gamma^{-2} - \chi + \theta^2 \right) \right] \right\} \quad (15'')$$

For $L \gg L_{abs}$ and integrating (15') over all angles one obtains

$$\frac{dN}{d\omega} = \left| \frac{dN}{d\omega} \right|_1 + \left| \frac{dN}{d\omega} \right|_2 \qquad (15''')$$

where

$$\left| \frac{dN}{d\omega} \right|_1 = \frac{\alpha c}{2\pi\omega} \left[\left(1 - \frac{2\chi'\gamma^{-2}}{\chi'^2 + \chi''^2} \right) \ln \frac{\left(\gamma^{-2} - \chi' \right)^2 + \chi''^2}{\gamma^{-4}} - 2 \right] \quad \text{and}$$

$$\left| \frac{dN}{d\omega} \right|_2 = \frac{\alpha c}{\pi\chi''} \left[\chi' - \gamma^{-2} \frac{\chi'^2 - \chi''^2}{\chi'^2 + \chi''^2} \right] \left[\frac{\pi}{2} + arctg \frac{\chi' - \gamma^{-2}}{\chi''} \right] \qquad (15'''')$$

In general, one can not identify $\left| \dfrac{dN}{d\omega} \right|_1$ and $\left| \dfrac{dN}{d\omega} \right|_2$ with XTR and XCR

contributions, however, the authors of [11] obtain a formula and show that in some cases such a separation is possible and at very high energies the XTR contribution increases $\sim \gamma$, while XCR contribution is constant.

Before proceeding let us discuss some characteristics of XCR using numerical results. Figure 3a shows XCR angular distribution calculated by formula (14) for Si at w=99.7 eV at various electron energies. The XTR and XCR peaks and their energy dependence are seen. The spectral distribution of XCR in Si at E=10 MeV calculated by formulae (7) and (10) are shown in Fig. 3b. Since XCR beams with energies in the water widow are important for some applications. In Fig. 4 shows the calculated [11] XCR spectra for low Z metals from which it is easy to make micron thick radiator. Monochromatic lines at higher energies are not practical since $(dN / d\omega) \sim 1/\omega$ and substances with higher values of Z have no $\varepsilon_1 > 0$.

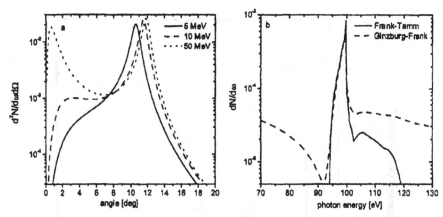

Figure 3. Angular distribution of XCR in Si at w=99.7 eV and various E a) and the spectral distribution of XCR in Si at E=10 MeV b) [11].

As it follows from (15) for the case $L \gg L_{abs}$ the XCR intensity is given by the same formula (15) in which χ' is replaced by $\chi = \chi' + i\chi''$. If also $Z_f^{XCR} \ll L_{abs}$ or $\chi' \gg \chi''$ and $\gamma \gg \gamma_{thr}^{XCR}$ then [15] for angles close to (10) one can neglect the second term of (15) and write

$$\frac{d^2 N}{d\omega d\Omega}\left(\theta \approx \theta^{XTR}\right) \approx \frac{\alpha}{4\pi^2 c} \frac{1}{\left|\theta - \theta^{XCR}\right|^2 + (\chi'' / 2\sqrt{\chi' - \gamma^{-2}})^2} \quad (16)$$

Therefore XCR with the given frequency takes place around θ^{XCR} in an angular interval with width

$$\Delta\theta \approx \chi''\left(\chi' - \gamma^{-2}\right)^{-1/2}, \quad (17)$$

while the spectral distribution of XCR has a Lorentz form with width [20]

$$\Gamma(\omega) = \frac{2\chi''(\omega)}{\left|d\chi'(\omega)/d\omega\right|}. \quad (18)$$

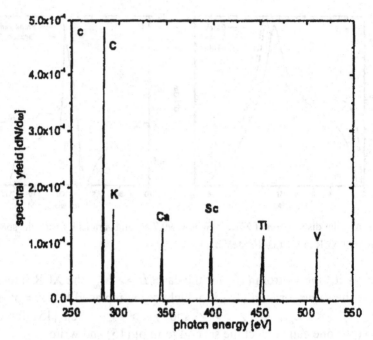

Figure 4. Narrow spectral lines of XCR in water window region [11].

2.4   Stimulated XCR

Since the amplification rate g per centimeter of the radiator of stimulated radiation produced by relativistic particles is determined by the derivatives of the line forms of the spontaneous radiation the narrow spectral and angular distributions of XCR give hope that one can expect intense stimulated XCR production when a sufficiently intense electron beam passes through a layer [15]. As the calculations [21] show the total gain, $G = gL$ achievable for an electron beam with density ρ_e is given by

$$G = \alpha \lambda^2 \rho_e \frac{\chi'^2}{\gamma \chi''^3} \left(1 - \frac{\gamma_{thr}}{\gamma}\right) \frac{d}{dx} \frac{1}{x^2 + 1/4}, \tag{19}$$

where $\gamma_{thr} = 1/\sqrt{\chi'}$, and $x = \left(\omega - \omega^{XCR}\right)/\Gamma$. If we requiring that the inhomogeneous widening due to multiple scattering on the radiator thickness $L \approx \lambda/(2\pi\chi'')$ be less than $\Gamma(\omega)$ we obtain a lower value of energy

$$\gamma > \gamma_{cr} = \left(\frac{2\lambda}{\alpha \chi''^2 L}\right)^{1/2}, \tag{20}$$

which decreases essentially G. Therefore the observation of stimulated XCR even for high ρ_e seems impossible in case of single layer radiator.

2.5 XCR Produced in Multilayers

The technology of preparation of multilayers consisting of alternating layers of two materials has been developed since the multilayers found wide application as X-ray mirrors, monochromators etc. After the works [22] and [23,24] devoted to the experimentally study of XTR and PXR production in thin multilayers first theoretically considered in [25] there is no doubt that the multilayers will serve as good radiators for production of X-ray beams. However to our knowledge no theoretical and experimental results on XCR produced in multilayers has been published. At present multilayers made of 10-100 pairs of layers of various materials with thickness 0.5-500 nm are available.

As it has been mentioned in the Introduction there are two types of calculations. First type are the results [17] obtained by the coherent summation method (CSM) which as in case of XTR does not take into account the reflections from the interfaces. The second type of results [18] are obtained using complicated quantum mechanical scattering theoretical methods (STM), and it is not clear are the reflections taken into consideration or not. In both cases taking one or both two materials with dielectric constant greater than 1 one can expect an intense narrow band XCR production in the same narrow region where $\varepsilon > 1$. Moreover according to [12] this parametric Cherenkov radiation cam be enhanced as in XTR. As the numerical results [18] show in the angular distribution of XCR there is satisfactory agreement between the CSM and STM results in case of some peaks and great differences in case of other peaks. No numerical calculations have been made for the XCR spectral distributions.

Recently we have derived [19] a CSM formula taking into account the reflections following the method described and used in [14] for a not realistic stack consisting of one type of layers with vacuum or air intervals. Our formula is valid also for optical CR and TR and it has been used in [26] for the study of optical parametric radiation.

Let the radiator consists of N pairs of plates with dielectric constants ε_1 and ε_2. According to lengthy calculations the spectral-angular distribution of the radiation in the second medium of the N-th pair produced by an ion with charge Z is given by the formula

$$\frac{d^2 N_{ph}}{dwd\theta} = \frac{2\beta^2}{137\pi} \frac{Z^2}{w} \frac{|\varepsilon_1 - \varepsilon_2|^2 u_2^2 \sin^3 \theta}{|1-u_1^2|^2 |1-u_2^2|^2} |F_N(w,\theta)|^2, \tag{21}$$

where

$$F_N(w,\theta) = B\frac{U_{N-1}(\xi)}{Q_N(\xi)} + \frac{B + D\exp[ib_0(1+u_2)]}{2Q_N(\xi)[\cos(a_0 + b_0) - \xi]} f_N(w,\theta), \tag{22}$$

$$f_N(w,\theta) = U_{N-2}(\xi) - U_{N-1}(\xi)\exp[-i(a_0 + b_0)] + \exp[-iN(a_0 + b_0)], \tag{23}$$

$$\xi = \cos(a_0 u_1)\cos(b_0 u_2) - \frac{1}{2}\left[\frac{u_1\varepsilon_2}{u_2\varepsilon_1} + \frac{u_2\varepsilon_1}{u_1\varepsilon_2}\right]\sin(a_0 u_1)\sin(b_0 u_2),$$

$$Q_N = Q_1 U_{N-1} - 4U_{N-2}\varepsilon_1\varepsilon_2 u_1 u_2 \exp[i(a_0 u_1 + b_0 u_2)],$$

$$Q_1 = (\varepsilon_1 u_2 + \varepsilon_2 u_1)^2 - (\varepsilon_1 u_2 - \varepsilon_2 u_1)^2 \exp(2ia_0 u_1),$$

$$B = 2u_1(1+u_2)(1-\beta^2\varepsilon_1 - u_2)\exp[-ia_0(1-u_1)] -$$
$$(\varepsilon_1 u_2 + \varepsilon_2 u_1)(1+u_1)(1-\beta^2\varepsilon_2 - u_1) +$$
$$(\varepsilon_1 u_2 - \varepsilon_2 u_1)(1-u_1)(1-\beta^2\varepsilon_2 + u_1)\exp(21a_0 u_1),$$

$$D = 2u_1(1+u_2)(1-\beta^2\varepsilon_1 - u_2)\exp[ia_0(1+u_1)] -$$
$$(\varepsilon_1 u_2 + \varepsilon_2 u_1)(1+u_1)(1-\beta^2\varepsilon_2 + u_1)\exp(2ia_0 u_1) +$$
$$(\varepsilon_1 u_2 - \varepsilon_2 u_1)(1-u_1)(1-\beta^2\varepsilon_2 + u_1)$$

and

$$a_0 = \frac{a\omega}{v} = \frac{2\pi a}{\lambda}, b_0 = \frac{b\omega}{v} = \frac{2\pi b}{\lambda}, \tag{24}$$

$$u_{1,2} = \beta(\varepsilon_{1,2} - \sin^2 \theta)^{1/2}. \tag{25}$$

In the above formulae $w = \hbar\omega$ is the energy of the emitted photon and $U_N(\xi)$ are second kind Chebishev polynomial of the argument ξ, satisfying the recurrent relations

$$U_N - 2\xi U_{N-1} + U_{N-2} = 0. \tag{26}$$

In expressions (21)-(26) $\varepsilon_{1,2} = \varepsilon_{1,2}' + i\varepsilon_{1,2}''$ are the complex dielectric constants of the layers. For various values of w the values of $\varepsilon_{1,2}'$ are given in tables, while $\varepsilon_{1,2}''$ can be determined with the help of relation $\mu_{1,2} = \omega\,\varepsilon_{1,2}''/c$, where $\mu_{1,2}$ are the linear absorption coefficients.

For the X-ray region $\omega > E_K$ where E_K is the energy of the K-edge the formulae (21)-(26) give the more practical formulae [13] for XTR, which are derived assuming $\varepsilon < 1$. However they are not valid formulae when $\varepsilon > 1$. The formulae (21)-(26) are valid for the X-ray regions close to the K,L, M... absorption edges where the dielectric constants can be greater than 1, and XCR can be produced. Nevertheless the CSM formulae (21)-(26) are sufficiently complicated and one can not make predictions on the XCR properties. Only following [12,14] one can say that XCR described by (21)-(26) has parametric nature due to the interferences between the radiation produced at various interfaces, and as it is seen the radiation intensity must be enhanced for such values of frequencies and angles for which the denominators in (22) are minimal. Some results from [19] on XCR and their comparison with the results of the works {17, 18] are shown in Fig 5. The place and magnitude of the second, third and very small fourth peaks are in

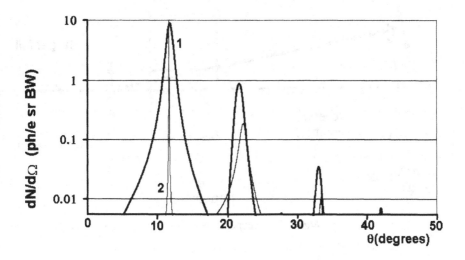

Figure 5. The angular distribution of XCR produced by 15 MeV electrons in a Si/Mo multilayer consisting of N=101 pairs with thicknesses a = b = 65 nm at w = 99.8 eVcalculated with the help of the formulae (21)-(26) (thick solid curves, 1) and CSM formulae (thin curves, 2)

3.   EXPERIMENTAL RESULTS [11]

3.1   The Experimental Arrangement

The experimental arrangement used in [11] is shown if Fig. 6.a). The electron beam from a medical linac has energy E =10 MeV $\Delta E / E \approx 4\%$, $I_{av} \approx 0.6$ nA, beam emittance 10 mm mrad, spot diameter ~1 mm. After deflection 90^0 the beam hits a carbon Faraday cup. 10 μ m thick Ti and V foils served as XCR targets. The CCD detectors were placed on a circle at 1m distance from the target after a thin aluminum (150nm) / carbon (27nm) filter to prevent CCD from softer background TR. The solid angle accepted by the detector was 2.2 10^{-4} sr. Only 10% of 80000 pixels were illuminated. The CCD efficiency was ~ 90 % in the region 0.28 – 15 KeV with spectral resolution ~ $1/\sqrt{w}$ and 165 eV (FWHM) at 1.5 keV. The angular interval under which the radiation angular-spectral distribution can be measured is from -20^0 up to $+20^0$. The energy calibration of the CCD camera was made with the help of K_α lines of the characteristic radiation from Al, Ti and V.

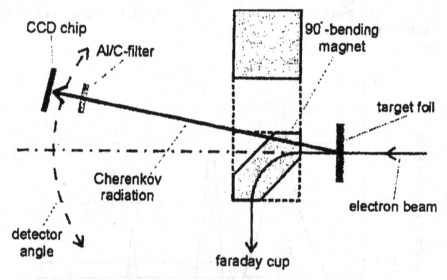

Figure 6. The experimental setup [11]

3.2   Spectral Distribution

The spectral distribution of the radiation measured under $\theta = 4^0$ is shown in Fig. 7a. The two peaks shown in the inserts b) an c) are the XCR and CR peaks, respectively, produced in the target by 10 MeV electrons in

Ti. The linearly pedestal decreasing with w low energy background is due TR, while the tail is background bremsstrahlung. After Gauss fitting of XCR the place of the peaks from Ti and V are at 459 ± 2 eV and 519 ± 3 eV, respectively when the corresponding L- edges are at 453 eV and 512 eV.

The procedure of the analysis is explained in Fig.7d. The doted curve is obtained integrating (15), while the dashed curve is obtained taking into account the absorption on Al/C filter. The solid curve is the convolution of the dashed curve with the spectral resolution and efficiency of CCD. It coincides with the measured XCR peak (Fig.7b). The shift of the peak's maximum value is explained by the fact that at a few eV below L-edge χ' becomes 0 when $\varepsilon = 1$ and there is no XTR and XCR (this is the explanation in [11]) and by the fact that the value of χ or μ has a jump at L-edge. Let us note that the experimental width of the XCR peak is ~100 eV, compared with the ~2 eV width of the theoretical peak. Therefore, indeed, XCR intensity is ~50 times higher that the observed one.

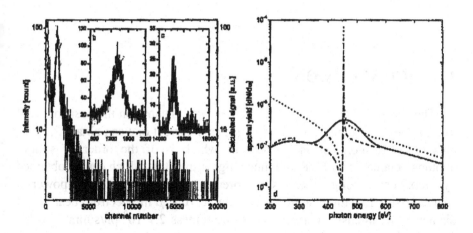

Figure 7. Measured a, b and c) and calculated spectral distribution of XCR [11].

3.3 Angular Distribution

The angular distribution measured for Ti and V and shown in Fig. 8a and b, respectively, as points is obtained by integrating curves as Fig.7b obtained for various angles and taking into account the efficiency of CCD and absorption in Al/C filter. Atypical ChR rather TR type behavior is seen. The theoretical XCR angular distributions shown as dotted curve are broadened due to multiple scattering as well as due to angular divergence of the electron beam.

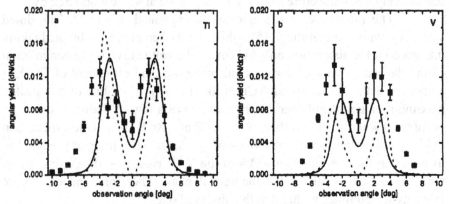

Figure 8. Measured (points) and calculated (solid and dotted curves) angular distribution of XCR [11].

4.   CONCLUSION

The authors [11] conclude that as their results show on the basis of XCR using a 10 MeV small accelerator with 1 mA current one can obtain 2.2 10^{12} ph/s or XCR monochromatic photons which is of the order of similar photons obtainable by PXR and much higher than that which can be obtained by monochromizing the usual bremsstrahlung beam or powerful characteristic radiation sources. Having a 100 μ m electron beam spot diameter this intensity corresponds to brightness 2.7 10^9 ph/s/mm²/sr/o.1% BW which is higher than the laser plasma and higher harmonic methods provide. The conclusion is: XCR is very suitable for application of soft X-ray photon beams, especially for the purpose of soft X-ray microscopy in the water window region.

References

1. P. Rullhusen, X. Artru and P. Dhez, Novel Radiation Sources Using Relativistic Electrons (World Scientific, Singapore, 1998).
2. K.A. Ispirian, On the Soft X-Ray Transition Radiation, Preprint YerPhI, EFI-6(72), 1972.
3. K.A. Ispirian and S.T. Kazandjian, Transition Radiation and Optical Properties in VUV region, *Fiz. Tverdogo Tela*, **15**, 1551 (1973).
4. M.A. Piestrup et al, Cerenkov Radiation as a Source of Ultaviolet Radiation, *J.Appl. Phys.* **44**, 5160 (1973).
5. M.A. Piestrup et al, Cerenkov Radiation as a Light Souirce for 2000-620 A° Spectral Range, *Appl. Phys. Lett.* **28**, 92 (1976).
6. V.A. Bazilev et al, Cherenkov radiation as intense X-ray source, *Pisma Zh. Eksp. Teor. Fiz.* **24** 406 (1976).
7. V.A. Bazilev et al, X-Ray Cherenkov Radiation. Theory and Experiment, *Zh. Eksp. Teor. Fiz.* **81**, 1664 (1981).
8. M.J. Moran, B.Chang, M.B. Schneider and Z.K. Maruyama, Grazing-Incidence Cherenkov X-Ray Generation, *Nucl. Instr. And Meth.* **B48**, 287 (1990).
9. W. Knulst, O.J. Luiten, M.J. van der Wiel and J. Verhoeven, Observation of Narrow-Band Si L-Edge Cherenkov Radiation Generated by 5 MeV Electrons, *Appl. Phys. Lett.* **79**, 2999 (2001).
10. W. Knulst, O.J. Luiten, M.J. van der Wiel and J. Verhoeven, High-Brightness, Compact Soft X-ray Source Based on Cherenkov Radiation, *Appl. Phys. Lett.* **83**, 4050 (2003).
11. W. Knulst, O.J. Luiten, M.J. van der Wiel and J. Verhoeven, High-Brightness, Compact Soft X-ray Source Based on Cherenkov Radiation, *Proc. SPIE,* **5196**, 393 (2004).
12. Ya.B. Fainberg and N.A. Khizhnyak, Energy Losses of Charged Particles Passing through layered Dielectric, *Zh. Eksper. Teor. Fiz.* **32**, 883 (1957).
13. M.L. Ter-Mikaelian, The Influence of the Medium on High Energy Processes at High Energies, Publishing House of Acamemy of Science of Armenia, Yerevan, 1969; High Energy Electromagnetic Processes in Condensed Media (Wiley Interscience, New York, 1972).
14. G.M. Garibian and Yan Shi, Rentgenovskoe Perekhodnoe Izluchenie, (Publ. of Academy of Science of Armenia, Yerevan, 1983).
15. V.A.Bazilev and N.K.Zhevago, Izluchenie Bistrikh Chastits v Veshchestve i vo Vneshnikh Polyakh (Moscow, Nauka ,1987).
16. V.L. Ginzburg and V.N. Tsitovich, Transition Radiation and Transition Scattering (Adam Higler, Bristol, 1990).
17. A.E. Kaplan, A.E.Law and P.L. Shkolnikov, X-Ray Narrow-Line TR source Based on Low Energy Electron Beams Traversing Multilayer Nanostructure, *Phys. Rev.* **E52** 6795 (1995).
18. B.Lastdrager, A.Tip and J.Verhoeven, Theory of Cerenkov and Transition Radiation from Layered Structures, *Phys. Rev.,* **E61**, 5767 (2000).
19. M.A. Aginian, L.A. Gevorgian, K.A. Ispirian, and I.A. Keropyan, X-Ray Transition Radiation Produced by High-Energy Particles in Multilayers and Their Applications, in Proc of this Workshop.
20. B.L. Henke, E.M. Gullikson and J.C. Davis, X-ray interactions: photoabsorption, scattering, transmission and reflection at E=50-30000 eV, Z=1-92, Atom. Data and Nucl. Data Tables, 54, 181,1993; see also http://www.cxro.lbl.gov/optical_constants.

21. V.A. Bazylev, N.K. Zhevago and M.A. Kumakhov, Theory of stimulated X-ray Cherenkov radiation, *Dokladi Akad Nauk SSSR, Fizika,* **263,** 855 (1982).

22. K.Yamada, T.Hosokawa and H. Takenaka, Observation of soft X-Rays of Single-Mode RTR from Multilayert Target with Submicron Period, *Phys.Rev.* **A59** 3673 (1999).

23. V.V. Kaplin et al, Observation of Bright Monochromatic X-Rays Generated by Relativistic Electrons Passing through a Multilayer Mirror, *Appl. Phys. Lett.* **76** 3647 (2000).

24. V. Kaplin et al, Tunable, Monochromatic X-Rays using internal Beam of a Betatron, *Appl. Phys. Lett.* **80** 3427 (2002).

25. N.K. Zhevago, Soft XTR at Inclined Incidence on Mutilayer Structure with Period of the order of the Wavelength, Proc. Intern.Symp. on TR of HEP (Yerevan,1977, p200).

26. M.A. Aginian, K.A. Ispirian, M.K. Ispiryan, V.G. Khachatryan, UV Radiation Produced in a Stack of Thin Plates and TR detectors, *Nucl. Instr. and Meth.* **A522,** 99 (2004).

ON X–RAY SOURCES BASED ON CHERENKOV AND QUASI-CHERENKOV EMISSION MECHANISMS

C. Gary[1], V. Kaplin[2], A. Kubankin[3], V. Likhachev[3], N. Nasonov[3], M. Piestrup[1], S. Uglov[2]

[1] *Adelphi Technology, Inc. 981-B Industrial Rd. San–Carlos, CA 94070, USA*

[2] *NPI Tomsk Polytechnic University, Lenin Ave. 2-A, Tomsk, 634050, Russia*

[3] *Laboratory of Radiation Physics, Belgorod State University, 14 Studencheskaya st., Belgorod, 308007, Russia*

Abstract   A variety of possible schemes of X–ray sources based on Cherenkov like emission mechanisms is considered theoretically. The possibility to increase substantially an angular density of parametric X–ray source under conditions of grazing incidence of emitting relativistic electrons on the reflecting crystallographic plane of a crystalline target is shown. The growth of Cherenkov X–ray angular density due to modification of the structure of Cherenkov cone in conditions of grazing incidence of an electron beam in the surface of a target is discussed as well as the peculiarities of Cherenkov X–ray generation from relativistic electrons crossing a multilayer nanostructure. The question of relative contributions of parametric X–ray and diffracted bremsstrahlung to total emission yield from relativistic electrons moving in a perfect crystal is elucidated. X–ray generation during multipasses of an electron beam through an internal target in circular accelerator is considered as well.

Keywords: Relativistic electron, Cherenkov X–ray radiation, Parametric X–ray radiation, Bremsstrahlung, X–ray source

1.   Introduction

Creation of an effective and inexpective X–ray source alternative to synchrotrons is a major focus of interest for the studies of coherent emission from relativistic electrons in dense media. A number of novel sources based on transition radiation [1], channelling radiation [2], parametric X–ray radiation [3], Cherenkov radiation [4] have been studied and considered as candidates for possible applications. However, the intensity

H. Wiedemann (ed.), Advanced Radiation Sources and Applications, 235–265.
© 2006 Springer. Printed in the Netherlands.

of these sources must be increased for them to be practical [5]. Because of this, the study of new possibilities to increase the spectral-angular density of such sources is a subject of much current interest.

The main attention in this paper is devoted to the analysis of Cherenkov like emission mechanisms. Substantial disadvantage of these mechanisms consists in the strongly non-uniform angular distribution of emitted photons close to a hollow cone. As a consequence one should collimate the emitted flux to obtain a uniform photon distribution on the surface of irradiated test specimen. A possibility to use the effect of emission cone modification in order to increase the part of collimated photon flux employed is studied in this paper. Parametric X–rays (PXR) from relativistic electrons moving in a crystal or multilayer nanostructure is considered in Sec.2. Substantial growth of PXR local angular density in conditions of small orientation angles of the emitting particle velocity relative to reflecting crystallographic plane is shown. Sec.3 is devoted to study of Cherenkov X–ray radiation in the vicinity of photoabsorption edge of the target's material [1, 6–20].The case of grazing incidence of emitting electrons on the surface of the target is the object of investigations.

Cherenkov X–ray radiation from relativistic electrons crossing a multilayer nanostructure is considered in Sec.4 of this paper. This scheme is of interest because of possible arrangement of the irradiated test specimen in the immediate vicinity of the multilayer due to large emission angles achievable on condition under consideration.

Relative contributions of PXR and diffracted bremsstrahlung to total emission yield from relativistic electrons crossing a perfect crystal are studied in Sec.5. This problem is of the great interest because the results of some recent experiments [11, 12] are in contradiction with the generally accepted opinion that PXR yield is determined in the main by the Bragg scattering of the fast particle Coulomb field (actually PXR emission mechanism). Performed studied show that the contribution of diffracted bremsstrahlung to total PXR yield may be very substantial in contrast with above opinion.

2. PXR in conditions of small orientation angles of emitting particle velocity relative to reflecting plane

Let us consider properties of PXR from relativistic electrons moving through a crystal aligned by reflecting crystallographic plane at the small angle $\varphi/2$ to the velocity of the emitting electron (see Fig.1). The well known kinematical formula for the spectral-angular distribution of emitted PXR energy [3, 13]

$$\frac{\mathrm{d}E}{\mathrm{d}t\,\mathrm{d}\omega\,\mathrm{d}\Omega} = \frac{8\pi Z^2 e^6 n_a^2 |S(\mathbf{g})|^2 e^{-g^2 u_T^2}}{m^2(1+g^2R^2)^2} \frac{(\epsilon\omega\mathbf{v}-\mathbf{g})^2 - [\epsilon\omega(\mathbf{n},\mathbf{v})-(\mathbf{n},\mathbf{g})]^2}{[g^2+2\omega\sqrt{\epsilon}(\mathbf{n},\mathbf{g})]^2}$$
$$\times\delta\left[\omega\left(1-\sqrt{\epsilon}(\mathbf{n},\mathbf{v})\right)-(\mathbf{g},\mathbf{v})\right] \qquad (1)$$

is used for our purposes. Here Z is the atomic number of the crystalline target, n_a is its atomic density, $S(\mathbf{g})$ is the structure factor of an elementary cell, $\mathbf{g}$ is the reciprocal lattice vector, u_T is the mean square amplitude of thermal vibrations of atoms, R is the screening radius in the Fermi-Thomas atom model, the simplest model with exponential screening is used, $\epsilon = 1 - \omega_0^2/\omega^2$, ω_0 is the plasma frequency, $\mathbf{n}$ is the unit vector to the direction of emitted photon propagation, $\mathbf{v}$ is the velocity of emitting electron.

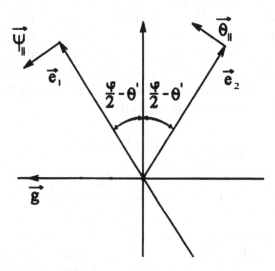

Figure 1. The geometry of PXR process. $\mathbf{e}_1$ is the axis of the emitting electron beam, $\mathbf{e}_2$ is the axis of the emitted photon flux, θ' is the orientation angle describing possible turning of the crystalline target by the goniometer, $\mathbf{g}$ is the reciprocal lattice vector, $\Theta_\parallel$ and $\Psi_\parallel$ are the components of the angular variables Θ and Ψ parallel to the plane determined by the vectors $\mathbf{e}_1$ and $\mathbf{e}_2$.

Introducing the angular variables Θ and Ψ in accordance with formulae

$$\mathbf{v} = \mathbf{e}_1\left(1 - \tfrac{1}{2}\gamma^{-2} - \tfrac{1}{2}\Psi^2\right) + \Psi, \qquad (\mathbf{e}_1, \Psi) = 0,$$
$$\mathbf{n} = \mathbf{e}_2\left(1 - \tfrac{1}{2}\Theta^2\right) + \Theta, \qquad (\mathbf{e}_2, \Theta) = 0, \qquad (\mathbf{e}_1, \mathbf{e}_2) = \cos\varphi, \quad (2)$$

one can reduce the formula (1) to more simple form in the case of small orientational angle $\varphi/2 \ll 1$ under consideration

$$\frac{\mathrm{d}E}{\mathrm{d}t\,\mathrm{d}\omega\,\mathrm{d}^2\Theta} \approx E_0\frac{g}{\omega}\frac{\frac{g}{\omega}(\frac{g}{\omega}-\varphi+2\theta'+2\Theta_{\parallel})-\frac{1}{\gamma^2}}{(\frac{g}{\omega}-\varphi+2\theta'+2\Theta_{\parallel})^2} \tag{3}$$

$$\times \delta\left[\frac{1}{\gamma^2}+(\Psi_\perp-\Theta_\perp)^2+(\varphi+\Psi_{\parallel}-\Theta_{\parallel})^2-\frac{g}{\omega}(\varphi+2\theta'+2\Psi_{\parallel})\right],$$

$$E_0 = \frac{16\pi Z^2 e^6 n_a^2 |S(\mathbf{g})|^2 e^{-g^2 u_T^2}}{m^2(1+g^2 R^2)^2 g^3},$$

where two-dimensional angles Ψ and Θ describe the angular spread in the beam of emitting electrons and the angular distribution of emitted photons respectively, $\Psi^2 = \Psi_{\parallel}^2+\Psi_\perp^2$, $\Theta^2 = \Theta_{\parallel}^2+\Theta_\perp^2$, the angle θ' describing the possible turning of the crystay by the goniometer is outlined in Fig.1, γ is the Lorentz-factor of emitting electron, the energy of emitted photon ω is assumed to be large as compared with the critical energy $\gamma\omega_0$ determining the manifestation of the density effect in PXR [14].

Let us consider the orientational dependence of strongly collimated PXR ($\Theta^2 \ll \gamma^{-2}$) from electron beam crossing a thin enough crystal (the multiple scattering angle Ψ_{tot} achievable at the exit of the target is assumed to be small as compared with γ^{-1}). The simple formula follows from (3) in conditions under consideration

$$\frac{\mathrm{d}E}{\mathrm{d}t\,\mathrm{d}^2\Theta} = E_0\gamma^2 g F_0(2\gamma\theta',\gamma\varphi), \tag{4}$$

$$F_0 = \frac{\gamma^2\varphi^2}{1+\gamma^2\varphi^2}\frac{\left(2\gamma\theta'-\frac{1}{\gamma\varphi}\right)^2}{(1+4\gamma^2\theta'^2)^2}$$

Obviously, in the limit $\gamma\varphi \to \infty$ the dependence $F_0(2\gamma\theta')$ is reduced to well known in PXR theory curve with two symmetrical maxima. On the other hand, amplitudes of these maxima differ essentially from each other in the range of small orientation angles, as may be seen from Fig.2. The discussed asymmetry implies that the most part of emitted PXR photons is concentrated in the narrow range of observation angles close to the left-hand maximum in PXR angular distribution. It is of first importance that the amplitude of this maximum increases with decreasing of the orientation angle $\varphi/2$. Thus, the possibility to increase substantially PXR angular density can be realized in the range of grazing incidence of emitting electrons on the reflecting crystallographic plane of the crystalline target.

In circumstances where $\varphi \ll 1$ special attention must be given to PXR spectral density. Integrating (3) over observation angles allows

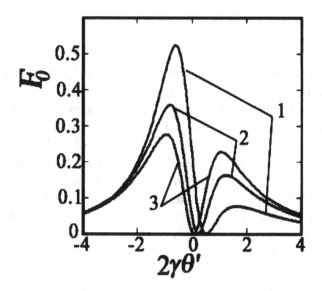

Figure 2. Asymmetry of PXR orientational dependence. The presented function $F_0(2\gamma\theta')$ is connected with PXR angular density by Eq.4. The curves 1,2,3 have been calculated for $\gamma\varphi = 2,\ 5,\ 20$ respectively.

one to obtain in the case of strongly collimated radiation $\Theta_d \ll \varphi$ (Θ_d is the photon collimator angular size) best suited for our purposes the following expression for the spectrum of emitted energy

$$\frac{dE}{dt\,d\omega} \approx E_0 \gamma^{-1} \frac{(2\gamma\theta' - \frac{1}{\gamma\varphi})^2}{(1 + 4\gamma^2\theta'^2)^2} \gamma^3 \varphi^3 \frac{\omega_B}{\omega}$$

$$\times \sqrt{\frac{\Theta_d^2}{\varphi^2} - \left(1 - \sqrt{\frac{\omega_B}{\omega}\left(1 + \frac{2\theta'}{\varphi}\right)} - \frac{1}{\gamma^2\varphi^2}\right)^2} \tag{5}$$

$$\times \sigma\left[\frac{\omega}{\omega_B} - \frac{1 + 2\theta'/\varphi}{\left(1 + \frac{\Theta_d}{\varphi}\right)^2 + \frac{1}{\gamma^2\varphi^2}}\right] \sigma\left[\frac{1 + 2\theta'/\varphi}{\left(1 - \frac{\Theta_d}{\varphi}\right)^2 + \frac{1}{\gamma^2\varphi^2}} - \frac{\omega}{\omega_B}\right]$$

where $\omega_B = g/\varphi$ is the Bragg frequency in the vicinity of which PXR spectrum is concentrated, $\sigma(x) = 1$ if $x > 0$ and $\sigma(x) = 0$ if $x < 0$.

As one would expect, amplitude of the spectrum (5) as a function of the angle θ' peaks at the value of θ' corresponding to maximum in the orientational dependence (4). In line with (5), the relative width of PXR spectral peak

$$\frac{\Delta\omega}{\omega} \approx 4\frac{\gamma^2\varphi^2}{1+\gamma^2\varphi^2}\frac{\Theta_d}{\varphi} \tag{6}$$

increases with decreasing of φ in the range $\varphi \gg \gamma^{-1}$, but this growth is returned if φ becomes comparable with γ^{-1}.

The width (6) is determined by the collimator size Θ_d. In the real conditions this width may be changed due to multiple scattering of emitting electrons. In order to estimate an influence of multiple scattering on PXR properties one should average the general expression (3) over scattering angles. It is interesting to note in this connection that the representation of PXR spectral-angular distribution in the form (3) is very convenient for further analysis because the scattering angle Ψ appears in the argument of δ-function in (3) only.

Using the general formula

$$\frac{dE}{d\omega\,d^2\Theta} = \int_0^L dt \int d^2\Psi f(t,\Psi)\frac{dE}{dt\,d\omega\,d^2\Theta} \tag{7}$$

and the distribution function

$$f(t,\Psi) = \frac{1}{\pi(\Psi_0^2 + \Psi_S^2 t)}\exp\left[-\frac{\Psi^2}{\Psi_0^2 + \Psi_S^2 t}\right], \tag{8}$$

where Ψ_0 is the initial angular spread of emitting electron beam, $\Psi_S = \frac{1}{\gamma}\sqrt{L_{Sc}}$ is the multiple scattering angle per unit length, $L_{Sc} \approx e^2 L_{Rad}/4\pi$, L_{Rad} is the radiation length, one can obtain the following formula for the emission spectral-angular distribution:

$$\frac{dE}{d\omega\,d^2\Theta} = E_0\frac{\gamma L_{Sc}}{2}\Phi\left[\frac{\omega}{\omega_B},\Theta,\theta',\gamma\varphi,\frac{L}{L_{Sc}}\right], \tag{9}$$

with

$$\begin{aligned}
\Phi =\ & \gamma\varphi\frac{\omega_B}{\omega}\frac{\frac{\omega_B}{\omega}\left[\frac{\omega_B}{\omega} - 1 + \frac{2}{\varphi}(\theta' + \Theta_\parallel)\right] - \frac{1}{\gamma^2\varphi^2}}{\left[\frac{\omega_B}{\omega} - 1 + \frac{2}{\varphi}(\theta' + \Theta_\parallel)\right]^2} \\
& \times\sigma\left[\frac{\omega_B}{\omega}\left(\frac{\omega_B}{\omega} - 1 + \frac{2}{\varphi}(\theta' + \Theta_\parallel)\right) - \frac{1}{\gamma^2\varphi^2}\right] \\
& \times\int_{t_-}^{t_+}\frac{dt}{\kappa(t)}\left[E_1\left(\frac{t^2 + (\kappa - \Theta_\perp/\varphi)^2}{\Psi_0^2/\varphi^2 + \gamma^2\varphi^2 L/L_{Sc}}\right)\right.
\end{aligned} \tag{10}$$

$$-E_1 \left(\frac{t^2 + (\kappa - \Theta_\perp/\varphi)^2}{\Psi_0^2/\varphi^2} \right)$$

$$+E_1 \left(\frac{t^2 + (\kappa + \Theta_\perp/\varphi)^2}{\Psi_0^2/\varphi^2 + \gamma^2\varphi^2 L/L_{\text{Sc}}} \right) - E_1 \left(\frac{t^2 + (\kappa + \Theta_\perp/\varphi)^2}{\Psi_0^2/\varphi^2} \right) \Bigg],$$

$$\kappa(t) = \sqrt{(t_+ - t)(t - t_-)},$$

$$t_\pm = \frac{\omega_B}{\omega} - 1 + \frac{\Theta_\parallel}{\varphi} \pm \sqrt{\frac{\omega_B}{\omega}\left(\frac{\omega_B}{\omega} - 1 + \frac{2}{\varphi}(\theta' + \Theta_\parallel)\right) - \frac{1}{\gamma^2\varphi^2}}$$

Let us use the result (9) to study the influence of multiple scattering on the orientational dependence of strongly collimated emission. It should be noted that the formula (9) does not take into account an influence of photoabsorption of emitted photons, but small values of the angle φ correspond to large values of the Bragg frequency $\omega_B = g/\varphi$, so that the emission of weakly absorbed hard X-rays is considered actually in this section.

PXR spectrum (the function $\Phi(\omega/\omega_B)$) calculated by (9) in the maximum of orientational dependence ($\Theta = 0$, $2\gamma\theta' = -\sqrt{1 + 1/\gamma^2\varphi^2} + 1/\gamma\varphi$) is illustrated by the curves presented in Fig.3 and Fig.4. The simplest case of an electron beam without initial angular spread ($\Psi_0 = 0$) is considered in this paper. The curves presented in Fig.3 illustrate the effect of substantial increase of PXR spectral width with decreasing the parameter $\gamma\varphi$. The effect of saturation of PXR spectral density due to an influence of multiple scattering with an increase in the thickness of the target is demonstrated by the curves presented in Fig.4.

An advantage of PXR emission mechanism discussed above consists in the possibility to generate X-rays in wide frequency range. On the other hand the intensity of PXR is not very high. More intensive source of X-rays can be created on the base of Cherenkov emission mechanism [8, 9] to be studied in the next section.

3. Cherenkov X-rays in conditions of grazing incidence of emitting electrons on the surface of a target

Cherenkov emission mechanism allows to generate soft X-rays in the vicinity of atomic absorption edges, where a medium refractive index may exceed unity [4]. This theoretical prediction has been experimentally confirmed [6–10]. Let us consider Cherenkov X-rays under special conditions of grazing incidence of emitting electrons on the surface of a target. We are interesting in the emission process in soft X-ray

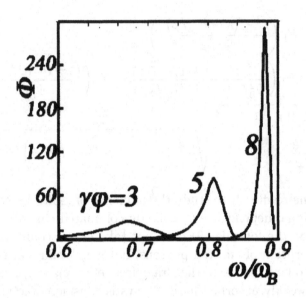

Figure 3. The growth of PXR spectral width with decreasing the incidence angle. The presented function Φ is connected with PXR spectral density by Eq.9. The curves have been calculated for fixed values of the parameters $\Theta = 0$, $\Psi_0 = 0$, $2\gamma\theta' = -\sqrt{1 + 1/\gamma^2\varphi^2} + 1/\gamma\varphi$, $L/L_{Sc} = 0.3$ and different values of the parameter $\gamma\varphi$.

range of the emitted photon energies ω, where an influence of a photoabsorption is very important. Assuming that the photoabsorption length $l_{ab} \sim 1/\omega\chi''(\omega)$ (χ'' is the imaginary part of the dielectric susceptibility) is less than the electron path in the target L/φ (L is the thickness of the target, φ is the grazing incidence angle, $\varphi \ll 1$) we are led to the simple model corresponding to the emission of a fast electron moving from semi-infinite absorbing target to a vacuum where the emitted photons are recorded in X–ray detector (see Fig.5). Since background in the small frequency range under study is determined in the main by transition radiation, we neglect the contribution of bremsstrahlung. In addition to this we consider the emission from electrons moving with uniform velocity $\mathbf{v}$, assuming that the value of multiple scattering angle achievable on the distance of the order of l_{ab}, $\Psi_{ms} \sim \sqrt{L_{ab}/\gamma^2 L_{Sc}}$ is small relative to characteristic angle of the Cherenkov cone $\sqrt{\chi'(\omega)}$ ($\chi'(\omega)$ is the real part of dielectric susceptibility).

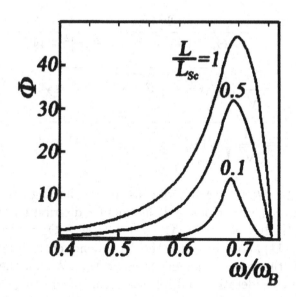

Figure 4. The saturation of PXR yield due to the influence of multiple scattering. The curves have been calculated for fixed values of the parameters $\gamma\varphi = 3$, $\Theta = 0$, $\Psi_0 = 0$, $2\gamma\theta' = -\sqrt{1 + 1/\gamma^2\varphi^2} + 1/\gamma\varphi$ and different values of the parameter L/L_{Sc}.

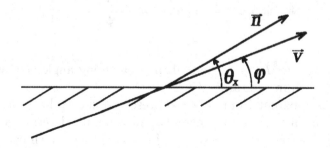

Figure 5. Geometry of the Cherenkov radiation process. **n** is the unit vector to the direction of emitted photon propagation, **v** is the emitting electron velocity, φ is the incidence angle.

Since the solution of the task under study is well known, we present the final result only

$$\frac{dE}{d\omega\, d^2\Theta} = \frac{16e^2}{\pi^2} \frac{\chi'^2 + \chi''^2}{(\Theta_x + \tau')^2 + \tau''^2} \tag{11}$$

$$\times \frac{\left[\gamma^{-2} + (\Psi_y - \Theta_y)^2 + \Theta_x^2 - \Psi_x^2\right]^2 + 4\Psi_x^2(\Psi_y - \Theta_y)^2}{\Omega_-^2 \Omega_+^2}$$

$$\times \frac{\Psi_x^2 \Theta_x^2}{[\gamma^{-2} + (\Psi_y - \Theta_y)^2 + \Theta_x^2 + \Psi_x^2 - 2\Psi_x \tau']^2 + 4\Psi_x^2 \tau''^2}$$

$$\tau' = \frac{1}{\sqrt{2}} \sqrt{\sqrt{(\Theta_x^2 + \chi')^2 + \chi''^2} + \Theta_x^2 + \chi'},$$

$$\tau'' = \frac{1}{\sqrt{2}} \sqrt{\sqrt{(\Theta_x^2 + \chi')^2 + \chi''^2} - \Theta_x^2 - \chi'},$$

where the angles Ψ and Θ describe as above the angular spread of the beam of emitting electrons and the angular distribution of emitted photons respectively (see Fig.5), $\Omega_\pm = \gamma^{-2} + (\Psi_y - \Theta_y)^2 + (\Psi_x \pm \Theta_x)^2$.

The formula (11) allows to search the dependence of the total emission distribution on the value of incidence angle Ψ_x. Within the range of small Ψ_x under consideration multiple scattering of emitting electrons in the target can influence substantially on emission properties. To account such influence one should average the expression (11) over Ψ_x and Ψ_y. We use in this work the simple distribution function

$$f(\Psi_x, \Psi_y) = \frac{1}{\pi \Psi_L^2} \exp\left[-\frac{\Psi_y^2 + (\Psi_x - \varphi)^2}{\Psi_L^2}\right], \qquad (12)$$

where $\Psi_L = \sqrt{L/\gamma^2 L_{Sc}\varphi}$ is the multiple scattering angle achievable at the electron path in the target L/φ.

As will be apparent from (11) the presented spectral-angular distribution contains a maximum (Cherenkov maximum), determined by the condition $\gamma^{-2} + (\Psi_y - \Theta_y)^2 + \Theta_x^2 + \Psi_x^2 - 2\Psi_x \tau' = 0$, which can be represented as

$$\gamma^{-2} - \chi' + \Theta_y^2 + \left(\varphi - \sqrt{\Theta_x^2 + \chi'}\right)^2 = 0 \qquad (13)$$

in the most interesting for us frequency range of anomalous dispersion before absorption edge where χ'' is usually much less than χ' (as this takes place $\tau' \approx \sqrt{\Theta_x^2 + \chi'}$, $\tau'' \approx \frac{1}{2}\chi'' / \sqrt{\Theta_x^2 + \chi'}$) and the Cherenkov radiation can be realized if $\chi' - \gamma^{-2} > 0$ in accordance with (13) (this is well known Cherenkov threshold in X–ray range). An influence of multiple scattering is neglected in (13) as well.

Let us consider the angular structure of the Cherenkov peak versus the orientation angle φ. From (11) for large enough $\varphi \gg \sqrt{\chi'} > \gamma^{-1}$ the emission angular distribution comprises two symmetric cones

$$\frac{dE_0}{d\omega\, d^2\Theta} = \frac{e^2}{\pi^2} \frac{\chi'^2}{\left[\gamma^{-2} - \chi' + \Theta_y^2 + (\varphi - \Theta_x)^2\right]^2 + \chi''^2}$$

$$\times \frac{\Theta_y^2 + (\varphi - \Theta_x)^2}{\left[\gamma^{-2} + \Theta_y^2 + (\varphi - \Theta_x)^2\right]^2} \tag{14}$$

The first of these radiation cones corresponding to the condition (13) represents Cherenkov radiation. Second one describes well known transition radiation. The structure of these cones is changed essentially when decreasing of the incidence angle φ. The distribution of emission intensity over azimuth angle on the Cherenkov cone becomes strongly non-uniform in contrast with (14). To show this let us compare the magnitudes of the distribution (11) in the plane $\Theta_y = 0$ at the points

$$\Theta_{x\,\text{max}}^{(\pm)} = \sqrt{\left(\varphi \pm \sqrt{\chi' - \gamma^{-2}}\right)^2 - \chi'}, \tag{15}$$

following from (13) and corresponding to the maximum of the Cherenkov radiation intensity. Such magnitudes follow from (11) and (15)

$$\left.\frac{dE_{\text{max}}^{(\pm)}}{d\omega\, d^2\Theta}\right|_{\Theta_y=0} = \frac{4e^2}{\pi^2} \frac{\chi' - \gamma^{-2}}{\chi''^2} \frac{1}{\varphi^2} \tag{16}$$

$$\times \left(\frac{1}{\varphi \pm \sqrt{\chi' - \gamma^{-2}}} + \frac{1}{\sqrt{\left(\varphi \pm \sqrt{\chi' - \gamma^{-2}}\right)^2 - \chi'}}\right)^{-2}$$

First of all it is necessary to note that two maxima (16) can be realized in the range of large enough values of $\varphi > \sqrt{\chi'} + \sqrt{\chi' - \gamma^{-2}}$ only, as it follows from (15). Obviously, the value of these maxima one the same for large enough $\varphi \gg \sqrt{\chi'}$ and coincide with that following from (14)

$$\left.\frac{dE_{\text{max}}^{(\pm)}}{d\omega\, d^2\Theta}\right|_{\Theta_y=0} = \frac{dE_{0\,\text{max}}}{d\omega\, d^2\Theta} = \frac{e^2}{\pi^2} \frac{\chi' - \gamma^{-2}}{\chi''^2}.$$

In accordance with (15) only the maximum $\frac{dE_{\text{max}}^{(+)}}{d\omega\, d^2\Theta}$ is realized in the range $\sqrt{\chi'} - \sqrt{\chi' - \gamma^{-2}} < \varphi < \sqrt{\chi'} + \sqrt{\chi' - \gamma^{-2}}$ (the Cherenkov cone begins to contact with the target's surface if $\varphi = \sqrt{\chi'} + \sqrt{\chi' - \gamma^{-2}}$). The magnitude of this maximum can exceed substantially the asymptotic value $\frac{e^2(\chi' - \gamma^{-2})}{\pi^2(\chi''^2)}$. The ratio $\left(\frac{dE_{\text{max}}^{(+)}}{d\omega\, d^2\Theta} \middle/ \frac{dE_{0\,\text{max}}}{d\omega\, d^2\Theta}\right)$, as the performed analysis of the expression (16) implies, depends strongly on the parameter $\gamma^2\chi'$. For example, this ratio has a value of about 10 if $\gamma^2\chi' = 5$.

The result obtained is of great importance for the creation of an effective Cherenkov based X–ray source. Indeed, Cherenkov emission mechanisms allow to generate very intensive soft X–ray beams (a yield of the order of 10^{-3} photons/electron has been obtained experimentally from a single foil [8, 9]). On the other hand, the angular density of Cherenkov radiation is not high since Cherenkov photons are emitted in a hollow cylindrical cone with a relatively large characteristic angle $\Theta_{Ch} \sim \sqrt{\chi'}$ to the electron trajectory. To obtain a uniform angular distribution of the emitted photons one must extract a small part of Cherenkov cone by the photon collimator. As a consequence, the useful part of the total emission yield is reduced substantially. In the range of small incidence angles $\varphi \ll 1$, the distribution of Cherenkov photons over azimuth angle becomes strongly non-uniform. Therefore the possibility to increase the used part of emission yield is opened by placing of photon collimator at the point corresponding to the maximum of the angular density of Cherenkov radiation. Such possibility is demonstrated by Fig.6, where the angular distribution of Cherenkov radiation calculated by the formula (11) for fixed photon energy is presented.

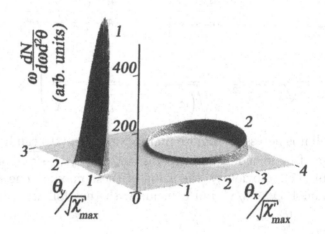

Figure 6. The dependence of Cherenkov cone structure on the incidence angle φ. The presented spectral-angular distributions of Cherenkov radiation have been calculated for Be target, $\omega = 111.6\ eV$ $1/\gamma\sqrt{\chi'_{max}} = 0.4$, $\chi'_{max} = 0.05$. Distribution 1 corresponds to the value of $\varphi = 0.17\sqrt{\chi'_{max}}$. Distribution 2 corresponds to $\varphi = 3\sqrt{\chi'_{max}}$.

As illustrated in Fig.6, there is not only the effect of non-uniform distribution of emitted photons over azimuth angle, but the effect of decreasing of Cherenkov emission angle when decreasing of the incidence angle φ as well. The effect in question have the simple geometrical interpretation [15]. As is clear from Fig.5 a photon emitted at the angle $\Theta_{x\,max}^{(+)} > \varphi$ has a shorter path L_{ph} in the target than that of L_{el} emitting electron. This effect is small for large orientational angles $\varphi \gg \sqrt{\chi'}$. The photon yield is formed in this case at the part of electron inside trajectory of the order of absorption length $l_{ab} \sim 1/\omega\chi''$. On the other hand in the range of small $\varphi < \sqrt{\chi'}$ the ratio $L_{el}/L_{ph} \sim L_{el}/L_{ab} \sim \Theta_{x\,max}^{(+)}/\varphi$ is increased substantially in accordance with (16), which is to say that the useful part of electron trajectory and consequently the photon yield are increased. Geometrical interpretation allows us to explain both azimuth non-uniformity of the angular distribution of emitted photons (photon path in the target is increased with increasing azimuth angle) and the effect of "Cherenkov angle" decreasing.

It should be noted that the degree of non-uniformity in the emission angular distribution over azimuth angle depends strongly on the value of incidence angle φ. The great importance of correct choice of φ is demonstrated by the curves presented in Fig.7. These curves describing the ratio $\left(\frac{dE_{max}^{(+)}}{d\omega\,d^2\Theta} \Big/ \frac{dE_{0\,max}}{d\omega\,d^2\Theta} \right)$ as a function of φ have been calculated by the formula (16) for different values of the parameter $\gamma^2\chi'$.

Cherenkov X-ray radiation yield from Be target has been calculated in this work by the use general formula (11) and dielectric susceptibilities $\chi'(\omega)$ and $\chi''(\omega)$ determined experimentally [16]. The curves presented Fig.8 describe the spectra of Cherenkov photons, emitted from the above mentioned target into the collimator with finite angular size. The collimator's center was placed in performed calculations at the point corresponding to maximum of the emission angular density. Its angular size was chosen so that the emission yield in such a collimator was close to saturation for small incidence angle φ when the emission angular distribution over azimuth angle was strongly non-uniform. The presented curves demonstrate the substantial growth of the emission yield when decreasing of the incidence angle φ.

It should be noted that the emission angular density increases very substantially when increasing emitting particle energy, but this growth is followed by decreasing of optimum value of incidence angle φ (see Fig.7). Along that the influence of multiple scattering of emitting electrons in the target increases as well. Obviously such influence must constrain the discussed growth of the emission angular density. We have calculated the spectral-angular distribution of Cherenkov radiation from Be target on

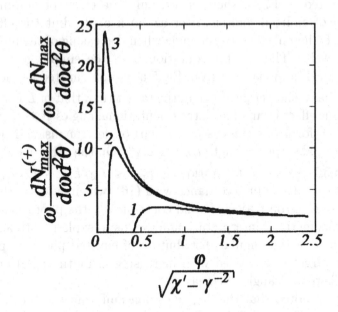

Figure 7. The amplification factor for Cherenkov angular density as a function of the incidence angle φ. The curves 1, 2 and 3 correspond to the value of the parameter $\gamma^2\chi' = 2$, 5 and 10 respectively.

the basis of the formula (11) averaged over beam spread at the exit of the target using the distribution function (12). The result of calculations, presented in Fig.9, shows a strong suppression of the angular density of Cherenkov radiation due to multiple scattering (this is because of very small angular width of Cherenkov cone proportional to χ'' as follows from (16)). On the other hand, the yield fixed by a photon collimator with finite angular size is not changed substantially as it is evident from the Fig.10.

Comparison of above considered X–ray sources based on PXR and Cherenkov mission mechanisms shows that PXR has an advantage over Cherenkov source consisting in the emission angle. This property allows to arrange an irradiated sample in the immediate vicinity of the source. The possibility to integrate properties of PXR and Cherenkov radiation sources into a single pattern is studied in the next section of the paper devoted to X–ray emission from relativistic electrons crossing a multi-layer nanostructure under conditions when Cherenkov radiation can be realized.

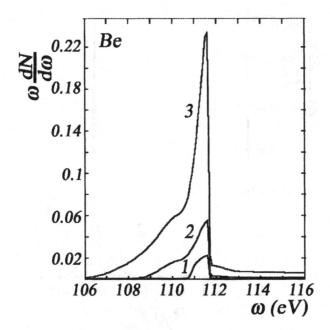

Figure 8. The spectrum of Cherenkov radiation from Be target as a function of the of the incidence angle φ. The curves have been calculated for *Be* target, $1/\gamma\sqrt{\chi'_{max}} = 0.1$, $\chi'_{max} = 0.05$, the collimator angular sizes $\Delta\Theta_x = 0.3\sqrt{\chi'_{max}}$, $\Delta\Theta_y = 0.3\sqrt{\chi'_{max}}$. Curves 1, 2 and 3 corresponds to $\varphi = 5\sqrt{\chi'_{max}}$, $0.5\sqrt{\chi'_{max}}$ and $0.05\sqrt{\chi'_{max}}$ respectively.

4. Cherenkov X–rays from relativistic electrons crossing a multilayer nanostructure

Consider X–ray emission from relativistic electrons moving in a medium with a periodic dielectric susceptibility $\chi(\omega, \mathbf{r}) = \chi_0(\omega) + \sum_{\mathbf{g}} {}'\chi_{\mathbf{g}}(\omega)\, e^{i(\mathbf{g},\mathbf{r})}$.

In the case of a one-dimensional structure consisting of alternative layers with thicknesses a and b and susceptibilities $\chi_a(\omega)$ and $\chi_b(\omega)$, respectively, the quantities $\chi_0(\omega)$ and $\chi_{\mathbf{g}}(\omega)$ are determined by the expressions

$$\chi_0(\omega) = \frac{a}{T}\chi_a + \frac{b}{T}\chi_b,$$

$$\chi_{\mathbf{g}}(\omega) = \frac{1 - e^{i(\mathbf{g},\mathbf{a})}}{igT}(\chi_a - \chi_b),$$
(17)

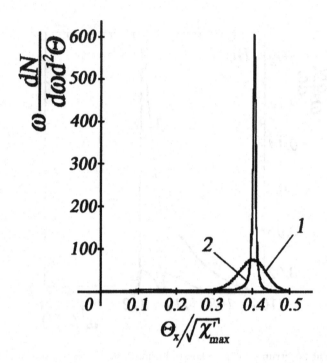

Figure 9. An influence of the multiple scattering on the Cherenkov radiation spectral-angular distribution. The curves 1 and 2 have been calculated with and without account of the multiple scattering respectively. The curves calculated for $1/\gamma\sqrt{\chi'_{\max}} = 0.04$, $\varphi = 0.08\sqrt{\chi'_{\max}}$.

where $T = a + b$ is the period of multilayer structure, $\mathbf{g} = \mathbf{e}_x g$, $g \equiv g_n = \frac{2\pi}{T}n$, $n = 0, \pm 1, \ldots$, $\mathbf{e}_x$ is the normal to the surface of a target (see Fig.11).

This task was under study in connection with the problem of X–ray source creation based on PXR and diffracted transition radiation from relativistic electrons crossing a multilayer nanostructure. The general solution obtained within the frame of dynamical diffraction theory can be found in [17], where an emission process in the range of hard X–rays far from photoabsorption edges has been considered. In contrast to this, soft X–ray generation in the vicinity of a photoabsorption edge is analyzed in this work. PXR properties can be changed substantially in this case due to the occurrence of Cherenkov radiation [18].

Embarking on a study of emission properties one should note that only Bragg scattering geometry can be realized in the case under consideration. Since soft X–rays are strongly absorbed in a dense medium,

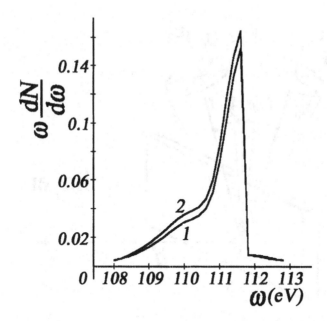

Figure 10. An influence of the multiple scattering on the Cherenkov radiation spectral distribution. The curves 1 and 2 have been calculated with and without account of the multiple scattering respectively. The curves have been calculated for *Be* target, $1/\gamma\sqrt{\chi'_{max}} = 0.11$, $\chi'_{max} = 0.05$, the collimator angular sizes $\Delta\Theta_x = 0.2\sqrt{\chi'_{max}}$, $\Delta\Theta_y = 0.3\sqrt{\chi'_{max}}$ and incidence angle $\varphi = 0.1\sqrt{\chi'_{max}}$.

the simple model of semi-infinite multilayer nanostructure can be used for calculations. Keeping in mind fundamental aspects of the discussed problem only, we shall restrict our consideration to the specific case of the emitting electron moving with a constant velocity $\mathbf{v} = \mathbf{e}_1(1 - \frac{1}{2}\gamma^{-2})$ (see formulae (2)). Determining the unit vector $\mathbf{n}$ to the direction of emitted photon propagation by the formula (2) and using general results [17, 19] one can obtain the following expression for the spectral-angular distribution of emitted energy

$$\frac{dE}{d\omega \, d^2\Theta} = \sum_{\lambda=1}^{2} |A_\lambda|^2, \tag{18}$$

where

$$A_\lambda = \frac{e}{\pi}\Theta_\lambda \frac{(\chi'_g + i\chi''_g)\alpha_\lambda}{\Delta - \chi'_0 + \delta'_\lambda - i(\chi''_0 + \delta''_\lambda)}$$

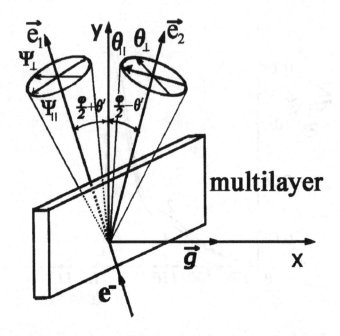

Figure 11. Cherenkov X–ray radiation from a multilayer X–mirror. Designations are the same as in Fig.1

$$\times \left[\frac{1}{\gamma^{-2} + \Theta^2 - \Delta - \delta'_\lambda + i\delta''_\lambda} - \frac{1}{\gamma^{-2} + \Theta^2} \right], \qquad (19)$$

$$\Delta = 2\sin^2\left(\frac{\varphi}{2}\right) \left[1 - \frac{\omega}{\omega_B} + (\theta' + \Theta_\parallel)\cot\frac{\varphi}{2} \right],$$

$$\alpha_1 = 1, \qquad \alpha_2 = \cos\varphi,$$

and

$$\delta'_\lambda = \frac{\operatorname{sign}(D_\lambda)}{\sqrt{2}} \sqrt{\sqrt{C_\lambda^2 + D_\lambda^2} + C_\lambda},$$

$$\delta''_\lambda = \frac{1}{\sqrt{2}} \sqrt{\sqrt{C_\lambda^2 + D_\lambda^2} - C_\lambda},$$

$$C_\lambda = (\Delta - \chi'_0)^2 - (\chi'^2_g - \chi''^2_g)\alpha_\lambda^2 - \chi''^2_0,$$

$$D_\lambda = 2[\chi''_0(\Delta - \chi'_0) + \chi'_g\chi''_g\alpha_\lambda^2],$$

with $\Theta_1 = \Theta_\perp$, $\Theta_2 = 2\theta' + \Theta_\parallel$, $\Theta^2 = \Theta_1^2 + \Theta_2^2$, and χ'_0, χ''_0 the real and imaginary part of the average dielectric susceptibility χ_0. The Bragg frequency is $\omega_B = g/2\sin(\varphi/2)$, and finally $\chi'_g = \left(\sin\left(\pi\frac{a}{t}\right)/\pi\right)(\chi'_a - \chi'_b)$, and $\chi''_g = \left(\sin\left(\pi\frac{a}{t}\right)/\pi\right)(\chi''_a - \chi''_b)$.

The total emission amplitude A_λ in (18) can be represented in the form

$$A_\lambda = A_\lambda^{\text{PXR}} + A_\lambda^{\text{DTR}},$$

$$A_\lambda^{\text{PXR}} = \frac{e}{\pi} \frac{\Theta_\lambda}{\gamma^{-2} + \Theta^2 - \chi_0' - i\chi_0''} \frac{(\chi_g' + i\chi_g'')\alpha_\lambda}{\gamma^{-2} + \Theta^2 - \Delta - \delta_\lambda' + i\delta_\lambda''},$$

$$A_\lambda^{\text{DTR}} = -\frac{e}{\pi}\Theta_\lambda \left[\frac{1}{\gamma^{-2} + \Theta^2} - \frac{1}{\gamma^{-2} + \Theta^2 - \chi_0' - i\chi_0''} \right] \tag{20}$$

$$\times \frac{(\chi_g' + i\chi_g'')\alpha_\lambda}{\Delta - \chi_0' + \delta_\lambda' - i(\chi_0'' + \delta_\lambda'')},$$

where A_λ^{PXR} describes the contribution of parametric X-rays, whereas A_λ^{DTR} is the amplitude of diffracted transition radiation [19]. It is clear that the Cherenkov like contribution to total emission yield is determined by the terms in (20) characterized by pole like singularity. Obviously, DTR amplitude has no poles (Cherenkov pole $\gamma^{-2} - \chi_0 + \Theta^2 = 0$ is spurious because it disappears due to an interference between DTR and PXR, the reflection coefficient described by the last factor in the formula for A_λ^{DTR} has no poles close to real axis, as is easy to see taking into account the structure of the coefficients δ_λ and C_λ from (18)). Thus, the possible contribution of diffracted Cherenkov radiation is determined by PXR emission amplitude only, because the equality $\gamma^{-2} + \Theta^2 - \Delta - \delta_\lambda = 0$ can be fulfilled.

Let us consider the field of existence of the maximum in PXR reflex neglecting initially the influence of photoabsorption ($\chi_0'' = \chi_g'' = 0$). The equation of PXR maximum realization

$$\Delta_0 - \Delta' - \text{sign}(\Delta')\sqrt{\Delta'^2 - \chi_g^2\alpha_\lambda^2} = 0$$

$$\Delta_0 = \gamma^{-2} - \chi_0 + \Theta^2, \qquad \Delta' = \Delta - \chi_0 \tag{21}$$

has the solution

$$\Delta' = \frac{\Delta_0^2 + \chi_g^2\alpha_\lambda^2}{2\Delta_0} \tag{22}$$

in two non-overlapping ranges of the values of the parameter Δ_0.

The first of them determined by the inequality

$$\Delta_0 > |\chi_g\alpha_\lambda| \tag{23}$$

corresponds to the branch of ordinary PXR. The second one determined by the inequality

$$\Delta_0 < -|\chi_g\alpha_\lambda| \tag{24}$$

corresponds to Cherenkov branch of PXR. This radiation can appear only with the proviso that the Cherenkov condition $\Delta_0 < 0$ is fulfilled.

Since $|\Delta'| > |\chi_g \alpha_\lambda|$ as it is obvious from (22), both branches are realized outside the region of anomalous dispersion. Radiation corresponding to these branches can appear inside the region of anomalous dispersion $|\Delta'| < |\chi_g \alpha_\lambda|$ having regard to photoabsorption, but the yield of this radiation is small, as the performed analysis has shown.

Let us estimate the greatest possible spectral angular density of the discussed emission mechanism. By assuming that $\chi_{0,g}'' \ll \chi_{0,g}'$, one can obtain from (18) the following simple formula

$$\left(\frac{\mathrm{d}N_\lambda}{\mathrm{d}\omega\, \mathrm{d}^2\Theta} \right)_{max} \approx \frac{e^2}{\pi^2} \frac{1}{\omega} \left(\frac{\chi_g' \alpha_\lambda}{\Delta_0} \right)^2 \left(\frac{\Theta_\lambda}{\delta_\lambda''} \right)^2, \tag{25}$$

$$\delta_\lambda'' = \chi_0'' \left| \frac{\Delta_0^2 + (\chi_g' \alpha_\lambda)^2}{\Delta_0^2 - (\chi_g' \alpha_\lambda)^2} \right| - \chi_0'' \alpha_\lambda \left| \frac{2\Delta_0 \chi_g' \alpha_\lambda}{\Delta_0^2 - (\chi_g' \alpha_\lambda)^2} \right|,$$

$$\Delta_0 < -\chi_g' \alpha_\lambda \sqrt{1 - \left(2 \frac{\chi_0'' - \chi_g'' \alpha_\lambda}{\chi_g' \alpha_\lambda} \right)^{\frac{2}{3}}} \approx -\chi_g' \alpha_\lambda,$$

describing the Cherenkov branch of PXR.

Obviously, the density (25) and that for ordinary Cherenkov radiation are of the same magnitude in the frequency range, where $|\Delta_0| \geq |\chi_g' \alpha_\lambda|$. Thus, the discussed emission mechanism based on self-diffracted Cherenkov radiation is of interest for X–ray source creation.

The structure of emitted photon flux calculated by the general formula (18) for $Be - C$ multilayer nanostructure is illustrated by the curves presented in Fig.12 and Fig.13. The period of nanostructure T and the orientation angle $\varphi/2$ were chosen so that the Bragg frequency ω_B and the frequency corresponding to maximum in real part of Be dielectric susceptibility were close to each other. Presented figures demonstrate the strong dependence of the emission angular distribution on the orientation angle. Distribution presented in Fig.12 has been calculated for fixed energies of emitting electrons and emitted photons. Large value of the angle $\varphi > \pi/2$ has been used in the performed calculations. On the other hand, distribution presented in Fig.13 has been calculated for the small values of the angle $\varphi < \pi/2$. It is crucial for the purposes of X–ray source creation that the presented distributions are strongly non-uniform. This property analogous to that discussed in the previous sections of the paper allows to increase the yield of strongly collimated radiation.

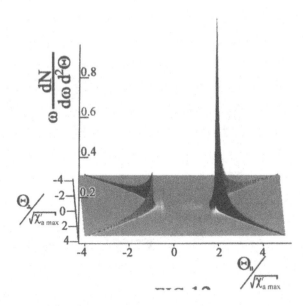

Figure 12. Spectral-angular distribution of Cherenkov radiation from multilayer X–mirror. The curve has been calculated for the fixed values of the parameters: $1/\gamma\sqrt{\chi'_{a\,max}} = 0.2$, $\theta'/\sqrt{\chi'_{a\,max}} = 0.53$, $\omega_B = \omega = 111.6\text{eV}$, $\varphi = 162°$, $a/T = 0.9$, $\chi'_{a\,max} = 0.05$ (Be).

5. Relative contribution of free and virtual photons to the formation of PXR yield

It should be noted that the model of PXR process used in above calculations does not take into account the contribution of real photons of diffracted bremsstrahlung to the formation of total emission yield since PXR cross-section was calculated for rectilinear trajectory of emitting electrons. As a consequence, only the contribution of actually parametric X–rays appearing due to the Bragg diffracted of virtual photons of the fast electron Coulomb field was taken into consideration in the performed analysis.

The discussed question concerning the relative contributions of indicated emission mechanisms to total emission yield is of interest for the problem of PXR based X–ray source creation because recent experiments [10, 11] pointed to a discrepancy between obtained data and the theory based on averaging of the ordinary PXR cross-section over multiple scattering angles. An exact statement of the discussed problem based on the kinetic equation approach was used in work [20]. Unfortunately, the expansions used in [20] allow to describe the case of thin enough target only, when the influence of multiple scattering is small. Currently the

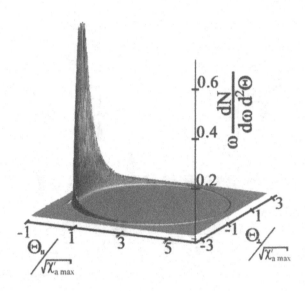

Figure 13.   The same but for $\theta'/\sqrt{\chi'_{a\,\max}} = 0.7$, $\varphi = 44°$.

exact approach was used to analyze the influence of multiple scattering on the PXR spectral width [21, 22], but the question providing the subject matter for the present section.

Let us consider PXR process in more detail as compared with that in previous section of this paper. Starting from well known equations of dynamical diffraction theory [23]

$$(k^2 - \omega^2(1 + \chi_0))\mathbf{E}_{\omega\mathbf{k}} - \mathbf{k}(\mathbf{k}, \mathbf{E}_{\omega\mathbf{k}}) - \omega^2 \sum_{\mathbf{g}} {}'\chi_{-\mathbf{g}}(\omega)\mathbf{E}_{\omega\mathbf{k}+\mathbf{g}} = 4\pi i\omega\mathbf{J}_{\omega\mathbf{k}}$$

(26)

where $\mathbf{E}_{\omega\mathbf{k}}$ is the Fourier-transform of the electric field, χ_0 and χ_g are the components of the crystalline dielectric susceptibility $\epsilon(\omega, \mathbf{r}) = 1 + \chi_0(\omega) + \sum_{\mathbf{g}} {}'\chi_{\mathbf{g}}e^{i(\mathbf{g}, \mathbf{r})}$, $\mathbf{J}_{\omega\mathbf{k}}$ is the Fourier-transform of the emitting electron current density, one can obtain on the basis of well known methods [23] the following expression for an emission field, propagating along the direction of Bragg scattering

$$\mathbf{E}_{\lambda\mathbf{k}+\mathbf{g}} = \frac{4\pi i\omega^3\chi_{\mathbf{g}}\alpha_\lambda}{D_\lambda}(\mathbf{e}_{\lambda\mathbf{k}}, \mathbf{J}_{\omega\mathbf{k}}),$$

(27)

$$D_\lambda = (k^2 - \omega^2(1 + \chi_0))((\mathbf{k} + \mathbf{g})^2 - \omega^2(1 + \chi_0)) - \omega^4\chi_{\mathbf{g}}\chi_{-\mathbf{g}}\alpha_\lambda^2,$$

where $\mathbf{E}_{\omega\mathbf{k}+\mathbf{g}} = \sum\limits_{\lambda=1}^{2} \mathbf{e}_{\lambda\mathbf{k}+\mathbf{g}}E_{\lambda\mathbf{k}+\mathbf{g}}$, $\mathbf{e}_{\lambda\mathbf{k}}$ and $\mathbf{e}_{\lambda\mathbf{k}+\mathbf{g}}$ are the polarization vectors, $(\mathbf{k}, \mathbf{e}_{\lambda\mathbf{k}}) = (\mathbf{k}+\mathbf{g}, \mathbf{e}_{\lambda\mathbf{k}+\mathbf{g}}) = 0$, $\mathbf{g}$ is the reciprocal lattice vector (see Fig.11).

To determine an emission spectral-angular distribution one should calculate Fourier-integral $E_\lambda^{Rad} = \int d^3k_g \; e^{i(\mathbf{k}_g,\mathbf{n})r}E_{\lambda\,\mathbf{k}+\mathbf{g}}$ in the wavezone by the stationary phase method (here $\mathbf{k}_g = \mathbf{k}+\mathbf{g}$, $\mathbf{n}$ is the unit vector to the direction of emitted photon propagation). The result of integration has the form

$$E_\lambda^{\text{Rad}} = \frac{4\pi^3 i\omega^3 \chi_{\mathbf{g}}\alpha_\lambda}{\sqrt{\Delta'^2 + \omega^2\chi_{\mathbf{g}}\chi_{-\mathbf{g}}\alpha_\lambda^2(1-(\mathbf{n},\mathbf{g})/\omega)}} \tag{28}$$

$$\times \left[(\mathbf{e}_{\lambda\mathbf{k}_+}, \mathbf{J}_{\omega\mathbf{k}_+})e^{i\xi_+ r} - (\mathbf{e}_{\lambda\mathbf{k}_-}, \mathbf{J}_{\omega\mathbf{k}_-})e^{i\xi_- r}\right]\frac{e^{i\omega r}}{r},$$

$$\xi_\pm = \frac{1}{2(1-(\mathbf{n},\mathbf{g})/\omega)}\left[-\delta' \pm \sqrt{\delta'^2 + \omega^2\chi_{\mathbf{g}}\chi_{-\mathbf{g}}\alpha_\lambda^2(1-(\mathbf{n},\mathbf{g})/\omega)}\right]$$

$$+ \frac{\omega}{2}\chi_0,$$

$$\delta' = \frac{g^2}{2\omega} - (\mathbf{n},\mathbf{g})\left(1 + \frac{1}{2}\chi_0\right), \quad \mathbf{k}_\pm = (\omega + \xi_\pm)\mathbf{n} - \mathbf{g}$$

For the further analysis it is very convenient to introduce the angular variables $\boldsymbol{\Theta}$ and $\boldsymbol{\Psi}_t$ by the formulae analogous to (2).

Using (2) and (28) one can obtain the following expression for the spectral-angular distribution of the number of emitted photons:

$$\frac{dN_\lambda}{d\omega \, d^2\Theta} = \frac{e^2\omega|\chi_g|^2\alpha_\lambda^2}{8\pi^2}\frac{1}{\delta^2 + \chi_g\chi_{-g}\alpha_\lambda^2\cos\varphi}$$

$$\times \text{Re}\left\langle \int dt \int\limits_0^\infty d\tau \, \Omega_{\lambda\,t}\Omega_{\lambda\,t+\tau}e^{-i\omega\tau} \right. \tag{29}$$

$$\left. \times \left[e^{i(\mathbf{k}_+,(\mathbf{r}_{t+\tau}-\mathbf{r}_t))} + e^{i(\mathbf{k}_-,(\mathbf{r}_{t+\tau}-\mathbf{r}_t))}\right]\right\rangle,$$

where $\delta = \delta'/\omega_{\text{B}}$, $\Omega_{1t} = \Theta_\perp - \Psi_{\perp t}$, $\Omega_{2t} = \Theta_\parallel + \Psi_{\parallel t} + 2\theta'$, θ' is the orientation angle (see Fig.11), the value $\theta' = 0$ corresponds to exact Bragg resonance orientation of the crystal relative to emitting electron velocity, the brackets $\langle\rangle$ mean the averaging over all possible trajectories of electrons in the target $\mathbf{r}_t \equiv \mathbf{r}(t)$, the angle $\boldsymbol{\Psi}_t$ is the time-dependent quantity because of multiple scattering. In accordance with (29) two branches of propagating in the crystal electromagnetic waves take the contribution to total emission yield within the frame of used dynamical diffraction approach.

Procedure of averaging of the expressions analogous to (29) is described in book [3], where an influence of multiple scattering on the ordinary bremsstrahlung from relativistic electrons, moving in amorphous medium, has been considered in detail. Using the corresponding results [3] one obtain from (29) the final expression for the total emission intensity

$$
\frac{dN_\lambda}{dt\,d\omega\,d^2\Theta} = \frac{e^2\omega|\chi_g|^2}{4\pi^2}\frac{\Omega_{\lambda t}^2\alpha_\lambda^2}{\sigma_\lambda^2}\mathrm{Re}\int_0^\infty d\tau\frac{\cos(\frac{\omega}{2}\sigma_\lambda\tau)}{\cosh^2(\sqrt{2i\omega q}\tau)} \tag{30}
$$
$$
\times\exp\left(-\tfrac{i\omega}{2}(\gamma^{-2}-\chi_0-\delta)\tau-\sqrt{\tfrac{i\omega}{8q}}\Omega_t^2\tanh(\sqrt{2i\omega q}\tau)\right),
$$

where $\sigma_\lambda^2 = \delta^2 + \chi_{\mathbf{g}}\chi_{-\mathbf{g}}\alpha_\lambda^2\cos\varphi$, $\quad q = 1/4L_{\mathrm{Sc}}\gamma^2$.

Let us use the general result (30) to elucidate the conditions such that the diffracted bremsstrahlung contribution can be substantial. The emission angular density

$$
\frac{dN}{d^2\Theta} = \int_0^L dt\int d^2\Psi_t\,f(t,\Psi_t)\int_0^\infty d\omega\sum_{\lambda=1}^2\frac{dN_\lambda}{dt\,d\omega\,d^2\Theta} \tag{31}
$$

in the most suitable characteristic for our purposes because this characteristic is very sensitive to the action of multiple scattering.

First of all let us integrate the intensity distribution (30) over emitted photon energies keeping in mind that the emission considered in this section is concentrated in the narrow vicinity of $\omega = \omega_{\mathrm{B}}$, because the Bragg diffraction process extracts this segment of initially wide spectra of both real photons of bremsstrahlung and virtual photons of the emitting electron Coulomb field. Taking into account that $\omega \approx \omega_{\mathrm{B}}$ in (30) except "fast variable" $\delta(\omega)$ (so-called resonance defect) one can perform the integration by the transformation of variables $d\omega = \left(\frac{d\delta}{d\omega}\right)^{-1}d\delta$. The result of integration has the following form

$$
\frac{dN_\lambda}{dt\,d^2\Theta} = -\frac{e^2\omega_{\mathrm{B}}^4|\chi_g|^2}{2\pi g^2}\frac{\Omega_{\lambda t}^2\alpha_\lambda^2}{\Omega_t^2\beta_\lambda}\mathrm{Im}\left\{\left[\beta_\lambda+i\left(\frac{1}{\gamma^2}-\chi_0\right)\right]\right.
$$
$$
\left.\times\int_0^\infty d\tau\exp\left[-\frac{\omega_{\mathrm{B}}}{2}\left[\beta_\lambda+i\left(\frac{1}{\gamma^2}-\chi_0\right)\right]\tau\right]\right. \tag{32}
$$

$$-\sqrt{\frac{i\omega_B}{8q}}\,\Omega_t^2\,\tanh\left(\sqrt{2i\omega_Bq\tau}\right)\Bigg]\Bigg\}\,,$$

where $\beta_\lambda^2 = \chi_{\mathbf{g}}\chi_{-\mathbf{g}}\alpha_\lambda^2\cos\varphi$ (we are considering the parametric X–rays for Laue geometry, so $\varphi < \pi/2$).

The emission intensity (32) takes into account both parametric X–rays and diffracted bremsstrahlung contribution. To estimate the relative contribution of diffracted bremsstrahlung let us compare the exact result for the total emission angular density $\frac{dN}{d^2\Theta}$, following from (31) and (32), with that, following from the general formula (31) and simplified formula (32) corresponding to the emission from a fast electron moving with a constant velocity. The last case corresponds to the limit $q \to 0$ in (32), when this formula can be reduced to the ordinary PXR angular distribution

$$\frac{dN_\lambda}{dt\,d^2\Theta} \to \frac{dN_{0\lambda}}{dt\,d^2\Theta} = \frac{e^2\omega_B^3|\chi_g|^2}{\pi g^2}\,\frac{\Omega_{\lambda\,t}^2\alpha_\lambda^2}{(\gamma^{-2}+\gamma_m^{-2}+\Omega_t^2)^2+\beta_\lambda^2}\,,\tag{33}$$

where $\gamma_m = \omega_B/\omega_0$ (ω_0 is the plasma frequency).

Calculating the quantity $\frac{dN}{d^2\Theta}$ we have restricted our selves to the case of small incidence angle $\varphi \ll 1$, when $\alpha_2 = \cos\varphi \approx \alpha_1 = 1$. Such conditions are most appropriate for the diffracted bremsstrahlung contribution to be substantial. Using the distribution function (8) and performing the integration in (31) over scattering angles $\mathbf{\Psi}_t$ one can obtain the following expression for $dN/d^2\Theta$:

$$\frac{dN}{d^2\Theta} = \frac{e^2\omega_B^3 L_{Sc}}{\pi g^2}F,\tag{34}$$

$$F = -\eta\,\mathrm{Im}\Bigg\{\left(1+\frac{\gamma^2}{\gamma_m^2}(1-i\eta)\right)\int\limits_0^\infty dt\,\coth(t)$$

$$\times\exp\left[-\sqrt{i}\left(1+\frac{\gamma^2}{\gamma_m^2}(1-i\eta)\right)\frac{\gamma_L}{\gamma}t\right]$$

$$\times\left[E_1\left(\frac{\sqrt{i}\gamma^2\Theta_0^2\frac{\gamma_L}{\gamma}\tanh(t)}{1+\sqrt{i}\left(\frac{L}{L_{Sc}}+\gamma^2\Psi_0^2\right)\frac{\gamma_L}{\gamma}\tanh(t)}\right)\right.$$

$$\left.-E_1\left(\frac{\sqrt{i}\gamma^2\Theta_0^2\frac{\gamma_L}{\gamma}\tanh(t)}{1+\sqrt{i}\gamma^2\Psi_0^2\frac{\gamma_L}{\gamma}\tanh(t)}\right)\right]\Bigg\}\,,$$

where $\gamma_L = \sqrt{\omega_B L_{Sc}/2}$, $\eta = |\chi_g/\chi_0| < 1$, $\Theta_0^2 = \Theta_\perp^2 + (2\theta' + \Theta_\parallel)^2$.

The function $F(\gamma\Theta_0)$ describing the contribution of both actually PXR and diffracted bremsstrahlung must be compared with analogous function

$$
F_0 = -\eta \operatorname{Im}\left\{\left(1 + \frac{\gamma^2}{\gamma_{\mathrm{m}}^2}(1 - i\eta)\right)\right.
$$

$$
\times \int_0^\infty \frac{dt}{\sqrt{\left(1 + \frac{\gamma^2}{\gamma_{\mathrm{m}}^2}(1 - i\eta) + \gamma^2\Theta_0^2 + t\right)^2 - 4\gamma^2\Theta_0^2 t}}
$$

$$
\left. \times \left[E_1\left(\frac{t}{\frac{L}{L_{\mathrm{Sc}}} + \gamma^2\Psi_0^2}\right) - E_1\left(\frac{t}{\gamma^2\Psi_0^2}\right)\right]\right\}, \tag{35}
$$

taking into account the contribution of PXR only and following from (37) and (33). The difference between these functions depends strongly on the parameters $\gamma/\gamma_{\mathrm{m}}$ and $\gamma/\gamma_{\mathrm{L}}$. The physical meaning of the parameter $\gamma/\gamma_{\mathrm{m}} = \gamma\omega_0/\omega_{\mathrm{B}}$ is very simple. This parameter describes an influence of the density effect on PXR and bremsstrahlung emission mechanisms. Screening of the Coulomb field of emitting relativistic electron due to the density effect occurring in the range $\gamma > \gamma_{\mathrm{m}}$ is responsible for PXR yield saturation as a function of the energy of emitting electron [14]. Changing of an emitted photon phase velocity due to the polarization of medium electrons is responsible for bremsstrahlung yield suppression (Ter-Mickaelian effect [3]) in the frequency range $\omega < \gamma\omega_0$ (in the case $\omega \approx \omega_{\mathrm{B}}$ under consideration this inequality is equivalent to $\gamma > \gamma_{\mathrm{m}}$).

The parameter $\gamma/\gamma_{\mathrm{L}}$ describes an influence of another classical electrodynamical effect in the physics of high energy particle bremsstrahlung known as Landau-Pomeranchuk-Migdal effect (see [3]). LPM effect arises with the proviso that the multiple scattering angle of emitting particle Ψ_{Coh} achievable at the distance of the order of so-called formation length $l_{\mathrm{Coh}} \approx 2\gamma^2/\omega$ ($l_{\mathrm{Coh}} \approx 2\gamma^2/\omega_{\mathrm{B}}$ in the case in question) exceeds the characteristic emission angle of emitting particle $\Psi_{\mathrm{em}} \approx \gamma^{-1}$. Obviously, $\Psi_{\mathrm{Coh}}^2 = \frac{1}{\gamma^2 L_{\mathrm{Sc}}}\frac{2\gamma^2}{\omega_{\mathrm{B}}} = \gamma_L^{-2}$, so that $\gamma/\gamma_{\mathrm{L}} = \Psi_{\mathrm{Coh}}/\Psi_{\mathrm{em}}$ and therefore the condition $\gamma > \gamma_{\mathrm{L}}$ means LPM effect manifestation.

Let us consider the distribution $F(\gamma\Theta_0)$ in the range of small emitting particle energies $\gamma \ll \gamma_{\mathrm{L}}$. Close inspersion of the integral (34) shows that the effective values of the variable in integration $t_{\mathrm{eff}} \ll 1$ in conditions under consideration independently of the parameter $\gamma/\gamma_{\mathrm{m}}$. Because of this, $\tanh(t) \approx t$ and the formula (34) can be reduced to more simple one

$$
F \;\to\; -\eta\,\mathrm{Im}\left\{\left(1+\frac{\gamma^2}{\gamma_{\mathrm{m}}^2}(1-\mathrm{i}\eta)\right)\int\limits_0^\infty \frac{\mathrm{d}t}{t}\,\exp\left[-\left(1+\frac{\gamma^2}{\gamma_{\mathrm{m}}^2}(1-\mathrm{i}\eta)\right)t\right]\right.
$$

$$
\left.\times\left[E_1\left(\frac{\gamma^2\Theta_0^2 t}{1+\left(\frac{L}{L_{\mathrm{Sc}}}+\gamma^2\Psi_0^2\right)t}\right)-E_1\left(\frac{\gamma^2\Theta_0^2 t}{1+\gamma^2\Psi_0^2 t}\right)\right]\right\} \qquad (36)
$$

The performed numerical analysis has shown that the function $F_0(\gamma\Theta_0)$ in (35) and the modified function $F(\gamma\Theta_0)$ in (36) coincide. By this means PXR characteristics are well described with the constraint $\gamma \ll \gamma_{\mathrm{L}}$ within the framework of ordinary PXR theory based on the calculation of PXR cross-section using the rectilinear trajectory of emitting particles. To explain this conclusion it is necessary to note that in conditions $\gamma \ll \gamma_{\mathrm{L}}$ under consideration the trajectory of emitting particle is close to stright line at the distance of the order of l_{Coh} for which the emitted photon is formed (the used expansion $\tanh(t) \approx t$ implies that the bend of electron's trajectory is neglected). As this takes place, the structure of the emitted electron electromagnetic field consisting of both virtual photons of the electron Coulomb and free photons of the bremsstrahlung is close to that for the electron moving along the rectilinear trajectory with a constant velocity. As a consequence, the structure of the diffracted by crystalline atomic planes electromagnetic field differs little from the ordinary PXR field.

In the opposite case $\gamma > \gamma_{\mathrm{L}}$ the velocity of emitting electron turns through the angle $\Psi_{\mathrm{Coh}} > \gamma^{-1}$ at the distance l_{Coh}. Since the value γ^{-1} is the scale if angular distribution of virtual photons associated with the electron's Coulomb field, the structure of total electromagnetic field of emitting electron differs substantially from that for the electron moving along the rectilinear trajectory. Because of this the structure of diffracted field is changed substantially as well. Relative contribution of diffracted bremsstrahlung depends in the case in question on the parameter γ/γ_m. This contribution is small if $\gamma > \gamma_m$ and bremsstrahlung is suppressed by Ter-Mikaelian effect. On the other hand, the contribution of diffracted bremsstrahlung can be very essential if $\gamma < \gamma_m$, but $\gamma > \gamma_{\mathrm{L}}$ (obviously, both these inequalities can be valid simultaneously with the proviso that $\gamma_L < \gamma_m$ only).

The discussed results are demonstrated by the Fig.14 and Fig.15, where the functions $F(\gamma\Theta_0)$ and $F_0(\gamma\Theta_0)$ calculated by the formula (34) and (35) respectively are presented. The curves presented in Fig.15 predict the dominant contribution of the diffracted bremsstrahlung to total emission yield under special conditions.

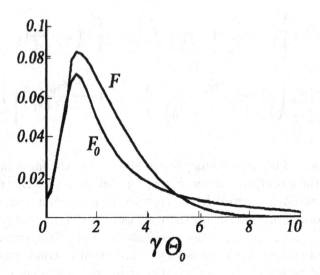

Figure 14.   The emission angular density with and without diffracted bremsstrahlung contribution. The presented functions $F(x)$ and $F_0(x)$ defined by (32) and (33) have been calculated for fixed values of the parameters $L/L_{sc} = 0.5$, $\eta = 0.8$, $\gamma_L/\gamma_m = 0.5$ and $\gamma_m/\gamma = 5$.

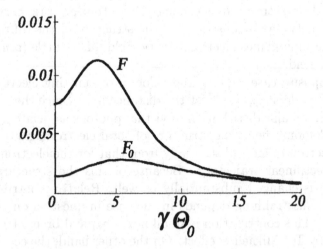

Figure 15.   The same but for $\gamma_m/\gamma = 0.3$.

6.   Conclusions

Performed analysis has shown that the angular density of X–rays produced by X–ray sources based on Cherenkov and quasi-Cherenkov emission mechanisms can be increased substantially in conditions of grazing incidence of emitting electrons at the surface of a radiator, when the angular distribution of emitted photons becomes strongly non-uniform. Owing to this fact, most part of emitted photons is concentrated in the small region of observation angles resulting in the high yield of collimated radiation.

In the case of Cherenkov X–ray source the discussed non-uniformity is caused by the strong dependence of a photoabsorption coefficient on the direction of emitted photon propagation. In accordance with performed calculations an increase in the emission angular density of the order of 5 − 10 and more possible in the range of incidence angles of the order of Cherenkov emission angle.

The analogous enhancement of the emission angular density is possible for PXR source (in the case being considered the anisotropic angular distribution of emitted photons is caused by the dependence of PXR reflection coefficient on the photon energy and strong connection of this energy with the observation angle given by the condition of Bragg diffraction). On the other hand, the growth of PXR angular density with decreasing the emitting electron incidence angle relative to reflecting crystallographic plane of the crystalline target is attended by substantial growth of the spectral width of emitted photon flux.

Cherenkov X–ray radiation is possible not only in homogeneous media but in multilayer periodic nanostructure as well. An advantage of the last scheme consists in the possibility to generate X–rays at large angles relative to emitting electron velocity and consequently to arrange an irradiated sample in the immediate vicinity of the radiator. The performed calculations have shown the possibility to obtain in this scheme the intensity and angular density of emitting photons close to that achievable in the ordinary scheme used the homogeneous target.

Analysing the relative contributions from diffracted bremsstrahlung and PXR to the total emission yield from relativistic electrons crossing a crystal shows that diffracted bremsstrahlung can dominate under special conditions as elucidated in this note.

Acknowledgments

This work was supported by RFBR (grants:02-02-16941, 03-02-16263) and Program "DOPFIN" Russian Ministry of Science and Education. One of the authors (A.K.) is grateful to Russian Ministry of Education and CRDF (grant VZ-010-0) for financial support.

References

[1] Ginzburg V. and Frank I. *Emission from uniformly moving appared when crossing a boundary between two different media.* J. Phys. USSR, **9** p.353, 1945.

[2] Kumakhov M. *On the theory of electromagnetic radiation of charged particles in a crystal.* Phys. Lett., **57**, p.17, 1976

[3] Ter-Mikaelian M. *High Energy Electromagnetic Processes in Condenced Media*, Wiley, New York, 1972.

[4] Bazylev V., Glebov V., Denisov E. et al. *JEPT Letters*, **24**, p.406, 1976.

[5] Rullhusen P., Artru X. and Dhez P. *Novel Radiation Sources Using Relativistic Electrons*, Word Scientific, Singapore, 1999.

[6] Bazylev V., Glebov V., Denisov E. et al. *JEPT*, **81**, p.1664, 1981.

[7] Moran M, Chang B., Schueider M. and Maruyama X. *Grazing-incidence Cherenkov X-ray generation.* Nucl. Instr. Meth., B48, p.287, 1990.

[8] Knulst W., Luiten O., van der Wiel M. and Verhoeven J. *Observation of narrow band Si L-edge Cherenkov radiation generated by 5 MeV electrons.* Appl. Phys. Lett., **79**, p.2999, 2001.

[9] Knulst W., van der Wiel M., Luiten O. and Verhoeven J. *High-brightness, narrow band and compact soft X-ray Cherenkov sources in the water window.* Appl. Phys. Lett., **83**, p.1050, 2003.

[10] Knulst W., van der Wiel M., Luiten O. and Verhoeven J. *High-brightness compact X-ray source based on Cherenkov radiation.* Proc. SPIE Int. Soc. Opt. Eng, **5196**, p.393, 2004.

[11] Chefonov O., Kalinin B., Padalko D. et al. *Experimental comparizon of parametric X-ray radiation and diffracted bremsstrahlung in a pyrolytic graphite crystal.* Nucl. Instr. Meth., B**173**, p.263, 2001.

[12] Bogomazova E., Kalinin B., Naumenko G. et al. *Diffraction of real and virtual photons in a pyrolitic graphite crystal as a source of intensive quasimonochromatic X-ray beam.* Nucl. Instr. Meth., B**201**, p.276, 2003.

[13] Feranchuk I. and Ivashin V. *Theoretical investigation of parametric X-ray features.* J. Physique, **46**, p.1981, 1985

[14] Nasonov N. and Safronov A. *Polarization bremsstrahlung of fast charged particles. Radiation of Relativistic Electrons in Periodic Structures*, NPI Tomsk Polytechnical University, Tomsk, p.134, 1993.

[15] Knulst W. *Soft X-ray Cherenkov radiation: towards a compact narrow band source.* PhD Thesis, Technic University Eindhoven, 2004.

[16] Henke B., Gullikson E. and Davis J. *Atomic data.* Nucl. Data Tables, **54**, p.18, 1993.

[17] Nasonov N., Kaplin V., Uglov S., Piestrup M. and Gary C. *X-rays from relativistic electrons in a maltilayer structure.* Phys. Rev., E**68**, p.03654, 2003.

[18] Nasonov N. and Safronov A. *Interference of Cherenkov and parametric radiation mechanisms of a fast charged particle.* Phys. Stat. Sol., B**168**, p.617.

[19] Caticha A. *Transition-diffracted radiation and the Cherenkov emission of X-rays.* Phys. Rev., A**40**, p.4322, 1989.

[20] Baryshevsky V., Grubich A. and Le Tien Hai. *Effect of multiple scattering on parametric X-radiation. JETP*, **94**, p.51, 1988.

[21] Shulga N. and Tabrizi M. *Influence of multiple scattering on the band width of parametric X-rays. JETP Lett.*, **76**, p.337, 2002.

[22] Shulga N. and Tabrizi M., *Method of functional integration in the problem of line width of parametric X-ray relativistic electron radiation in a crystal Phys. Lett.*, A**308**, p.467, 2003.

[23] Pinsker Z. *Method of functional integration in the problem of line width of parametric X-ray relativistic electron radiation in a crystal. Phys. Lett.*, A**308**, p.467, 2003.

INVESTIGATION OF FAR-INFRARED SMITH-PURCELL RADIATION AT THE 3.41 MEV ELECTRON INJECTOR LINAC OF THE MAINZ MICROTRON MAMI

H. Backe, W. Lauth, H. Mannweiler, H. Rochholz, K. Aulenbacher, R. Barday,
H. Euteneuer, K.-H. Kaiser, G. Kube, F. Schwellnus, V. Tioukine
Institut für Kernphysik, Universität Mainz, D-55099 Mainz, Germany

Abstract   An experiment has been set up at the injector LINAC of the Mainz Microtron MAMI to investigate far-infrared Smith-Purcell radiation in the THz gap (30 μm $\leq \lambda \leq 300 \mu$m). The essential components are a superconductive magnet with a magnetic induction of 5 Tesla in which 200 mm long gratings of various periods between 1.4 mm and 14 mm are located, and a composite silicon bolometer as radiation detector. First experiments were performed in the wavelength region between 100 μm and 1 mm with a bunched 1.44 MeV electron beam.

Keywords: Smith-Purcell radiation, FEL, SASE

Introduction

Smith-Purcell (SP) radiation is generated when a beam of charged particles passes close to the surface of a periodic structure, i.e., a diffraction grating. The radiation mechanism was predicted by Frank in 1942 [Frank, 1942] and observed in the visible for the first time by Smith and Purcell [Smith and Purcell, 1953] with an 250-300 keV electron beam. In a number of subsequent experiments the results were confirmed, for references see [Kube et al., 2002]. Soon after the discovery, potential applications of the SP effect became the topic of interest. In a number of theoretical and experimental studies the SP effect has been discussed as a basis for free electron lasers, for particle acceleration, or for particle beam diagnostics, for references see also [Kube et al., 2002]. In particular we mention the work of Urata et al. [Urata et al., 1998] in which superradiant Smith-Purcell Emission was observed for the first time at a wavelength $\lambda = 491$ μm with a 35 keV electron beam from a scanning electron microscope.

H. Wiedemann (ed.), Advanced Radiation Sources and Applications, 267–282.
© 2006 *Springer. Printed in the Netherlands.*

The optical radiation emission from diffraction gratings has been investigated at the Mainz Microtron MAMI using the low-emittance 855 MeV electron beam [Kube et al., 2002]. A general feature of SP radiation from optical diffraction gratings at ultrarelativistic beam energies is the weakness of the radiation. The reason was found in the smallness of the radiation factors $\mid R_n \mid^2$. According to Van den Berg's theory [Van den Berg, 1973a; Van den Berg, 1973b; Van den Berg, 1974; Van den Berg, 1971], these factors become only large at low electron beam energies.

In this contribution we report on first experiments with a 1.44 MeV beam at the injector of the Mainz Microtron MAMI to produce Smith-Purcell radiation in the wavelength region between 30-300 μm. Such radiation is of particular interest since in this so-called THz gap no compact and efficient radiation sources like lasers or electronic devices are currently available. In the following section, first some basic theoretical considerations underlying our experiments are described.

1.   Basic Theoretical Background

According to the approach of di Francia [Di Francia, 1960] the radiation mechanism can be understood as the diffraction of the field of the electron by the grating. It is diffracted in radiating and non-radiating orders. The latter are evanescent surface waves with a phase velocity which might coincide with the velocity of the electron. In this case the electron can transfer energy to the wave and indirectly drive the radiating orders. The angular distribution of the number of photons dN per electron radiated into the nth order into the solid angle $d\Omega$ is

$$\frac{dN}{d\Omega} = \alpha \mid n \mid N_W \frac{\sin^2 \Theta \sin^2 \Phi}{(1/\beta - \cos \Theta)^2} \mid R_n \mid^2 e^{-\frac{d}{h_{int}}\xi(\Theta,\Phi)} \tag{1}$$

$$\xi(\Theta, \Phi) = \sqrt{1 + (\beta\gamma \sin \Theta \cos \Phi)^2} \tag{2}$$

where α is the fine-structure constant, N_W the number of grating periods, $\beta = v/c$ the reduced electron velocity, Θ, Φ are the emission angles as introduced in Fig. 1. The interaction length

$$h_{int} = \frac{\beta\gamma}{4\pi}\lambda, \tag{3}$$

where $\gamma = (1 - \beta^2)^{-1/2}$ is the Lorentz factor, describes the characteristic finite range of the virtual photons emitted and reabsorbed by the electrons and

$$\lambda = \frac{D}{\mid n \mid}(1/\beta - \cos \Theta),$$

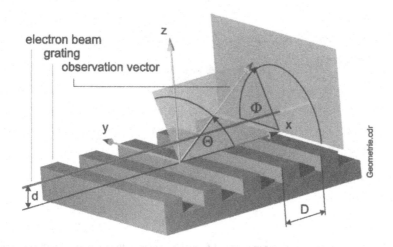

Figure 1. Definition of geometry. The electron moves with constant reduced velocity $\beta=v/c$ at a distance d above the grating surface in x direction. The groves, oriented in the y direction, repeat with the grating period D. The photon wave-vector **k** makes the polar angle Θ with the positive x axis and the azimuthal angle Φ with the positive y axis in the plane spanned by the y and x axis

the wavelength of the emitted radiation. The quantity D is the grating period.

According to Eq. (1) the intensity decreases exponentially with increasing distance d between electron and grating surface. As shown by Fig. 2 (a), the normalized radial emittance ε_N^r of the electron beam restricts the minimum distance of the beam axis to the value

$$d_{min} = \sqrt{L\frac{\varepsilon_N^r/\pi}{\beta\gamma}}. \tag{4}$$

As has been discussed in detail in Ref. [Kube et al., 2002] the exponential coupling factor in Eq. (1) imposes restrictions on the wavelength and the grating length, see also Fig. 2 (a). These are relaxed to a great extent if the electron beam is guided in a strong magnetic field B, see Fig. 2 (b). The radius $R = 2\rho + r$ of an electron beam in the magnetic field can be described by the sum of the cyclotron radius $\rho = (p/(eB)) \cdot \vartheta$ and the radius r of the electron beam. The latter is connected to the transverse beam emittances ε_y and ε_z by the relation $\sqrt{\varepsilon_y\varepsilon_z/2} = \varepsilon_r = \pi \cdot \vartheta \cdot r$. A minimization with respect to the angle ϑ yields

$$\vartheta_{opt} = \sqrt{\frac{eB}{2m_ec}\frac{\varepsilon_N^r/\pi}{(\beta\gamma)^2}}$$

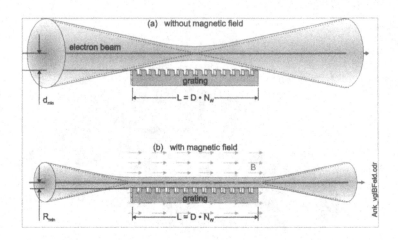

Figure 2. Coupling of an electron beam of finite emittance to the grating (a) without magnetic guiding field, and (b) with magnetic guiding field

and

$$R_{opt} = \sqrt{8 \frac{m_e c}{eB} \frac{\varepsilon_N^r}{\pi}}.$$

With $h_{int} = \kappa \cdot R_{opt}$, and $\kappa \geq 1$ a coupling parameter, a critical wave length

$$\lambda_{crit} = \frac{8\sqrt{2}\pi\kappa}{\beta\gamma} \sqrt{\frac{m_e c}{eB} \frac{\varepsilon_N^r}{\pi}}.$$

can be defined which has no functional dependence on the length of the grating anymore. The meaning of this quantity is that for operational wavelengths $\lambda \leq \lambda_{crit}$ the radiation production mechanism becomes inefficient. In the above equations m_e is the rest mass of the electron.

The intensity $dN/d\Omega$ of the emitted radiation, Eq. (1), maximizes for $\Phi_{opt} = 90°$ and

$$\cos \Theta_{opt} = \beta = \sqrt{1 - 1/\gamma^2}. \tag{5}$$

Eqns. (3), (4), and (5) reduce to the optimum grating period

$$D_{opt} = \beta\gamma^2 \lambda_{crit}. \tag{6}$$

Financial restrictions fixed the length of the grating and the magnetic field to $L = 200$ mm and $B = 5$ Tesla, respectively. With these parameters settled the additional restriction that the grating should have at least $N_W \geq 100$ periods defines the region of operation, see Fig. 3.

As will be pointed out in the next subsection in more detail the Self Amplified Spontaneous Emission (SASE) amplification mechanism is driven by micro bunching of the electron beam. The relevant quantity to achieve micro bunching is the acceleration imposed on the electrons by the evanescent electric field. At constant force the acceleration scales as γ^{-3}. As a consequence, an electron beam energy as small as possible should be chosen. However, Fig. 3 tells that the THz gap can be accessed only by electron beams with an energy greater than about 1 MeV provided the normalized radial emittance is small enough, $\varepsilon_N^r \leq 2 \cdot 10^{-7}\pi$ m rad. The lowest energy at which the MAMI injector can be operated in a stable mode was found to be 1.44 MeV. With a normalized emittance $\varepsilon_N^r = 66 \cdot 10^{-9}\pi$ m rad and $\kappa = 2$ we obtain $\Theta_{opt} = 15.18°$, $h_{int} = 29.3$ μm, $\lambda_{crit} = 91.5$ μm, $D_{opt} = 1.121$ mm, and at $\vartheta_{opt} = 2.67$ mrad a $R_{opt} = 13.4$ μm. For the sake of simplicity we have chosen as design parameters $\lambda = 100$ μm, $\Theta_{opt} = 15.18°$, and a grating period $D = 1.4068$ mm.

Finally, the radiation factors $\mid R_n \mid^2$ in Eq. (1) remains to be discussed. First of all a suitable shape of the grating structure must be chosen. We decided to use a lamellar, i.e. a rectangular, shape because it can easily be manufactured. The relevant parameters of such a grating are the ratio h/D of the groove depth to the grating period D, and the ratio a/D of the lamella width, i.e. the remaining material between the grooves, both taken with respect to the grating period D. The optimization was performed numerically by a modal expansion of the integral representation of the field inside the grooves using a Green's function formalism as described by Van den Berg [Van den Berg, 1974]. The procedure yielded $a/D = 0.505$ and $h/D = 0.086$ with a mean radiation factor $\langle\mid R_1 \mid^2\rangle_\Theta = 0.23$ [Rochholz, 2002].

These considerations define the experimental parameters and the experimental setup to be described in the following section.

Coherent Enhancement and Self Amplified Spontaneous Emission

For our research program to investigate radiation production mechanisms in the wavelength region between 30-300 μm special emphasis was put on amplification processes, in particular the coherent enhancement of Smith-Purcell radiation by a bunched beam and the Self Amplified Spontaneous Emission (SASE).

The radiated number of photons $\langle dN/d\Omega\rangle_{coh}$ from a bunch of N_e electrons, distributed according to Gaussians with a standard deviations σ_x, σ_y, and σ_z in the longitudinal x and transverse y and z directions, enhances with respect to Eq. (1) by a factor $N_e(S_{inc} + (N_e - 1)S_{coh})$. The incoherent S_{inc} and coherent S_{coh} form factors are given by [Doria et al., 2002]

$$S_{inc} = \frac{1}{\sqrt{2\pi}\sigma_z} \int_0^\infty e^{-z\,\xi(\Theta,\Phi)/h_{int}} e^{-(z-d)^2/2\sigma_z^2} dz \tag{7}$$

$$S_{coh} = \left| \frac{1}{\sqrt{2\pi}\sigma_z} \int_0^\infty e^{-z\,\xi(\Theta,\Phi)/2h_{int}} e^{-(z-d)^2/2\sigma_z^2} dz \right|^2$$

$$\times \left| \frac{1}{\sqrt{2\pi}\sigma_x} \int_{-\infty}^\infty e^{-ik_x x} e^{-x^2/2\sigma_x^2} dx \right|^2$$

$$\times \left| \frac{1}{\sqrt{2\pi}\sigma_y} \int_{-\infty}^\infty e^{-ik_y y} e^{-y^2/2\sigma_y^2} dy \right|^2. \tag{8}$$

The integrals in the longitudinal x and transverse y directions have been evaluated for $k_x = k/\beta = \omega/(c\beta) = 2\pi/(\beta\lambda)$ and $k_y = 0$ yielding $S_{coh,x} = \exp(-(2\pi\sigma_x/(\beta\lambda))^2)$ and $S_{coh,y} = 1$, respectively. The quantity d denotes the center of gravity of the Gaussian in z direction.

SASE occurs if the electron beam bunch will additionally be micro-bunched while the passage over the grating. This process is driven by longitudinal forces originating from evanescent surface fields which travel synchronously with the electron bunch across the grating. Gain formulas were derived by Wachtel [Wachtel, 1979], Schächter and Ron [Schächter and Ron, 1989], as well as by Kim and Song [Kim and Song, 2001] in a two-dimensional model with continuous sheet currents. The gains according to Schächter and Ron, G_{SR}, and Kim and Song, G_{KS}, which were somehow modified by us to meet our experimental situation, are

$$G_{SR} = \frac{\sqrt{3}}{2(\beta\gamma)^2} \left[\frac{\beta c \bar{I}}{f_b F_{duty}\sqrt{2\pi}\sigma_x} \frac{4\pi\beta}{I_A\sqrt{2\pi}\sigma_z} \left(\frac{2\pi}{\lambda}\right)^2 e^{-d/h_{int}} F(\Delta_z) \right]^{1/3} \tag{9}$$

$$G_{KS} = \frac{\sqrt{2\pi}}{(\beta\gamma)^2} \left[\frac{\beta c \bar{I}}{f_b F_{duty}\sqrt{2\pi}\sigma_x} \frac{e_{00}}{I_A\sqrt{2\pi}\sigma_z} \cdot \frac{2\pi}{\lambda} e^{-d/h_{int}} F(\Delta_z) \right]^{1/2}, \tag{10}$$

respectively. The weighting function

$$F(\Delta_z) = \frac{\sinh(\Delta_z/2h_{int})}{\Delta_z/2h_{int}}$$

takes into account that in our experiment the transverse beam profile is in good approximation a Gaussian with standard deviation σ_z while in Ref. [Schächter and Ron, 1989] and in Ref. [Kim and Song, 2001] a rectangular shape of the transverse beam profile with a width Δ_z and a thin current sheet along the grating surface were assumed, respectively. Both quantities are connected by $\Delta_z = \sqrt{2\pi}\sigma_z$. Further on, $I_A = ec/r_e = 17045.09$ A is the Alfvén current, with r_e the classical electron radius, and e_{00} the reflection coefficient to the $n = 0$ mode of the grating [Kim and Song, 2001].

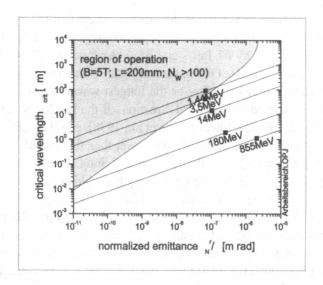

Figure 3. Critical wavelength λ_{crit} in first order $\mid n \mid$ = 1 as function of the normalized radial emittance ε_N^r for different beam energies. The region of operation is indicated for a grating with length L = 200 mm, number of periods $N_W \geq 100$, and magnetic field B = 5 Tesla.

The emitted power d^2P per unit solid angle $d\Omega$ and unit path length dx at a mean beam current $\bar{I}$ of duty cycle F_{duty}, and bunch repetition rate f_b is given by

$$
\frac{d^2P}{d\Omega\,dx} = \frac{2\pi\hbar c}{\lambda} \cdot \frac{\bar{I}}{e} \cdot \frac{dN}{d\Omega} \cdot \frac{1}{L} \left[e^{2G\cdot x} S_{inc} + \left(\frac{\bar{I}}{ef_bF_{duty}} - 1 \right) S_{coh} \right]
$$

$$
= 2\pi c m_e c^2 \frac{\bar{I}}{I_A} n^2 \frac{1}{D^2} \frac{\sin^2\Theta \sin^2\Phi}{(1/\beta - \cos\Theta)^3} \mid R_n \mid^2
$$

$$
\times \left[e^{2G\cdot x} S_{inc} + \left(\frac{\bar{I}}{ef_bF_{duty}} - 1 \right) S_{coh} \right]. \tag{11}
$$

The total emitted power dP per unit solid angle $d\Omega$ is obtained by integrating over the grating length coordinate x:

$$
\frac{dP}{d\Omega} = 2\pi c m_e c^2 \frac{\bar{I}}{I_A} n^2 \frac{L}{D^2} \frac{\sin^2\Theta \sin^2\Phi}{(1/\beta - \cos\Theta)^3} \mid R_n \mid^2
$$

$$
\times \left[\frac{e^{2G\cdot L} - 1}{2G\cdot L} S_{inc} + \left(\frac{\bar{I}}{ef_bF_{duty}} - 1 \right) S_{coh} \right]. \tag{12}
$$

A few remarks seem to be appropriate concerning the validity of Eqns. (9), (10), (11), and (12). These equations only hold if the bunch is sufficiently long to allow a micro-bunching of the electron plasma in the synchronously travelling evanescent surface mode. Such plasma waves have recently been investigated theoretically by Andrews and Brau [Andrews and Brau, 2004] who also derived a gain formula. One important result is that the evanescent mode has a wavelength somewhat larger as the longest wavelength of SP radiation which in fact is emitted in backward direction. If the results, which were derived for a special SP experiment [Urata et al., 1998], could be generalized the gain of Ref. [Andrews and Brau, 2004] is lower than the formula of Schächter and Ron predicts but higher as that of Kim and Song. Further on, it must be concluded from this work that the bunch length should be at least of the order of the wavelength of the backward emitted SP radiation $D(1/\beta + 1)$, see Eq. (4), otherwise the gain is expected to be lower as the gain formulas predict, i.e. for an already bunched beam the gain is a function of the bunch length itself, or the coherent bunch form factor S_{coh}. In the limiting case of an already coherently emitting bunch it is reasonably that the gain must be negligible small. Inspecting Eq. (12) we see that SASE can be neglected if $GL \ll 1$, i.e. for a grating of length $L = 0.2$ m, as in our experiment, for $G \ll 5/$m.

Let us finally discuss the ansatz in the square brackets of Eq. (11). Since the gain formulas were derived for initially incoherently emitting electrons we assumed that the gain factor acts only on the incoherent part. This ansatz should be a good approximation as long as the coherent form factor is small in comparison to the incoherent one, i.e. for $S_{coh} \ll S_{inc}$. For an already nearly coherently emitting bunch the gain formulas must be modified, as already mentioned above. However, our ansatz in the square brackets might still be right if such modified gain expressions would be used.

In Fig. 4 the gains of Ref. [Schächter and Ron, 1989; Kim and Song, 2001] are shown as function of the mean electrical beam current $\bar{I}$ for our experimental design parameters. The gain of Kim and Song is more than a factor 10 smaller than that of Schächter and Ron, and SASE for the former is expected to be negligible small. However, it must be stressed that in view of the discussion of the last paragraph the gain of Schächter and Ron constitutes most probably only an upper limit.

The emitted intensity as function of mean beam current $\bar{I}$ and distance d of the beam to the grating are shown in Fig. 5 and 6, respectively. The effect of coherent emission from the electron bunch does not exceed about 30 %. On the one hand, the large gain according to Schächter and Ron results for a constant d in a strong nonlinear increase of the emitted intensity as function of $\bar{I}$. On the other hand, at constant $\bar{I}$ the intensity drops off faster as the interaction length h_{int} predicts for spontaneous emission. Both predictions can experimentally be scrutinized. It was one of the goals of the experiments described in the next subsection to shed light on this rather perturbing theoretical situation.

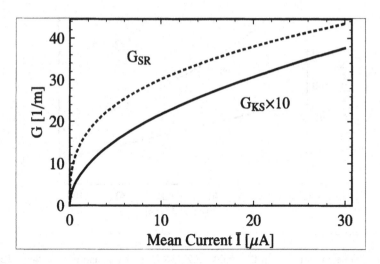

Figure 4. Gain G_{SR}, and G_{KS} according to Schächter and Ron [Schächter and Ron, 1989], and Kim and Song [Kim and Song, 2001], respectively, as function of the mean electron beam current $\bar{I}$. Parameters: kinetic energy of the electron beam 1.44 MeV, duty factor $F_{duty} = 0.5$, bunch repetition rate $f_b = 2.45$ GHz, standard deviation of bunch length $\sigma_x = 256.9$ μm, standard deviation of transverse beam size $\sigma_z = 15$ μm, radiation factor $| e_{00} | = 1$, magnetic field $B = 5$ Tesla, distance of beam to the grating $d = 30$ μm, observation angle $\Theta = 15.2°$, wave length $\lambda = 100$ μm

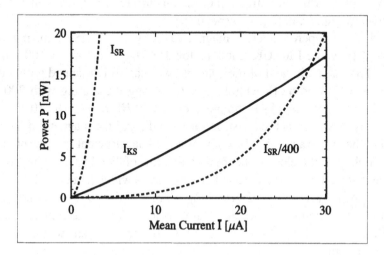

Figure 5. Integrated power in the limits $12.2° \leq \Theta \leq 18.2°$ and $60° \leq \Phi \leq 120°$ as function of the mean electron beam current $\bar{I}$. Shown are calculations with the gain of Kim and Song, labelled I_{KS}, and of Schächter and Ron, labelled I_{SR}. Radiation factor $|R_1|^2 = 0.23$, distance of the beam to the grating $d = 30$ μm, all other parameters as for Fig. 4

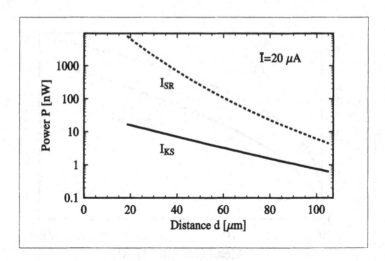

Figure 6.   Integrated power in the limits $12.2° \leq \Theta \leq 18.2°$ and $60° \leq \Phi \leq 120°$ as function of the beam distance d from the grating. Shown are I_{KS} and I_{SR} according to the gain of Kim and Song and of Schächter and Ron, respectively. Radiation factor $|R_1|^2 = 0.23$, mean electron beam current $\overline{I} = 20\ \mu A$, all other parameters as quoted in Fig. 4

2.   Experimental

Beam line and experimental setup at the 3.41 MeV injector of the Mainz Microtron MAMI for the investigation of the far-infrared Smith-Purcell radiation are shown in Fig. 7 and Fig. 8, respectively.

The 5 Tesla superconductive magnet (5T Cryomagnet System, manufactured by Cryogenic Ltd, UK London, Job J1879) has a bore of 100 mm diameter. The magnetic field of the solenoid was carefully mapped with a three dimensional Hall probe [Rochholz, 2002]. Along the grating with 200 mm length the magnetic field has a homogeneity of $|\Delta B|/B = 1.15 \cdot 10^{-3}$. More importantly, however, is the question for a bending of the magnetic field lines. Already a bending radius of $R_0 = 390$ m would be sufficient to prevent an optimal coupling of the electron beam with a radius of $R = 15\ \mu m$ to the grating. The relevant quantity is the inverse bending radius which in a two dimensional model is given by $1/R_0 = (\partial B/\partial z)/B_0$. Here z is the coordinate in the horizontal plane perpendicular to the symmetry axis. The measurement yielded $(\partial B/\partial z)/B_0 \leq 2.6 \cdot 10^{-5}$ cm^{-1} and fulfills the requirement within a region $\Delta z = \pm 1.4$ mm.

The liquid helium cooled composite silicon bolometer (Mod. QSIB/3, manufactured by Composite Bolometer System, QMC Instruments Ltd, UK West Sussex) served as radiation detector. A low-pass and a high-pass filter restrict the sensitive spectral range to $33\ \mu m \leq \lambda \leq 2$ mm. The f/3.5 Winston cone with an entrance aperture of 11 mm accepts angles $\theta \leq 8.1°$. The optical re-

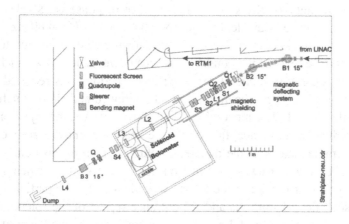

Figure 7.   Setup of the Smith-Purcell experiment at the injector LINAC A of the Mainz Microtron MAMI. The 1.44 MeV electron beam is deflected just in front of the 14 MeV Race Track Microtron RTM1 by an angle of totally 30°. An achromatic and isochronous system has been employed, consisting of two bending magnets (B1, B2) and six quadrupoles. The beam is properly coupled into the superconductive solenoid by means of a quadrupole doublet (Q1, Q2) and steerer magnets (S1, S2, S3, S4). Stray fields are symmetrized in this region by a magnetic mu-metal shielding. Downstream the solenoid the beam is deflected downwards by an angle of 15° (bending magnet B3) and finally dumped. Three zinc sulfide fluorescent screens (L1, L2, L3, L4) serve for beam diagnostic purposes.

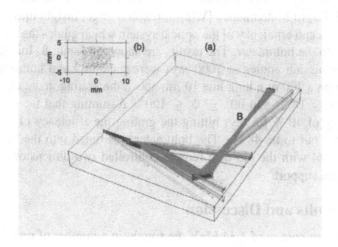

Figure 8.   (a)View onto the optical system. Far-infrared radiation produced by the grating G of 200 mm length is vertically focused by a cylindrical mirror M1 deflected by a plane mirror M2 onto a horizontally focusing mirror M3 and detected by a bolometer B. A cylindrical mirror above the grating focuses the radiation back to the electron beam. All components are plated with a layer of 5 μm of gold. Shown are also ray trace simulations. (b) Ray distribution perpendicular to the nominal ray direction at the bolometer position B.

sponsivity amounts to 10.7 kV/W , the speed of response is 285 Hz, and the system optical Noise Equivalent Power (NEP) 4.2 pW/$\sqrt{Hz}$ at 80 Hz. In combination with a lock-in amplifier the bolometer is sensitive enough to detect radiation under experimental conditions with a power as low as about 100 pW.

A special operating mode of the three linac sections was required to prepare a 1.44 MeV electron beam with good emittance and short bunch length [Mannweiler, 2004]. The most critical part, however, was the injection of the beam into the solenoid since, first of all, the beam axis must exactly coincide with the symmetry axis of the magnetic field. To achieve this goal any field distortions, even as far away as 2 m from the entrance of the solenoid, must be avoided. The iron joke of the RTM1 was identified to be particularly sensitive to such field distortions and the entrance region of the solenoid had to be shielded over a length of 1.5 m by mu metal. In case the beam axis does not exactly coincide with the symmetry axis of the solenoidal magnetic field the beam spot describes a circle at the fluorescent screen behind the solenoid if the magnetic field is varied. The injection of the beam can this way be controlled and optimized. Secondly, the beam must be focussed in such a manner to accomplish in the solenoid the required optimal angle ϑ_{opt}, Eq. (5), which is correlated to the beam radius R_{opt}. For this purpose the beam spot size was measured in the solenoid by the secondary electron signal from a scanner wire of 40 μm thickness.

Finally, ray tracing calculations were performed with the add-on Optica [Barnart, 2003] to Mathematica [Wolfram, 2003] to get insight into the focusing properties and efficiency of the optical system which guides the far-infrared radiation into the bolometer. The results are shown in figure 8. Initial coordinates and emission angles of 1000 rays were randomly and homogeneously distributed on a 200 mm long line 10 μm above the grating in angular regions $12.2° \leq \Theta \leq 18.2°$ and $60° \leq \Phi \leq 120°$. Assuming that the grating has a reflectivity of 40 % for rays hitting the grating the efficiency of the optical system turns out to be 48 %. The bolometer was tuned into the spot shown in figure 8 (b) with the aid of a remote controlled two-dimensional parallel displacement support.

3.   Results and Discussion

At the beam energy of 1.44 MeV we sought in a number of runs for spontaneous Smith-Purcell radiation with a wavelength of about 100 μm and, in particular, for the predicted SASE. Parameters and experimental setup were chosen as described above. Finally, only an upper limit $P \leq 0.085$ nW for the Smith-Purcell signal could be determined at a mean beam current $\bar{I} = 16$ μA and a magnetic field $B = 4.8$ Tesla. The only reasonable explanation for this rather disappointing result is that the beam spot was larger than expected and,

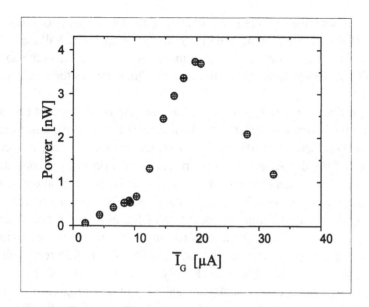

Figure 9. Bolometer signal as function of the mean beam current on the grating. Grating period 14.3 mm, wavelength $\lambda = 1.0$ mm at observation angle $\Theta = (15.2 \pm 3)°$, accepted wavelength region of the bolometer 33 μm $\leq \lambda \leq$ 2 mm.

therefore, the beam could not be coupled optimally to the grating. One or a combination of the following reasons may be responsible: (i) the non-standard operation mode of the three MAMI Linac sections for the 1.44 MeV beam deteriorated the transverse emittance and, in addition, also the bunch length, (ii) the beam could not be injected into the solenoid in such a manner that the optimal beam radius was achieved, and (iii) grating and electron beam in the solenoid could not be aligned properly. Indeed, measurements in the solenoid yielded a projected spot size in z direction with standard deviation of $\sigma_z \approx$ 100 μm. Also bunch length measurements were performed with the longitudinal phase space diagnostics of the injector linac [Euteneuer et al., 1988] which yielded $\sigma_x = 540$ μm. Assuming a distance $d = 2.5$ σ_z of the beam to the grating, even with the optimistic gain formula of Schächter and Ron an intensity of only $P = 0.05$ nW follows from the formulas presented in subsection 2 in accord with the experimental result.

To get further insight into the signal generation an experiment with a grating of the large period of 14.3 mm, $a/D = 0.51$, $h/D = 0.09$, $|R_1|^2 = 0.24$ was performed. The wavelength of the Smith-Purcell radiation at an emission angle of 15.2° amounts to $\lambda = 1.0$ mm. The interaction length $h_{int} = 298$ μm is large enough to couple even a beam with the experimentally determined large spot size of $\sigma_z = 105$ μm optimally to the grating. The intensity characteristics as function of the distance d of the electron beam from the grating was found to be consistent with the expected interaction length. The intensity of radiation as function of the grating current, see Fig. 9, was measured with half the electron beam current on the grating. The grating current, therefore, is the same as the current of the electrons above the grating which creates the Smith-Purcell radiation.

The non-linear growing of the signal as function of the current can be explained by the bunch coherence, as calculations with the above described formalism suggest, see figure 10 (a). The maximum of the intensity at a current of 20 μA and the decrease at larger currents is most probably a consequence of increasing bunch lengths, see figure 10 (b). It is worth mentioning that a gradual increase of the bunch length by only about 20 % explains already the experimentally observed dramatic decrease of the intensity at beam currents $I \geq 20$ μA. These results are in accord with experiments to determine the shape of an electron bunch with the aid of coherent Smith-Purcell radiation, see e.g. Ref. [Doria et al., 2002], or [Korbly and A.S. Kesar, 2003].

The just described experiment with the 14.3 mm grating corroborate the conjecture that the required settings of the accelerator and beam line for an 1.44 MeV electron beam with low transverse spot size and short longitudinal bunch length in the solenoid were not found, or can not be achieved at all. Therefore, further experiments were performed with the 3.41 MeV standard beam of good quality the results of which will be reported elsewhere [Mannweiler, 2004].

4.   Conclusions

Smith-Purcell (SP) radiation in the wave length region between 100 μm and 1000 μm has been investigated at the injector LINAC of the Mainz Microtron MAMI with a pulsed electron beam of 1.44 MeV energy. The beam was kept focused in a superconductive 5 Tesla magnet and guided above metal gratings with a length of 200 mm. The predicted large intensity of $\lambda = 100$ μm radiation at could not be observed. The most probable explanation for this fact is that (i) the beam spot size was too big and, therefore, the beam could not be coupled optimally to the grating, and (ii) gain formulas may not be applicable because the electron beam bunches in our experiment are shorter than the wavelength of the evanescent mode. At a wavelength of $\lambda = 1$ mm the observed intensity as function of the beam current can be explained by coherent emission from about 600 μm (rms) long electron bunches which slightly lengthen as the bunch charge increases.

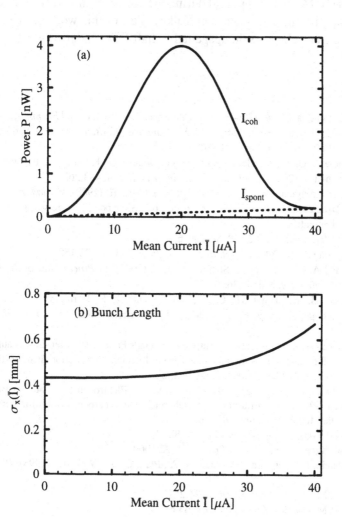

Figure 10. (a) Integrated intensity in the limits $12.2° \leq \Theta \leq 18.2°$ and $60° \leq \Phi \leq 120°$ as function of the mean electron beam current $\overline{I}$ for a grating with a period of 14.3 mm. Wave length $\lambda = 1.0$ mm at observation angle $\Theta = 15.2°$. A rectangular beam profile of width $\Delta_z = 131.6$ μm in z direction with a distance to the grating $d = 65.8$ μm was assumed. (b) Adapted standard deviation of the bunch length σ_x as function of the mean beam current $\overline{I}$. Other parameters: kinetic energy of the electron beam 1.44 MeV, duty factor $F_{duty} = 0.5$, bunch repetition rate $f_b = 2.45$ GHz, $\Delta_y = 263.2$ μm, radiation factor $|R_1| = 0.23$, magnetic field $B = 4.8$ Tesla.

Acknowledgments

One of us (H. M.) was supported during the course of this work by a stipend of the Studienstiftung des Deutschen Volkes. Part of this work has been supported by the Deutsche Forschungsgemeinschaft DFG under contract Ba1336/1-3.

References

Andrews, H. L. and Brau, C. A. (2004). *Phys. Rev. Spec. Topics - Acc. and Beams*, 7:070701.

Barnart, D. (2003) *Optica: Design Optics with Mathematica*. Wolfram Research. Version 1.3.0.

Di Francia, T. (1960). *Nuovo Cimento*, 16:61.

Doria, A., Gallerano, G., Giovenale, E., Messina, G., Doucas, G., Kimmitt, M. F., Andrews, H. L., and Brownell, J. (2002). *Nucl. Instr. Meth. in Phys. Res. A*, 483:263.

Euteneuer, H., Herminghaus, H., Nilles, K. W., and Schöler, H. (1988). In Tazzari, S., editor, *European Particle Accelerator Conference*, page 1149, Singapore, New Jersey, London,Hong Kong, World Scientific.

Frank, I. (1942). *Izv. Akad, Nauk SSSR, Ser Fiz*, 6:3.

Kim, K. J. and Song, S. B. (2001). *Nucl. Instr. Meth. Phys. Res. A*, 475:158.

Korbly, S. E. and A.S. Kesar, M. A. Shapiro, R. J. T. (2003). In *Proceedings of the Particle Accelerator Conference*, page 2536.

Kube, G., Backe, H., Euteneuer, H., Grendel, A., Haenbuck, F., Hartmann, H., Kaiser, K., Lauth, W., Schöpe, H., Wagner, G., Walcher, T., and Kretzschmar, M. (2002). *Phys. Rev. E*, 65:056501.

Mannweiler, H. (2004). Experimente zur Entwicklung eines Freie-Elektronen-Lasers auf der Basis des Smith-Purcell-Effektes im infraroten Spektralbereich, Dissertation, Institut für Kernphysik der Universität Mainz, 2004, in preparation.

Rochholz, H. (2002). Entwicklungsarbeiten für einen Freie-Elektronen-Laser auf der Basis des Smith-Purcell-Effektes im infraroten Spektralbereich, Diplomarbeit, Institut für Kernphysik der Universität Mainz, 2002, unpublished.

Schächter, L. and Ron, A. (1989). *Phys. Rev. A*, 40:876.

Smith, S. J. and Purcell, E. M. (1953). *Phys. Rev.*, 92:1069.

Urata, J., Goldstein, M., Kimmitt, M., Naumov, A., Platt, C., and Walsh, J. (1998). *Phys. Rev. Lett.*, 80:516.

Van den Berg, P. M. (1971). *Appl. Sci. Res.*, 24:261.

Van den Berg, P. M. (1973a). *J. Opt. Soc. Am*, 63:689.

Van den Berg, P. M. (1973b). *J. Opt. Soc. Am*, 64:1588.

Van den Berg, P. M. (1974). *J. Opt. Soc. Am*, 64:325.

Wachtel, J. M. (1979). *J. Appl. Phys.*, 50:49.

Wolfram, S. (2003). *Mathematica*. Wolfram Research. Version 5.0.

RADIATION BY RELATIVISTIC ELECTRONS IN THIN TARGETS AND AT COLLISIONS WITH SHORT BUNCHES OF RELATIVISTIC PARTICLES

N. F. Shul'ga [1,2], D. N. Tyutyunnik [1]
[1] *ITP NSC KIPT, Akademicheskaya 1, Kharkov 61108, Ukraine*
[2] *Belgorod State University, Belgorod 308007, Russia*
e-mail: shulga@kipt.kharkov.ua

Abstract. The process of radiation at collision of relativistic electron with short and long bunch of relativistic particles is considered. The analogy between this process and the process of radiation by relativistic electrons in oriented crystals is mentioned. Conditions, under which the coherent effect in radiation and the effect of suppression of coherent radiation take place, are obtained. Physical causes for arising of these effects are discussed.

Key words: amorphous media, coherent effect, Landau-Pomeranchuck effect, suppression of radiation, relativistic electron, long-range potential.

PACS number: 13.10.+q, 14.60.Ef, 41.75.Ht

1. Introduction

There are possible some different coherent and interference effects in radiation when a high energy electron moves in crystal close to one of the crystalline axes. Due to this effect the spectral density of radiation contains sharp peaks with high intensity of radiation [1-3].

Landau and Pomeranchuck shows [4] that the multiple scattering of high energy electrons in amorphous media can lead to suppression of radiation in the region of small radiated frequencies in comparison with the correspondent Bethe and Heitler result [5] under calculation of which was assumption that the radiation of the electron on different atoms of media has independent character.

In [6,7] was shown, that it is possible the analogous effects in radiation of electrons not only in thick but also in thin layers of crystalline and amorphous media. It is important that in last case the whole target plays as a single object interacted with the electron.

It was paid attention in 90-th [8] on the possibility of significant coherent effect in radiation under head-on collision of relativistic electron with a short bunch of relativistic particles. This effect takes place in the region of

H. Wiedemann (ed.), Advanced Radiation Sources and Applications, 283–292.
© 2006 *Springer. Printed in the Netherlands.*

frequencies for which the coherent length of radiation process is large in comparison with longitudinal length of the bunch.

In the present paper we pay attention that all of these effects have a general physical nature connected with features of the process of an electron radiation in the region of the radiation length. This allows to consider all of noted effects in radiation from the united viewpoint. In particular it is shows that the problem of radiation at collision of relativistic electron with a short and a long bunch of relativistic particles is similar in many parts to the problem of radiation of the electron in oriented crystals. It is possible both the coherent effect in radiation and the effect of suppression of coherent radiation in both cases. This allows to state the limiting values of the efficiency of radiation at collision of an electron with the bunch of the particles.

2.   The Coherence Length of Radiation Process

The process of radiation by relativistic electron is developed in large space region along the particle motion direction which shoots up with a particle energy increasing and decreasing of the frequency of radiated photon [9, 10]. The length of this region is called the coherent radiation length. Different coherent and interference effects are essential in the range of this length. Such effects take place for example at the motion of relativistic electron in a crystal near one of the crystal axes. The correlations between consistent collisions of the electron with lattice atoms are essential in this case. Because to these effects the spectral density of the electron radiation in oriented crystal can highly exceed the spectral density of radiation in nonoriented crystal (amorphous media).

In an amorphous media correlations between consistent collisions of the electron with atoms of media are absent. This denotes that wave phases of emitted waves at collision of electron with different atoms are accidental. Because of that the intensity of radiation of the electron in amorphous media is proportional to number of atoms with which the electron is collided. Landau and Pomeranchuck however showed [4] that this result is correct if the middle value of square of the multiple scattering angle of an electron in the range of the formation length of radiation is small in comparison with the square of typical radiation angle by relativistic electron. But if opposite condition is hold the effect of suppression of radiation in comparison with correspondent result of Bethe and Heitler [5] must takes place [4]. What is the physical reason for this effect? Toward this let us consider in detail the question how the relativistic electron radiates at collisions with atoms.

The electron is the charge, surrounded by own (coulomb) field of the particle. The lines of equal potentials of this field are pressed ellipsoids. As a result of interaction with atom the excitation of this field takes place. After that the excited part of the field comes off from the electron and then it is reorganized in the radiation field [10]. However, that takes place not at once for relativistic electron. This is connected with the fact that the radiation by relativistic electron occurs mainly in the direction of an electron motion. The excited part of the field can not quickly to be turning off from the electron. Really, the relative velocity of recession of the excited part of field and the electron is $v_r = c - v$, where c - is the velocity of light and v - is the electron velocity. It is possible to consider the Fourier component with the length λ of the stimulated part of field as a free field from the electron if this part of the field comes off from the electron on the distance λ (if it is not true then the interference between the own electron field and the stimulation of this field is very essential). Such division of the own electron field and his stimulated part takes place during the time $\Delta t = \lambda \big/ v_r \approx 2\gamma^2 \lambda \big/ c$. The length passed by the electron during this time $l_c = v \Delta t$ coincides with the coherent length of the radiation process [9,10].

Let us consider now the question about radiation intensity. If the electron collides with number atoms over the coherence length then the excitation of his field is increased. But if number of collisions of the electron with atoms over the coherence length will big enough that all Fourier components with the length λ of his field take a significant excitation, then it will be only re-excitation of this part of field. At that there do not arise additional stimulation. Therefore the additional radiation do not arise in this case. This is the physical reason for Landau-Pomeranchuck effect of suppression of radiation by a fast electron in media.

However in the case of Landau-Pomeranchuck effect the picture of the radiation process turns out some more complicated because this effect relates to the media the thickness of which is much bigger than the coherent length. There is some balance in this case between the excitation of the particle field in a media and transformation of this field in radiation. So the essential interest has the case for electron radiation in a thin media for which the coherent length of the radiation process is bigger than the thickness of a target. It was shown in [6, 7] that the effect of radiation suppression takes place also in this case. However this effect has some differences from Landau-Pomeranchuck effect. Namely in considered case the spectral density of the radiation is not depends from the target thickness L, starting from some thickness. The number of collisions increases in this case but the radiation do not increase with increasing of the target thickness. Such effect takes place when the particle passes through thin layers of an amorphous or a crystalline media. It will show further that the analogous effect takes place also at collision of relativistic electron with a short bunch of relativistic particles.

3.   Radiation at Collisions of Fast Electron with a Short Bunch of Relativistic Particles

The problem of coherent radiation by relativistic electron at collision with a short bunch of relativistic particles is similar to the problem of coherent radiation by relativistic electron in oriented crystals [10]. That is connected with the fact that the particles of a bunch are charged particles having a long-distant potential. For this case, as well known [5, 10], both the longitudinal as well as orthogonal distances, which are important for radiation, are equal to the coherence length $l_c = \dfrac{2\gamma^2}{\omega}$ on the order of value. If the orthogonal size of a bunch is smaller, than the length l_c, then correlations between sequential collisions of the electron with particles of the bunch are essential. As a result of this correlation the essential coherent effect in radiation is possible which is analogous to the coherent effect in radiation at collision of relativistic electron with oriented crystal. Let us show that at first for frontal collision of an electron with a short bunch of relativistic particles when the orthogonal size $L_\perp$ of the bunch is less than his longitudinal size L.

Taking into account the noted analogy the description of radiation by relativistic electron in the field of a bunch can be made on the basis of methods proved for description of the coherent radiation process by electrons in oriented crystal. Let us consider the simplest case of present problem, when the coherence length l_c is large in comparison with longitudinal size of the bunch. All particles of the bunch in this case take part in interaction as one particle with the effective charge $Q = N|e|$ where N is the particle number in the bunch and e is the electron charge. Than the effective constant of interaction of the electron with particles of bunch is determined by the expression $\dfrac{Ne^2}{\hbar c}$. This parameter is big in comparison with unit for high N values. The radiation by electron in the field of such bunch can be considered in the frame of classical theory of radiation.

When the condition $l_c \gg L$ holds the spectral density of radiation is defines by the expression [11]

$$\frac{dE}{d\omega} = \frac{2e^2}{\pi}\left\{ \frac{2\xi^2 + 1}{\xi\sqrt{\xi^2 + 1}}\ln\left(\xi + \sqrt{\xi^2 + 1}\right) - 1 \right\}, \tag{1}$$

where $\xi = \xi_N = \dfrac{\gamma \vartheta_N}{2}$, γ is the Lorenz factor of the electron and ϑ_N is the electron scattering angle. For the simplest case when the electron passes near the bunch at the distance $\rho > L_\perp$ from the bunch axes we have

$$\xi_N = N\frac{e^2}{m\rho}, \tag{2}$$

where m is the electron mass. If the condition $\xi_N \ll 1$ holds we obtain from (1) the following expression for spectral density

$$\frac{dE}{d\omega} \approx \frac{2e^2}{3\pi}\gamma^2\vartheta_N^2 = N^2\frac{8e^6}{3\pi m^2\rho^2}. \tag{3}$$

The inequality $\xi_N \ll 1$ is quickly destroyed with increasing of N. If, for example, $\rho \approx 10^{-2}\,cm$, then this inequality is broken for $N \approx 10^{10}$. But if the condition $\xi_N \gg 1$ is hold, we have in accordance with (1)

$$\frac{dE}{d\omega} \approx \frac{4e^2}{\pi}\ln 2\xi_N = \frac{4e^2}{\pi}\ln\frac{2e^2N}{m\rho}. \tag{4}$$

Thus, when the condition $\xi_N \ll 1$ is holds, the coherent effect in radiation of an electron in the field of bunch takes place according to which the intensity of radiation is proportional to N^2. However, if the condition $\xi_N \gg 1$ is hold, the effect of suppression of coherent radiation takes place and quadratic dependence of the intensity from N substitutes by more weak logarithmical dependence from N. The physical reason for the last effect is the same as the reason of suppression of high-energy electrons radiation in thin layer of media.

4. Coherent Radiation at Collisions of Fast Electrons with a Long Bunch of Relativistic Particles

Let us consider now the radiation of low-energy photons arising at falling of relativistic electron on a long bunch of charged particles under small angle ψ relative his axes. For this case in dipole approximation of classical theory of radiation we have [10]

$$\frac{dE}{d\omega} = \frac{e^2\omega}{2\pi}\int_\delta^\infty\frac{dq}{q^2}\left(1 - \frac{2\delta}{q}\left(1 - \frac{\delta}{q}\right)\right)\left|\vec{W}(q)\right|^2, \tag{5}$$

where $\delta = \omega/2\gamma^2$ and $\vec{W}(q)$ is the Fourier transformation of the orthogonal component of the particle acceleration

$$\vec{W}(q) = \int_{-\infty}^{+\infty} dt \dot{\vec{v}}_\perp (t) e^{iqt} . \tag{6}$$

(Here we use the system of units, where the light velocity c is equal to unit.)

In the present case the orthogonal component of electron acceleration is defined by the following equation of motion (see eq. (9) in [11])

$$\dot{\vec{v}}_\perp (t) = -\frac{2e}{\varepsilon} \frac{\partial U(\vec{\rho}(t))}{\partial \vec{\rho}(t)}, \tag{7}$$

where ε is the initial electron energy, $U(\vec{\rho}(t))$ is a full scalar potential created by particles of flying bunch in the point $\vec{\rho}(t)$ where the electron is situated. (The presence of factor two in this expression is a result, that scalar and vector potentials in relativistic case take an equal contribution in equation of motion [11].) The formation length of small radiated frequencies ω is large in comparison with the middle distance a between particles in the bunch. The microscopic structure of the bunch in longitudinal direction can be neglected for this case. This allows, as in the case of radiation by electron in crystals [10], to substitute the potential of the particle of bunch $U(\vec{\rho}(t))$ in (7) by the potential which is averaged on z coordinate along the bunch axes. This expression must be also averaged on the particle distribution of the flying bunch in orthogonal plane. For the most experiments this distribution has Gaussian form and the expression (6) must be averaged with the function

$$f(\vec{\rho}_1) = \frac{1}{\pi u^2} e^{-\frac{\vec{\rho}_1^2}{u^2}},$$

where $\sqrt{u^2}$ is the orthogonal size of the bunch and $\vec{\rho}_1$ is the fixed orthogonal coordinate of the particle in the bunch. So the average potential of the bunch is defined by the expression

$$\langle U(\vec{\rho}(t))\rangle = \frac{Q}{a}\int f(\vec{\rho_1})d^2\rho_1 \int_{-\infty}^{+\infty}\frac{dz}{\sqrt{z^2+\dfrac{(\vec{\rho}(t)-\vec{\rho_1})^2}{\gamma^2}}}. \tag{8}$$

Substituting (8) in (7), we can take the next expression for $\left|\vec{W}(q)\right|^2$:

$$\left|\vec{W}(q)\right|^2 = \chi e^{-\frac{u^2 q^2}{2\psi^2}}\left(A^2+B^2\right), \tag{9}$$

where $\chi = \dfrac{16e^2 Q^2}{\varepsilon^2 a^2 \psi^2}$ and

$$A = \int_{-\infty}^{+\infty}\frac{x\,dx}{x^2+\dfrac{q^2\overline{u^2}}{4\psi^2}}e^{-x^2}\sin\left(\frac{2b}{\sqrt{\overline{u^2}}}x\right), \tag{10}$$

$$B = \frac{q\sqrt{\overline{u^2}}}{2\psi}\int_{-\infty}^{+\infty}\frac{dx}{x^2+\dfrac{q^2\overline{u^2}}{4\psi^2}}e^{-x^2}\cos\left(\frac{2b}{\sqrt{\overline{u^2}}}x\right). \tag{11}$$

Here b is a value of the impact parameter of the electron relative to the bunch axis.

Let us consider some limit cases of obtained formulas.

If the condition $b \gg \sqrt{\overline{u^2}}$ is hold, then $A = B = \pi e^{-\frac{qb}{\psi}}$. For this case the spectral density of radiation (5) in the region of small radiated frequencies $\omega \ll \dfrac{\psi\gamma^2}{b}$ is defined by the next expression:

$$\frac{dE}{d\omega} = \frac{4\pi}{3}e^2\gamma^2\chi. \tag{12}$$

But if the particle flies near the axes of the bunch symmetry, then $b \approx 0$ and for small radiated frequencies integrals (10) and (11) take the form

$$A = 0, \quad B = \pi e^{\frac{q^2 \overline{u^2}}{4\psi^2}}.$$

The square of module of Fourier components (6) in this case comes to a constant $\left|\vec{W}(q)\right|^2 = \chi\pi^2$, and the spectral density of radiation (5) turns out the constant also, independent from the radiated frequency:

$$\frac{dE}{d\omega} = \frac{2\pi}{3} e^2 \gamma^2 \chi. \tag{13}$$

However, this spectral density of radiation is not equal to zero. Distinction from zero of (13) is a special feature of the long-range coulomb potential for charged particles (8).

For real experiments we takes up the radiation from the flux of charged particles. In corresponds with that the spectral density of radiation (5) must be averaged on the particle distribution in radiated bunch. This bunch has usually the Gaussian distribution. However, for the sake of simplicity, let's assume that orthogonal size of radiated bunch is bigger than the orthogonal size of flying bunch $\sqrt{\overline{u^2}}$. It is possible to consider in this case, that the bunch has almost the uniform distribution function in this case.

Usually, when such questions rates, it is used a conception of the radiation efficiency [10]:

$$\frac{dK}{d\omega} = \int \frac{dE}{d\omega} d^2\rho.$$

For the uniform bunch with the length L this expression can be written in the form

$$\frac{dK}{d\omega} = L\psi \int db \frac{dE}{d\omega}. \tag{14}$$

After averaging on impact parameters b we find that the square of the module of Fourier components (6) is defined by equation

$$\left\langle \left|\vec{W}(q)\right|^2 \right\rangle = 2\pi\chi e^{-\frac{q^2 \overline{u^2}}{2\psi^2}} \int_{-\infty}^{+\infty} \frac{dx}{x^2 + \frac{q^2}{\psi^2}} e^{-\frac{\overline{u^2}}{2}x^2}.$$

The last expression takes the next form for small radiated frequencies $\omega << \dfrac{2\sqrt{2}\psi\gamma^2}{\sqrt{u^2}}$,

$$\left\langle \left| \vec{W}(q) \right|^2 \right\rangle \approx \frac{2\pi^2 \chi\psi}{q}.$$

The radiation efficiency in this case is determined by the following formulae

$$\frac{dK}{d\omega} \approx \frac{32}{3}\frac{L}{a}\frac{e^4 Q^2}{m^2}\frac{1}{a\delta}, \tag{15}$$

when m is the electron mass. It is possible to see, that this expression do not depend on the incident angle ψ. This is an important property of the long-range potential (8).

5.   Conclusion

The radiation process by relativistic particles is developed at large spatial range along the particle motion direction. If an electron is collided with big number of atoms in this range the electron interaction with these atoms is carried out differently then with an isolated atom. It is possible both the increasing and the suppression of interaction efficiency of an electron with atoms of media in this case.

The electron interaction with atoms in amorphous media is carried out as with isolated atoms if the distances between atoms are far enough. In dense amorphous media the Landau-Pomeranchuck effect of suppression of radiation is possible which is connected with influence on radiation of the multiple electrons scattering in the region of coherent length.

The coherent effect in radiation is possible if relativistic electron is moving near one of the crystal axis. Due to this effect the intensity of radiation by electron in crystal can be much bigger than the intensity of radiation in nonoriented crystal (amorphous media). The effect of suppression of coherent radiation is possible for this case also. This effect is close to the Landau-Pomeranchuck effect of suppression of radiation by high energy electrons in amorphous media.

The analogous effects, as it is shown in present paper, are possible also at collisions of high energy electrons with the bunch of relativistic particles. The description of radiation process by electron can be fulfilled in this case on the basis of methods which were used for description of the radiation process in crystal. That allowed to make clear general properties and differences between radiation processes in these cases. It is general that the effect of coherent radiation and the effect of suppression of coherent radiation are possible in these cases. The differences are connected with types of potential with which the particle interact. The potential of particle bunch is the long-distant potential. But the potential of each crystal atomic string has a screening at distances on the order of value of Thomas-Fermi radius. Due to that, both longitudinal and transversal distances which are important for radiation at collision of fast electron with the bunch of relativistic particles have the macroscopic size.

The effect of coherent radiation at collision of relativistic electron with short and narrow bunch of relativistic particles can be used for monitoring of colliding beams of high energy particles.

Acknowledgments

This work is supported in part by Russian Foundation for Basic Research (project 03-02-16263).

References

1. Ferretti B. *Sulla bremsstrahlung nei cristalli*, (1950) *Nuovo Cim.* 7, 118.
2. Ter-Mikaelian M. L. *Interference radiation by relativistic electrons*, (1953) *Zh. Exp. Teor. Fiz.* 25, 296.
3. Überall H. *High-energy interference effect of bremsstrahlung and pair production*, (1956) *Phys. Rev.* 103, 1055.
4. Landau L. D. and Pomeranchuck I. Ya. *Electron-Shower processes at high energy*, (1953) *Dokl. Akad. Nauk SSSR* 92, 735.
5. Bethe H. and Heitler W. *On the stopping of fast particles and on the creation of passive electrons*, (1934) *Proc. Roy. Soc.* 146, 83.
6. Shul`ga N. F. and Fomin S. P. *On suppression radiation in an amorphous media and in a crystal*, (1978) *Lett. Zh. Exp. Teor. Fiz.* 27, 126.
7. Shul`ga N. F. and Fomin S. P. *On the space-time evolution of the process of ultrarelativistic electron radiation in a thin layer of substance*, (1986) *Phys. Lett.* A114, 148.
8. Ginzburg I. F., Kotkin G. L., Polityko S. I. and Serbo V. G. *Coherent bremsstrahlung at colliding beams. Method of calculation*, (1992) *Sov. J. Nucl. Phys.* 55, 1847; 55, 3324.
9. Ter-Mikaelian M. L. *High Energy Electrodynamical Processes in Condensed Media*; Wiley-Interscience, New-York (1972).
10. Akhiezer A. I. and Shul`ga N. F. *High Energy Electrodynamics in Matter*; Gordon and Breach, Amsterdam (1996).
11. Shul`ga N. F. and Tyutyunnik D. N. *On the radiation in collisions of short bunches of relativistic particles*, (2003) *JETP Lett.* 78, 700.

RING X-RAY TRANSITION RADIATION DETECTORS ON THE BASIS OF CHARGE COUPLED DEVICE ARRAY

V. G. Khachatryan
CANDLE-Center for the Advancement of Natural Discoveries using Light Emission
Avan, Acharian, 31,Yerevan, 375040, Armenia

Abstract.   The idea of creation of the ring imaging detectors for the high Z relativistic heavy ion detection applying charge coupled device (CCD) arrays as a X-ray photon detectors is discussed taking into account availability of CCD arrays and wide experience of their scientific applications in the field of experimental physics for the registration of the X ray photons. The results of the Monte Carlo calculations are presented.

1.   Introduction

Since the intensity of the forward X-ray transition radiation (XTR) depends particle Lorenz factor γ, XTR can be used for the measurement of the γ [1]. Recently it has been proposed to use also the angular distribution of XTR in order to improve the resolution of detector by constructing Ring Transition Radiation Detectors (RTRD) [2,3]. The construction of the RTRD where Xenon filled multiwire proportional chamber is applied for X-ray photon detection is described in the paper [2]. X-ray RTRD construction and principles of the operation have similarities with those of proximity focusing RICH detectors. No focusing elements are necessary but relatively large lever arm is required.

In the present work CCD array is proposed for X-ray detection because of its obvious advantages as position sensitive photon detector, compared with multiwire proportional chambers. Good spatial resolution and relatively high efficiency in the energy region of the photons 0.1 – 10 keV make CCD array a good choice for the measurement of angular distribution of the photons emitted by relativistic charged particle. Better spatial resolution of CCD detector compared with multiwire chambers enable one to choose shorter

293

H. Wiedemann(ed.), Advanced Radiation Sources and Applications, 293–297.
© 2006 *Springer. Printed in the Netherlands.*

lever arm and thus more compact construction of the RTRD. Another argument supporting CCD choice as a photon detector is that the technological advance continuously improves their parameters and makes them more affordable.

Monte Carlo simulations support the idea that it is possible to get ring images of X-ray transition radiation of the relativistic heavy ions in the multilayer radiator within the aperture of the photon detector and thus find out energy (or more precisely Lorenz factor) of the ion. Below a possible construction of the RTRD utilizing CCD array as a photon detector is described and results of Monte-Carlo simulation carried out to reveal principles of RTRD operation are presented.

2.   RTRD Construction

Figure 1 illustrates RTRD construction. The radiator is consisted of 22.5 microns polyethylene layers spaced with 500 microns gaps. Radiator overall thickness is limited to 2 cm reducing number of the radiated photons, but allowing neglect the distances between the production points of the photons. CCD stands 2m away from the radiator letting photons to fly away from the axis of incident charged particle direction of motion. The limit on detectable charged particle Lorenz factor is a distance of 2 m. That distance can be changed by tuning RTRD to different γ regions. However, longer distances result in higher radiation losses and scattering of X-ray photons in the air. A helium filled bag can be placed between the radiator and the CCD detector to reduce such losses.

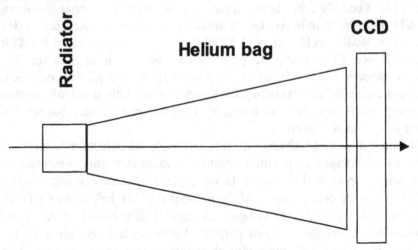

Figure 1. Schematic view of the ring imaging transition radiation detector.

The overall radiator thickness is 2 cm. A small size allows considering the X-ray photons produced by a point source. The CCD array should be positioned at some distance from the radiator. A 2 m distance is enough to get at least 10 bins of X-ray photons polar angle distribution for the incident charged particles, even with $\gamma = 5 \times 10^4$ provided that CCD detector has 50μm spatial resolution. A Helium filled bag may be used to reduce loss of X-ray photons due to scattering between the radiator and the CCD detector.

It is intended to use the measured angular distribution of the X-ray transition radiation of the charged particle crossing the multilayer radiator. In this respect it is advantageous to apply CCD arrays as a photon detectors because of their much better spatial resolution in comparison with those of the xenon multiwire proportional chamber. We assume that CCD array has relatively modest parameters that listed in the Table 1. The main requirement is relatively high sensitivity, i. e. capability to discriminate single X-ray photons. Usually it is easy to get good quantum efficiency in the region of 10 keV applying GdO_2S_2 phosphor screen that converts the incoming X-ray image into an optical image.

Table 1. CCD photon detector parameters

Sensitive area	1 cm x 1 cm
Pixel Size	12 μm
Pixel numbers	800 × 800
Spatial resolution	50 mm

It is possible to design a scientific grade CCD detector with relatively high spatial resolution and high sensitivity capable to discriminate single photons [4]. While wide dynamic range is not required for RTRD photon detector the high sensitivity is the most vital feature required for the CCD detector applicable in the RTRD.

3. Numerical Calculations

We have chosen for numerical calculations the quantum efficiency curve of the Advanced CCD Imaging Spectrometer (ACIS) onboard the Chandra X-Ray observatory ASIC experiment [5,6]. The probability of the X-ray photon registration in the region of the photon energies 0.2 keV to 12 keV is rather high (see Fig. 2).

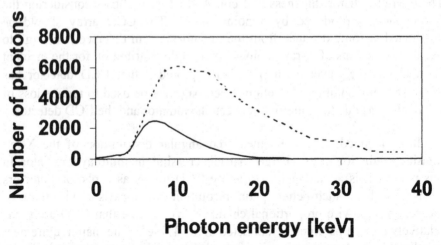

Figure 2. Monte Carlo simulations. The energy spectrum of the X-ray photons produced in multiplayer radiator by the charged particle (Z=82) within the interval of the polar angles from 0 to $5/\gamma$, where Lorenz-factor $\gamma = 3000$ (dashed line). Solid line shows the same spectrum taking into account quantum efficiency and thus cutting off 80% of photons.

Monte Carlo simulations of the X-ray photons produced by relativistic heavy ion with charge Z=82 and $\gamma = 3000$ have been done. Angular photon distributions relative to the direction of the motion of the incident particle are shown in the Fig. 3.

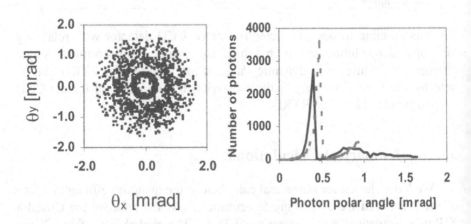

Figure 3. Monte-Carlo simulations. Left - Scatter plot θ_x vs. θ_y (Polar angles in the horizontal and vertical planes respectively) of the photons. This plot shows that the wide rings in the photon detector plane can be observed. Right –Polar angle distribution of photons. Particle charge is Z = 82. The solid line corresponds to $\gamma = 3000$, and the dashed line to $\gamma = 5000$.

Taking advantage of the interference between transition radiation coming from evenly positioned plates of the radiator it is possible to get a sharp maximum in angular distribution of photons. Obtaining that distribution from the measurement enables one to find out from the position of the distribution maximum and distribution shape the Lorenz factor of the incident charged particle in accordance with procedure described in the works [2,3].

4. Summary

Numerical calculations show that the creation of RTRD using photon detectors based on CCD arrays with affordable (and rapidly improving along with advance of technology) parameters appears feasible. That type of detectors can be used to find out the Lorenz factor of relativistic heavy ions in the range of $\gamma = 10^3 - 10^4$.

References

1. Garibian, G.M. and Yan Shi, Xray Transition Radiation, Publ. House of Academy of Sciences of Armenia, Yerevan, 1983.
2. Akopov, N.Z., Ispiryian, K.A., Ispirian, M.K., Khachatryan, V.G., Ring transtion radiation detectors in frascati Physics Series Vol. XXV (2001), pp.59-74.
3. Aginian,M.A., Ispiryian, K.A., Ispirian, M.K., Khachatryan, V.G., nucl. Instr. And Meth. A 522 (2004) 112.
4. Phillips, Walter C., et al, J. Synchrotron Rad. (2002). 9, 36-43
5. Sahai, Raghvendra,et al., Astrophys.J. 599 (2003) L87-L90.
6. Cristiani, S. et al, , Astrophys.J. 600 (2004) L119-L122.

HIGH ANALIZING POWER GAMMA POLARIMETER

K.R. Dallakyan
Yerevan Physics Institute, Brot. Alikhanyan Str. 2, 375036, Yerevan, Armenia

Abstract: A new approach for high-energy photon beam polarimetry was found using appropriate ranges of e^+e^- -pair emission polar and azimuthal angles. The pair particle energy and angular distributions are calculated by the Monte-Carlo method. On the basis of obtained results a design of polarimeter is given.

Keywords: Polarization, polarimetry, pair production, cross section asymmetry

1.   INTRODUCTION

The interest towards to photon beam polarimetry is connected to the planned in the large research centers (JLAB, CERN) experiments on photoproduction processes by polarized photon beams [1,2]. As photon beam polarization will be measured in parallel with the basic experiment, it lays out conditions on polarization measurement methods: the polarimeter should be sufficiently simple and inexpensive, it should not have an essential influence on the photon beam intensity (less than 1%), the errors should be less than 2-5%, the asymmetry has to be sufficiently high in order to diminish the influence of systematic errors, etc.

The comparison of photon beam linear polarization measurement methods1 shows that the most suitable methods in case of 4-10 GeV photon energies are the measurement of the asymmetry in the processes of triplet photoproduction in the field of atomic electrons and e+e--pair production in amorphous and crystalline media. The first process is more suitable from the

E-mail:kdallak@mail.yerphi.am

H. Wiedemann (ed.), Advanced Radiation Sources and Applications, 299–304.

standpoint of the measurement technique - one needs to measure the recoil electron asymmetry at large angles (larger than 15°). However, the analyzing power is low (about 10%) and the yield of reaction is relatively small. In case of pair production in crystals, the asymmetry grows with photon energy, becoming essentially high at photon energies $10 > 10$ GeV. Although this method faces some technical difficulties connected with the usage of a goniometric system, a magnetic spectrometer, etc., nevertheless, it has an essential advantage - the analyzing power does not depend on multiple scattering of pair particles. It should be pointed out that this method is the best at high photon energies of order of 10 GeV or higher.

The method to measure the asymmetry in the process of e^+e^--pair production in amorphous media seems more appropriate at JLAB experiments. The results of this method are discussed in the Proceedings [1] from Workshop on "Polarized Photon Polarimetry" and in the paper [3]. In paper[3] the asymmetry of nearly coplanar pairs with respect to the polarization plane is considered when one of pair particles is emergent within or beyond the small azimuthal angle interval of $2\Delta\Phi$ (the so-called wedge method). The asymmetry was calculated when pair particles energies are not recorded and for pair particles of equal energies after integration of cross section over polar angles. The maximum analyzing power made 18%.

2. RESULTS OF CALCULATIONS AND DISCUSSION

We proceed from the five dimensional differential cross section formula of e^+e^--pair photoproduction in the Coulomb field by polarized photons in high energy and small angle approximation, depending on pair particles energy y, and polar $\vartheta_\pm$ and azimuthal $\varphi_\pm$ angles [4,5].

The calculations are carried out for 4 GeV photons and a copper target. We repeated the calculations of the wedge method [3] in order to verify the Monte-Carlo simulation program's results. The results are shown in Fig. 1 for the asymmetry ratio $R = d\sigma(\varphi_+ = 0)/d\sigma(\varphi_+ = \pi/2)$ as a function of the angle $\Delta\Phi/\beta$ in the case of equal energies y=0.5 after integration over polar angles $0 \leq \gamma\vartheta_\pm \leq 28$. Curves a and b correspond to the cases when one pair particle is emergent within and beyond the small angular interval of $2\Delta\Phi$, respectively, and coincide with curves a and b of Fig.4 in [3]. Curves a' and b' represent the same dependencies as *a* and *b*, but now the first particle observes also emergent in the small angular range of $2\Delta\varphi$ ($\Delta\varphi/\beta$=1.98; β^{-1} =

$121Z^{-1/3}$). It occurs that at $\Delta\Phi \geq \Delta\varphi$ the influence of $\Delta\varphi$ is negligibly small and becomes essential at $\Delta\Phi < \Delta\varphi$.

The results of calculations show that high analyzing power can be attained by combining the two methods, selecting particles by azimuthal (the wedge method) and polar (the ring method) angles. The results of such selection are illustrated in Fig. 2. The curves *a* and *b* in Fig. 3 are the same as curves a and b in Fig. 1, but now particles are selected also by polar angles within the range $1.56 \leq \gamma\vartheta_\pm \leq 3.13$ where $\gamma = /m$ and the analyzing power is large, ~44 %. These results are obtained for pair particles of equal energies y = 0.5 in the ideal case when one of particles has azimuthal angle $\varphi = 0$ or $\varphi = \pi/2$ (i.e. $\Delta\varphi = 0$).

The curves c and d are calculated taking into account $\Delta\varphi$ ($\Delta\varphi = 0.05$ rad or $\Delta\varphi/\beta = 1.98$; $\beta^{-1} = 121Z^{-1/3}$). Curve c represents the case when the pairs are taken within the angular range of $1.56 \leq \gamma\vartheta_\pm \leq 3.13$ and the particles energies are not recorded $0 < y < 1$. The analyzing power is ~27%. Curve d is calculated for almost symmetrical pairs $0.475 < y < 0.525$ in the angular ranges $1.56 \leq \gamma\vartheta_\pm \leq 3.13$. The analyzing power is high, ~43%. In the case of c the analyzing power is low, than in the case of d but the pairs yield is high and there is no need in a pair spectrometer.

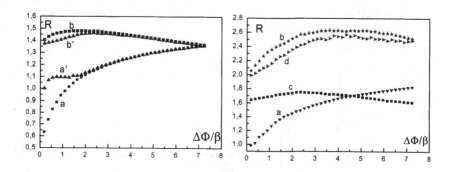

Figure 1. The asymmetry ratio as a function of $\Delta\Phi/\beta$.

Figure 2. The asymmetry ratio as a function of $\Delta\Phi/\beta$.

When it is difficult to reconstruct the kinematics of e^+e^- reaction (especially the photon propagation direction), we propose a new and more appropriate method for asymmetry measurement (the so-called coordinate method). It suggests to measure the electron and positron energies $y = \varepsilon_\pm/\omega$ and their coordinates in detection plane parallel (x) and perpendicular (y) to the polarization plane. The analyzing power is defined as A =

$(N_{\|0} - N_{\perp\flat})/(N_{0\|0} + N_\perp)$, where $N_{\|00}$ and $N_{\perp\flat}$ are the number of pairs with Δx $\geq k\Delta y$ and $\Delta y \geq k\Delta x$, respectively, where $\Delta x = |x_+ - x_-|$, $\Delta y = |y_+ - y_-|$ and k is a chosen constant quantity. Selecting particles in different energies and angular ranges one attain 15-20 % for analyzing power by this method.

The results obtained in this work can serve as a theoretical basis for creation of a polarimeter measuring the photon beam polarization degree at photon energies of a few GeV. Several different methods for the measurement of pair particle production cross section asymmetry are considered in this article. Each of these methods has both advantages and disadvantages. But it is possible to design a polarimeter using the potentialities of different methods simultaneously and thus decrease the systematic error.

3.   DESIGN OF POLARIMETER

For low energies and at large bases can bee constructed simple polarimeter. In particular simple and optimal polarimeter is designed for the -2 photon beam of the Yerevan electron accelerator based on the ring method. The wedge method is not appropriate in this case since the angles are very small and practically it is impossible to register such angles by simple detectors.

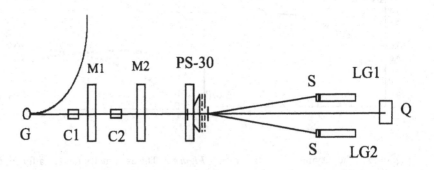

Figure 3. The layout of the polarimeter.

The layout of the polarimeter is shown in the Fig. 3. Photons created in diamond target, which is mounted in the goniometer placed in synchrotron ring, pass through collimators, cleaning lenses and magnet systems and exit to experimental hall. The photon beam will have intensity 10^8-10^9 photon/sec at peak energy of photon spectrum and will be collimated with 10^{-4} rad collimator. Polarimeter will be located downstream the collimator and will work in two regimes: with and without pair spectrometer. In the first case photon spectrum and polarization will be measured simultaneously. The basis of polarimeter is 18 m. In the second case the basis of the polarimeter makes up 25 m. For photon spectra measurement the polarimeter will be disconnected and pair spectrometer will be connected.

The polarimeter consists of scintillation amorphous target for pair production (calculations are carried out for copper target) and two pairs of scintillation and total absorption $PbWO_4$ counters. The scintillation and total absorption counters can be one pair but in this case they would have rotating possibility over the center of the beam. In this paper we present results of Monte-Carlo calculations for the case of 18 m base with pair spectrometer. The calculations have been done for photons with peak energy of 500 MeV.

Scintillation counters have 3x3 cm^2 sizes and will bee placed from the beam center at distances 5 cm (the centers) up-down and left-right. The total absorption counters will have 5x5 cm^2 sizes in order to avoid shower loss. It is natural that counters can bee chosen with small sizes but in that case the results will bee recalculated. For the chosen geography for a target with 10^{-4} radiation length thickness the analyzing power make up 21.84%, without taking account of multiple scattering in target the analyzing power will be 28.73%. Because of multiple scattering the analyzing power decreases to 5.6% for the target with 10^{-3} radiation length thickness.

4. ACKNOWLEDGEMENTS

The author is thankful to Prof. R. Avakian, H. Vardapetyan, A. Sirunyan, H. Akopyan and S. Darbinyan for usful discussions. The work was supported in part by ISTC grant A099 and in part by CRDF grant AP2-2305-YE-02.

References

1. Proceedings from JLAB/USC/NCCU/GWU Workshop on "Polarized Photon Polarimetry", (G. Washington University, June 2-3, 1998).
2. A. Apyan, R. O. Avakian, S. Bellestrero et al., Proposal to Study the Use a Crystal as a 'Quarter-Wave Plate' to Produce High Energy Circularly Polarized Photons, CERN/SPSC 98-17, SPSC/P308, July 13, 1998.
3. L.C. Maximon and H. Olsen, Measurement of linear photon beam polarization by pair production, *Phys. Rev.* **126**, 310 (1962).
4. H. Olsen and L.C. Maximon, Photon and electron polarization in high-energy bremsstrahlung and pair production with screening, *Phys. Rev.* **114**, 887 (1959).
5. R.Avakian, K. Dallakyan and S. Darbinyan, The search of large cross section asymmetries in the pair production process by polarized photons, hep-ex/9908048 v2 31, Aug 1999.

PROPOSAL FOR A PHOTON DETECTOR WITH PICOSECOND TIME RESOLUTION

R. Carlini[a], N. Grigoryan[b], O. Hashimoto[c], S. Knyazyan[b],
S. Majewski[a], A. Margaryan[b], G. Marikyan[b], L. Parlakyan[b],
V. Popov[a], L. Tang[a], H. Vardanyan[b], C.Yan[a]

a) *Thomas Jefferson National Accelerator Facility, Newport News, VA 23606, USA,*
b) *Yerevan Physics Institute, 2 Alikhanian Bros. Str., Yerevan-36, Armenia,*
c) *Tohoku University, Sendai, 98-77, Japan*

Abstract: A picosecond photon detector and a new timing concept are proposed here. Factors limiting time resolution of the technique are briefly discussed. Some R&D work is also reported.

Keywords: photon detector, picosecond, rf timing.

1.   INTRODUCTION

At present the detection of low light levels is carried out with solid state Charge Coupled Devices – CCDs, and vacuum photomultiplier tube -PMTs or hybrid photon detector -HPDs. Except CCDs, these instruments also provide fast time information used in different fields of science and engineering. In the high energy particle and nuclear physics experiments (see e.g. [1]), the time precision limit for the current systems consisting of particle detectors based on PMTs or HPDs, timing discriminators and time to digital converters is about 100 ps (FWHM).

However, it is well known that timing systems based on RF fields can provide ps precision. Streak cameras, based on similar principles, are used routinely to obtain picosecond time resolution of

H. Wiedemann (ed.), Advanced Radiation Sources and Applications, 305–311.

the radiation from pulsed sources [2-4]. The best temporal resolution is achieved with a single shot mode. Using continuous circular image scanning with 10^{-10} sec period, a time resolution of 5×10^{-13} sec for single photoelectrons and 5×10^{-12} sec for pulses consisting of ~17 secondary electrons, induced by heavy ions, was achieved in early seventies [5]. With a streak camera operating in the repetitive mode known as "synchroscan", typically, a temporal resolution of 2 ps (FWHM) can be reached for a long time exposure (more than one hour) by means of proper calibration [6]. But, streak cameras are expensive and their operation is complicated and they did not find wide application in the past, including the elementary particle and nuclear physics experiments.

The basic principle of a RF timing technique or a streak camera is the conversion of the information in the time domain to a spatial domain by means of ultra-high frequency RF deflector. Recent R&D work (ISTC Project A-372, [7]) has resulted in an effective 500 MHz RF circular sweep deflector for keV energy electrons. The sensitivity of this new compact RF deflector is about V/mm and is an order of magnitude higher than sensitivity of the RF deflectors used previously. The essential progress in this field can be reached by a combination of a new RF deflector created at YerPhI with new technologies for position sensors e.g. position sensitive electron multipliers from Burle Industries Inc., USA [8]. We are proposing a new concept for a compact and easy-to-use photon detector with picosecond time resolution based on the above combination.

2.   PICOSECOND PHOTON DETECTOR

The principle of the picosecond photon detector is to couple in vacuum the position sensitive technique to input window with photocathode evaporated on its inner surface (Fig. 1). A light photon incident on the photocathode generates a photoelectron, which is subsequently accelerated up to several keV, focused, and transported to the RF resonator deflector. Passing through the RF deflector, the electrons change their direction as a function of the relative phase of electromagnetic field oscillations in the RF cavity, thereby transforming time information into deflection angle. In this project, we propose to use the oscilloscopic method of RF timing, which will result in a circular pattern on the detector plane. Coordinates of the secondary electrons are measured with the help of a position sensitive detector located at some distance behind the RF deflecting system. The architecture of the technique is very similar to the design of the HPDs [9-11]. In contrast to ordinary PMTs, where a photoelectron initiates a multi-step amplification process in a dynode system, in the HPDs a photoelectron undergoes single-step acceleration from the photocathode towards the position sensitive detector where electron multiplication process happens in the solid state anode target material.

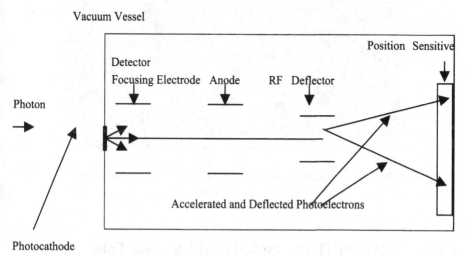

Figure 1. Schematic of the picosecond photon detector.

Recently at YerPhI, we have successfully employed a dedicated RF deflecting system at a frequency of about f = 500 MHz and obtained a circular scan of a 2.5 keV electron beam on the phosphor screen (see Fig. 2). This compact, simple and sensitive RF deflecting system is very suitable for high frequency RF analyzing of few-keV electrons.

The time resolution of the system presented in Fig. 1 depends on several factors [12, 13]:

2.1 Physical Time Resolution of Photocathode

General quantum-mechanical considerations show that the inherent time of the photoelectron emission should be much shorter than 10^{-14} sec. However, a delay and time spread of the electron signal - $\Delta\tau_p$ can be caused by the finite thickness of the photocathode Δl and the energy spread of the secondary electrons $\Delta\varepsilon$. For $\Delta l = 10$nm and $\Delta\varepsilon = 1$ eV we obtain $\Delta\tau_p \approx 10^{-13}$ sec.

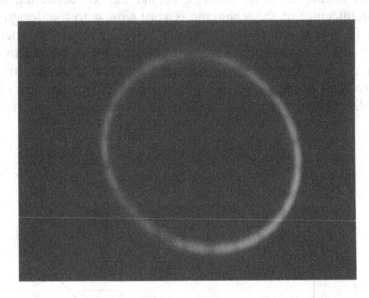

Figure 2. Image of a circularly scanned 2.5 keV electron beam on the phosphor screen.

2.2   Physical Time Resolution of Electron Tube

The minimal physical time dispersion of the electron optics is determined by chromatic aberration due to the electron initial energy spread-$\Delta\varepsilon$. The time spread $\Delta\tau_t$ due to this effect in the case of a uniform accelerating electric field near the photocathode plane- E in V/cm is $\Delta\tau_t = 2.34 \times 10^{-8} (\Delta\varepsilon)^{1/2}/E$ sec. For $\Delta\varepsilon = 1$ eV and $E = 10$ kV/cm we obtain $\Delta\tau_t \approx 2$ ps.

2.3   Technical Time Resolution of Electron Tube

Electron tube is a device with precise electron focusing in the effective electron transit time equalization. The time precision limit for the whole system consisting of photo-cathode, accelerating, focusing and transporting parts in the carefully designed system is estimated to be on the order of 0.1 ps (FWHM) for point like photocathode [14] and 10 ps (FWHM) for active photocathode diameter of 4 cm [15] or even larger [16].

2.4 Technical Time Resolution of the RF Deflector

By definition, the technical time resolution is $\Delta \tau_d = d / v$, where d is the size of the electron beam spot or the position resolution of the secondary electron detector if the electron beam spot is smaller, while v is the scanning speed: $v = 2\pi R / T$, here T is the period of rotation of the field, R is the radius of the circular sweep on the position sensitive detector. For example, if $T = 2 \times 10^{-9}$ sec (f = 500 MHz), $R = 2$ cm, and $d = 0.5$ mm, we have $v \geq 0.5 \times 10^{10} cm / \sec$ and $\Delta \tau_d \leq 10 \times 10^{-12}$ sec.

3. POSITION SENSITIVE DETECTOR

Position sensitive detectors for the RF timing technique should have the following properties:
- High detection efficiency for 2.5-10 keV single electrons;
- Position sensitivity, for example, $\sigma_x \leq$ of 0.5 mm;
- High rate capability.

These three requirements are to a different degree realized by several types of detectors. They include vacuum- based devices such as the Multi-anode PMT [17,18], Micro Channel Plates (MCP) [19] as well as an array of Si PIN [20] and Avalanche Photodiodes (APD) [21] or APD working in a Geiger mode (SiPM) [22].

However, the best approach is the development of a dedicated multi-anode PMT with circular anode structure.

Position determination can be performed in two basic architectures:

Direct readout: array of small pixels, with one readout channel per pixel, such as available with avalanche Si diodes. Position resolution in this case is about or better than the size of readout cell.

Interpolating readout: position sensor is designed in such a way that measurement of several signals (amplitudes or/and times) on neighboring electrodes defines event position.

Position resolution limit for both cases is $\Delta x/x \sim 10^{-3}$ [23].

In our case, we propose that a rather simple interpolating readout scheme can be applied and 2 ps technical time resolution can be reached using a f = 500 MHz RF deflector.

4.   SUMMARY OF THE TECHNIQUE

From the temporal resolution budget discussed above one can conclude that the expected parameters of the technique are:
   a) Internal time resolution for each photo-electron of about 2 ps;
   b) Technical time resolution in the range of $(2\text{-}10)\times10^{-12}$ sec for f = 500 MHz RF deflector;
   c) Absolute calibration of the system better than 10^{-12} sec is possible, which is an order of magnitude better than can be provided by a regular timing technique.

After analysis of the available technical solutions for electron multipliers, the best solution in our opinion is the development of a dedicated PMT based on MCP with circular anode structure and interpolating readout schemes for position determination with relative precision close to 10^{-3}.

This new photon detector and timing system can be operated at f = 100-1500 MHz frequency range and can find applications in:

- High energy elementary particle and nuclear physics experiments, e.g. in Cherenkov time-of-flight and time-of-propagation technique;
- Diffuse optical tomography applications, e.g. in detection of breast cancer;
- High energy gamma astronomy to separate showers generated by gamma particles and hadrons;
- Lunar laser ranging technique to reach millimeter-precision range information.

This work is supported in part by International Science and Technology Center- ISTC, Project A-372.

References

1.) William B. Atwood, Time-Of-Flight Measurements, SLAC-PUB-2620 (1980).

2.) E. K. Zavoisky and S. D. Fanchenko, Physical principles of electron-optical chronography, Sov. Phys. Doklady 1, 285 (1956).

3.) D. J. Bradly, Ultrashot Light Pulses, Picosecond Techniques and Applications, in Topics in Applied Physics, V. 18, 17 (1977).

4.) A. M. Prokhorov and M. Ya. Schelev, Recent research and development in electron image tubes/cameras/systems. SPIE Vol. 1358, International Congress on High-Speed Photography and Photonics, 280 (1990).

5.) M. M. Butslov, B. A. Demidov, S. D. Fanchenko et al., Observation of picosecond processes by electron-optical chronography, Sov. Phys. Dokl. Vol. 18, No.4, 238 (1973).

6.) Wilfried Uhring et al., "Very long-term stability synchroscan streak camera", Review of Scientific Instruments, Vol. **74**, N.5 p.2646 (2003).

7.) A. Margaryan, "Development of a RF picosecond timing technique based on secondary electron emission", ISTC Project No A372.

8.) Burle Industries Inc., http://www.burle.com

9.) R. Kalibjan, A phototube using a diode as a multiplier element, IEEE Trans. Nucl. Sci. NS-12 (4) p.367 (1965) and NS-13 (3) p.54 (1966).

10.) T. Gys, The pixel hybrid photon detectors for the LHCb-rich project, Nucl. Instrum. And Methods A **465**, 240 (2001).

11.) R. DeSalvo et al., First results on the hybrid photodiode tube, Nucl. Instrum. And Methds A **315**, 375 (1992).

12.) E. K. Zavoisky and S. D. Fanchenko, Image converter high-speed photography with $10^{-12} - 10^{-14}$ sec time resolution, Appl. Optics, **4**, n.9, 1155 (1965).

13.) R. Kalibjian et al., A circular streak camera tube, Rev. Sci. Instrum. **45**, n.6, 776 (1974).

14.) K. Kinoshita et al., Femtosecond streak tube, 19th International Congress on High-speed Photography and Photonics, SPIE Vol. **1358**, 490 (1990).

15.) Boris E. Dashevsky, New electron optic for high-speed single-frame and streak image intensifier tubes, SPIE Vol. **1358**, 561 (1990).

16.) Daniel Ferenc, Imaging hybrid detectors with minimized dead area and protection against positive ion. arXiv: physics/9812025 v1 (1998).

17.) S. Korpar et al., Multianode photomultipliers as position-sensitive detectors of single photons, Nucl. Instrum. and Methods A **442** 316 (2000).

18.) Jr. Beetz et al., Micro-dynode integrated electron multiplier, US Patent #6,384,519, May 7 (2002).

19.) A. S. Tremsin et al., Microchannel plate operation at high count rates: new results, Nucl. Instrum. And Methods A **379**, 139 (1996).

20.) C. B. Johnson et al., Circular-scan streak tube with solid-state readout, Applied Optics, Vol. **19**, No. 20, 3491 (1980).

21.) M. Suayama et al., Fundamental investigation of vacuum PD tube, IEEE Trans. Nucl. Science, V. **41**, No 4, 719 (1994).

22.) G. Bondarenko et al., Limited Geiger-mode microcell silicone photodiode: new results, Nucl. Instrum. And Methods A **442**, 187 (2000).

23.) Helmuth Spieler, Introduction to Radiation Detectors and Electronics, VII. Position-Sensitive Detectors, http://www-physics.lbl.gov/~spieler

FEMTOSECOND DEFLECTION OF ELECTRON BEAMS IN LASER FIELDS AND FEMTOSECOND OSCILLOSCOPES

E.D.Gazazyan, K.A.Ispirian, M.K.Ispirian, D.K.Kalantaryan* and
D.A.Zakaryan**
Yerevan Physics Institute, Br. Alikhanians 2, 375036, Yerevan, Armenia
CANDLE, Acharyan 31, 375040, Yerevan, Armenia
**Yerevan State University, 375049, Alex Manoogian 1, Armenia Yerevan*

Abstract: The interaction between the moving electron and circularly polarized laser
             photon beams in a finite length interaction region is considered. Then the
             detection of the electrons after a long lever arm allows to scan circularly the
             electrons with femtosecond period of the laser light. The principles of the
             construction of laser femtosecond oscilloscopes are considered.

Keywords: laser, electron, beams, oscilloscope, femtosecond pulses

As it is well known (see [1]) it is difficult to accelerate the charged particles by laser beams, especially in vacuum, because the fields are perpendicular to the direction of propagation of photons. Nevertheless, in [2] it has been shown that replacing the deflecting RF fields inside cavities by intense laser fields one can construct a femtosecond oscilloscope.

In this work, after solving the equations of motion of electrons in monochromatic, traveling wave with circular polarization taking into account the primary conditions some problems of construction and optimization of femtosecond oscilloscopes are considered. The results are in agreement with the well-known results obtained by the Hamilton-Jacobi [3,4] and other methods [5].

Solving the equations of motion of electron with momentum **p** in the electric **E** and magnetic field **H** of a wave have in the case of circularly polarized laser photon beam propagating along the axis OX in parallel with the electron beam with initial conditions

H. Wiedemann (ed.), Advanced Radiation Sources and Applications, 313–318.
© 2006 Springer. Printed in the Netherlands.

$$x_0 = 0, \ y_0 = 0, \ z_0 = 0, \ p_{x0} \neq 0, \ p_{y0} = 0, \ p_{z0} = 0, \tag{1}$$

one obtains the following solutions

$$x = \frac{2B}{k}\left(p_{x0} + \frac{2Be^2E_0^2}{\omega^2}\right) \cdot \eta - \frac{4B^2e^2E_0^2}{k\omega^2} \cdot \sin\eta, \tag{2a}$$

$$y = -\frac{2BeE_0 \sin\varphi_0}{k\omega}\eta - \frac{2BeE_0}{k\omega}\left[\cos(\eta+\varphi_0) - \cos\varphi_0\right], \tag{2b}$$

$$z = -\frac{2BeE_0 \cos\varphi_0}{k\omega}\eta + \frac{2BeE_0}{k\omega}\left[\sin(\eta+\varphi_0) - \sin\varphi_0\right], \tag{2c}$$

$$p_x = p_{x0} + \frac{2Be^2E_0^2}{\omega^2}(1-\cos\eta), \tag{2d}$$

$$p_y = \frac{eE_0}{\omega}\left[\sin(\eta+\varphi_0) - \sin\varphi_0\right], \tag{2e}$$

$$p_z = \frac{eE_0}{\omega}\left[\cos(\eta+\varphi_0) - \cos\varphi_0\right], \tag{2f}$$

where

$$B = \frac{\sqrt{m^2c^2 + p_{x0}^2} + p_{x0}}{2m^2c^2}, \tag{3}$$

$k = \omega/c$ and E_o are the wave number and amplitude of the laser field, $\eta = \omega t - kx$ and φ_0 is the phase at the entering of the electron into the interaction region. In agreement with [3-4] the motion described by (2a)-(3) is helical. However, the transversal displacement of the electrons in L_{int} is too small even when the laser fields are of the order of the critical fields.

In the work [2] it has been proposed a method for the enhancement of the transversal displacement. After the interaction region with length L_{int} which is determined by the parallel mirrors 1 and 2, the electrons fly a field free region before they reach and can be detected on a screen placed at a distance L (see Fig.1).

Since in the case of circular scanning there is no need of synchronization such scanning is preferable and has more application than in the cases of deflection in the same direction that is used in oscilloscopes with linear scanning. One can show that the transversal displacements can be enough large after L because the electrons coming out from the interaction region have large angles.

Indeed, from (2a) one obtains a useful relation for L_{int}

$$L_{int.} = \frac{2B}{k}\left(P_{x0} + \frac{2Be^2E_0^2}{\omega^2}\right) \cdot \Delta\eta - \frac{4B^2e^2E_0^2}{k\omega^2} \cdot \sin\Delta\eta, \quad (4)$$

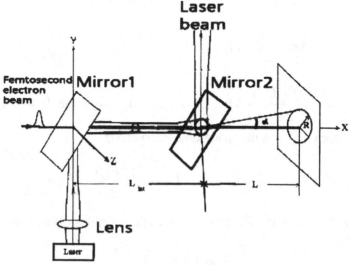

Figure 1. The principles and arrangement (schematically) for the measurement of the length and particle distribution in a femtosecond process.

where $\Delta\eta = \omega\tau - kL_{int}$ is the space-time interval between two mirrors (τ is the time of flight of electron within the interaction region $L_{int.}$, $\tau = L_{int}/v$).

The coordinates of the electrons on the screen are determined by the relations

$$y_2 = y_1 + L\frac{dy}{dx}\Big|_{x=L_-}, \quad z_2 = z_1 + L\frac{dz}{dx}\Big|_{x=L_-}, \quad (5)$$

where y_1, z_1 are the coordinates in the beginning of the drift L.

The transversal displacement of the electrons from the axis OX is determined by $R = \sqrt{y_2^2 + z_2^2}$ and using (2b), (2c) and (5) one can write

$$R = \frac{2eE_0}{k\omega} \sqrt{B^2 \Delta\eta^2 + \left(4B^2 + \frac{L^2 k^2}{p_{x1}^2} + \frac{2LBk\Delta\eta}{p_{x1}}\right) \cdot \sin^2 \frac{\Delta\eta}{2} - 2B^2 \Delta\eta \sin \Delta\eta}$$

(6)

where

$$p_{x1} = p_{x0} + \frac{2Be^2 E_0^2}{\omega^2}(1 - \cos\Delta\eta),$$

(7)

$$\Delta\eta = \omega\tau - kL_{\text{int}} = \omega\frac{L_{\text{int}}}{v} - \frac{\omega}{c}L_{\text{int}} = \frac{2\pi L_{\text{int}}}{\lambda\gamma^2 \beta(\beta+1)},$$

(8)

$L_{\text{int}} = \dfrac{\lambda\gamma^2 \beta(\beta+1)}{2\pi}\Delta\eta$. Then one may write the following relationship

between L_{int} and the space-time interval $\Delta\eta$

$$L_{\text{int}} = \frac{\lambda\gamma^2 \beta(\beta+1)}{2\pi}\Delta\eta.$$

(9)

For the relativistic electrons ($\beta \sim 1$) the formulae (8) and (9) take the form

$$\Delta\eta = \frac{\pi L_{\text{int}}}{\lambda\gamma^2} \quad \text{and} \quad L_{\text{int}} = \lambda\gamma^2 \frac{\Delta\eta}{\pi}.$$

(10)

Figure 2 shows the trajectories of 50 MeV electrons with various initial phases for E=2 10^8 V/cm (laser CO_2 with W=10^{18} W/cm^2), L_{int}=5cm, L= 100cm.

For the optimal choice of the parameters of the femtosecond oscilloscope we have calculated for the fixed other parameters the dependence of the electron image radius upon electron energy, R(E), which is decreasing function of E. We have calculated also the dependence R upon L_{int} which is a periodic function of L_{int} with maximal, R_{max}, and minimal, R_{min} values of R. The period, DL, of the last dependence which must be larger than the difference of the electrons' paths due to $\Delta\eta$, the accuracy of the mirror installation, as well as due to the spot sizes of the electron beam, is an increasing function of E in contrast to R(E).

If the length of the bunch t_x under measurement is less than the period of the laser light T, then only a part of the circle with length L_x will be obtained on the screen, measuring which one can determine by $t_x = L_x T /(2\pi R)$. The relative error of the measurement of the time,

$\Delta t_x / t_x \approx \Delta L_r / L_r$, is smaller than the errors inherent for such measurements [6,7] which together with the position sensitive detectors will be considered in our next publications.

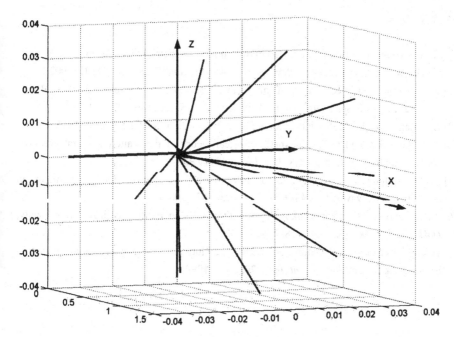

Figure 2. Trajectories of electrons. The coordinates are given in meters.

Table 1 shows the values of R_{max}, R_{min} and DL for three values of electron energy.

E_{kin}	R_{min}[cm]	R_{max}[cm]	DL
50keV	0.13	5.3	1.3237 μ m
1MeV	0.0218	0.864	30.059 μ m
20MeV	0.00159	0.06	6.0583 mm

One can conclude that as the above given results show, in principle, it is possible to construct a tabletop femtosecond oscilloscope. As a first step we propose to perform the following experiment, which is the laser analog of the RF experiment [7]. After splitting the two laser beams with certain delay are given to two interaction regions. In the first of which after the first scanning the electron beam femtosecond chopping takes place after a narrow slit. The second scanning will allow to test the proposed method of femtosecond time measurement.

Acknowlegement

The work is supported by the ISTC project A-372.

References

[1] T. Tajima and G. Mourou, Zettawatt-etawatt lasers and their applications in ultrastrong-field physics, *Phys. Rev. Spec. Topics, Acc. and Beams*, **5**, 031301 (2002).

[2] K.A. Ispirian and M.K. Ispiryan, Femtosecond Transversal Deflection of Electron Beams with the Help of Laser Beams and Its Applications, ArXiv: hep-ex/0303044, 2003.

[3] L.D. Landau and E.M. Lifshits, Teorya Polya, 5-th Edition, Nauka, Moscow, 1967.

[4] E.S. Saranchik and G.T. Schappert, Classical theory of the scattering of intense laser radiation by free electrons, *Phys. Rev. D*1, 2738 (1970).

[5] B.M. Bolotovski and A.V. Serov, Peculiarities of the motion of particles in electromagnetic waves, *Uspekhi Fiz. Nauk* ,**173,** 667, (2003)

[6] E.K. Zavoiski S.D. Fanchenko, Physical Principles of electron-optical chronography, *Dokl. Akad. Nauk SSSR,* **108,** 218, 1956.

[7] A.V. Aleksandrov et al, Setting up and time-resolution measurement of radio-frequency-based streak camera, *Rev. Scient. Instr.* **70,** 2622 (1999).

NOVEL FAST-ACTING SCINTILLATION DETECTORS FOR WIDE ENERGY RANGE APPLICATIONS

A. Fedorov, M. Korzhik, A. Lobko, and O. Missevitch

Institute for Nuclear Problems, Belarus State University, 11 Bobruiskaya Str., Minsk 220050, Belarus

Abstract: Latest developments in complex structure oxide crystalline scintillating compounds are reported. Particularly, lead tungstate (PWO) crystal with increased light yield is described. Study of improved PWO with avalanche photodiode (APD) readout allows confidently predict its successful future application in wide energy range, starting from comparatively low energies through tens of GeV. Besides, APD applications in direct detection mode have demonstrated good energy resolution in 1-10 keV energy range. Our experience with yttrium aluminate crystal and prospects of its usage at high counting rates at photon energies up to 500 keV are discussed. Late developments of scintillators with heavy elements in a crystal matrix for application in ~1...10 MeV energy range are presented. Methods and equipment of a detector monitoring and stabilization for studies, where long-term stability is important, are also discussed.

Keywords: Fast and heavy crystal scintillators; scintillation detectors for wide energy range application; scintillation detector monitoring and stabilization.

1.   INTRODUCTION

The record of scintillation materials research and development lasts for about a century and it is strongly connected with the steady progress in nuclear and material science. More than ninety various scintillation materials have been introduced to date, however growing demand for a scintillator applications in research, industry, and medical imaging stimulates scientists to put more efforts to their development.

319

H. Wiedemann (ed.), Advanced Radiation Sources and Applications, 319–328.
© 2006 Springer. Printed in the Netherlands.

Let us mention basic properties of inorganic scintillating crystals that make them good competitors among whole diversity of ionizing radiation detectors.

1. **High density.** Synthetic crystals based on complex structure oxides may have high density up to 8 g/cm^3 and high effective atomic number Z>70. Thus, they can provide good stopping power and, therefore, detectors based on them will be more space saving. That is a matter of especial significance in array detectors, where transversal dimensions of a detector cells determine its spatial resolution;

2. **Short scintillation decay time.** Crystal scintillators have a variety of decay constants and they may be very small, e.g. as low as 600 ps for BaF$_2$. Hence one can use them for gamma-quanta detection in conditions of high counting rate and also in detecting systems with high timing resolution. Generally, combination of high stopping power and short response time, which can be obtained simultaneously only in detectors based on a crystal scintillators, gives unique opportunities to the variety of applications;

3. **Radiation hardness.** Crystal scintillators can survive in intensive radiation fields. This property is important for many applications from high-energy physics through well lodging to space missions, as well as in intensive synchrotron and X-ray beams. High stability of crystal parameters under ionizing radiation is caused by sophisticated technology, used in modern crystal production facilities. Thus, high-tech growth approach can provide crystals with low level of non-controlled structure defects.

List above is not completed and crystal scintillators have other attractive features (e.g. signal linearity, weak temperature dependence, high light yield, possibility to grow large volume crystals, etc.). That is why new and new customers are demanding scintillators with advanced properties and stimulating solid-state physicists and crystal technologists to greater efforts.

As recent example, several high-energy physics experiments, like CMS, ALICE, and BTeV have chosen lead tungstate (PWO) crystal for construction of their electromagnetic calorimeters. This choice stimulated un-precedent number of R&D efforts devoted to PWO scintillation physics and growth technology. Joint effort of international community yielded in mass production of high quality PWO crystals in about decade, though in early nineties PWO was just one of a series of potentially prospective materials. Inspired by the result, many teams continue well-directed efforts in a scintillating crystal research. Some recent results are described below.

2.   IMPROVED PWO

The main goal of the lead tungstate optimization was the consistency of matching specification, which was developed foreseeing PWO application in experiments at CERN. For these applications the most crucial parameters are radiation hardness and fast scintillation kinetics. Optimized scintillation elements of CMS type have scintillation yield of fast component ~20 phe/MeV for small (several cm^3) samples. Corresponding full size scintillation cell has total light yield of ~10-12 phe/MeV.

However, for PWO applications at energies lower than in LHC or Tevatron, its scintillation light yield should be considerably higher. Such increase can be reached by two methods: 1) increase of the structural perfection of the crystal; 2) activation of the crystal with luminescent impurity centers, which have large cross-section of electron capture from conducting band and relatively short time of their following radiative recombination. Several efforts had been carried out to increase PWO light yield by an additional doping [1,2]. However, scintillation time increase occurs when crystal was doped with many of investigated impurity centers. Besides crystal activation with impurities centers, authors of [3] investigated the possibility to redistribute electronic density of states near the bottom of conducting band by the change of ligands contained in the crystal.

Another possibility to eliminate point structure defects in crystals is their doping with Y, La, Lu, which will suppresses the appearance of oxygen and cat ion vacancies in the crystal. Required concentration of activators is ~100 ppm. However, increase of scintillation yield by 30-50% was observed in crystals at small concentration of La (~50 ppm and less) only. Such dopant concentration is not enough for complete suppression of vacancies. Additional increase of the crystal structural perfection can be reached by combining the improved control of the melt stochiometric composition and activation of crystal with impurity of mentioned elements with concentration less than 50 ppm. In this case both systematic and fluctuating origins of vacancies in the crystal will disappear.

2.1   Properties of improved PWO samples

2.1.1   Light yield

Samples of several cm^3 volume cut from crystals grown in nitrogen atmosphere from corrected melt, with La concentration of about 40 ppm showed light yield about 41 phe/MeV at room temperature. Full size crystals of 23 cm length (CMS type), cut from the same boule have light yield not less than 20 phe/MeV at 18°C.

Taking into account, that temperature dependence of lead tungstate scintillation yield is -2%/°C [4], there is an additional possibility to increase scintillation light yield by crystal temperature decrease as seen from the Table 1.

Table 1. Improved PWO light yield as function of temperature

Sample temperature, °K	318	296	278	258
Light yield, phe/MeV	33	41	53	66

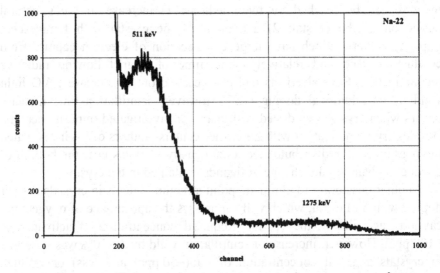

Figure 1. Pulse-height ^{22}Na spectrum recorded by 5x5x5 mm^3 PWO and 5x5 mm^2 APD at temperature T=-20±5°C.

Figure 1 presents pulse-height spectrum of ^{22}Na recorded using improved PWO sample of 5x5x5 mm^3 dimensions and 5x5 mm^2 avalanche photo-diode (APD) at temperature T=-20±5°C. One can see here quite acceptable view with fully observable peaks at 511 and 1275 keV. This spectrum shows applicability of the improved PWO for sufficiently lower energy range than it was used before.

For entirely low energy (~several keV) x-ray detection, an APD itself can be used in direct detection mode as a semiconductor detector with internal gain. Due to this internal gain, the signal-to-noise ratio is better than that of a PIN diode of similar sensitive area. We have tested HAMAMATSU S8664-55 5x5 mm^2 APD under ^{57}Co 6.4 keV x-rays. This APD has internal gain M=50 at 365 V bias voltage, 80 pF capacity, and 3 nA typical dark current.

Using charge-sensitive preamplifier with 2SK162 input JFET and quasi-Gauss shaping amplifier with shaping time constant 250 ns we have obtained 12% energy resolution for 6.4 keV at room temperature.

2.1.2 Scintillation kinetics

Analysis of scintillation kinetics of the improved PWO crystal (Fig. 2) shows that contribution of centers created on a basis of irregular groups, which cause components with $\tau\sim30$ ns, is less than 3%. Due to suppression of irregular centers in the crystal, scintillation kinetics remains rather fast with temperature decrease.

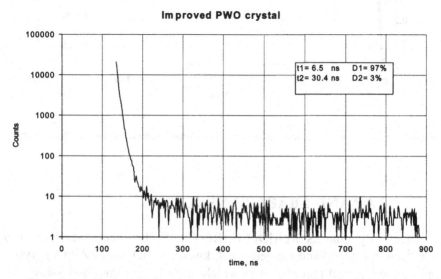

Figure 2. Scintillation kinetics of crystal grown from corrected melt with 20 ppm La concentration, T=300°K.

Figure 3 shows time dependence of scintillation pulse integral (light collected within specified time gate) measured at temperatures varying from –6 to +14°C. One can see only limited slowdown of scintillation kinetics constrained in first 50 ns time. No slow components appear in scintillation kinetics. So, such crystals can be successfully used to build fast detectors without risk to meet significant signal pile-up problem.

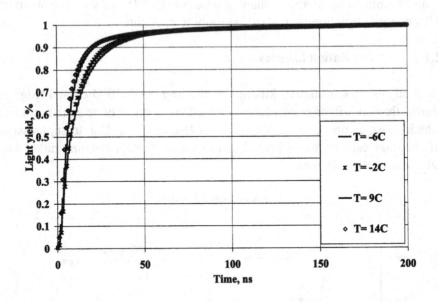

Figure 3. PWO scintillation integrals measured at various temperatures as function of time.

3.   YAP AND LuAP CRYSTALS

Since early nineties ReAlO₃:Ce (Re=Gd, Y, Lu) scintillation crystals have been developed and studied. First, known YAlO₃:Ce (YAP:Ce) crystal was well developed and now it is widely used for gamma- and x-ray, and alpha-particle measurements [5-7]. The radical improvement (multiple times) in throughput of transmission Mossbauer spectroscopy measurements by use YAP:Ce fast detectors opens a real opportunity for wide application of Mossbauer spectroscopy in industry and substantially widens an area of fundamental structure analytical problems related to small resonance effects observation which can be solved [8].

Due to the fact, that YAP:Ce crystal has relatively small effective atomic number ($Z_{eff}=32$) and photo-electric fraction, dense and much effective ($Z_{eff}=65$) LuAlO₃:Ce (LuAP:Ce) crystal has been proposed, grown and studied [9].

Whereas growth technology of large volume LuAP:Ce single crystals is complicated and still under development, mixed yttrium-lutetium LuYAP:Ce crystal was proposed and its Czokhralski growth technology developed. It was found that yttrium doping of "pure" LuAP:Ce crystal makes its growth technology more reliable, so good production yield of $(Lu_x-Y_{1-x})AlO_3$:Ce crystals can be obtained. It was also established that variation of Y/Lu ratio x in $(Lu_x-Y_{1-x})AlO_3$:Ce compounds allows to obtain scintillation materials with pre-defined detection efficiency, see Fig. 4. Respectively, "pure" YAP:Ce crystals, as well as LuYAP:Ce with 10-20% Lu content can be successfully used in gamma cameras designed to obtain images of human organs, using radio-isotope emission with energies between 50 keV and 190 keV. LuYAP:Ce with 60-80% Lu content have superior to Gd_2SiO_5:Ce stopping power at 511 keV and can be successfully used in positron emission tomography (PET). $(Lu_x-Y_{1-x})AlO_3$:Ce crystals with different Lu content have differing scintillation decay times. Due to this reason, multi-layer phoswich detectors based on $(Lu_x-Y_{1-x})AlO_3$:Ce with pulse-shape discrimination of hit layer can be developed for use in PET detectors with depth-of-interaction capability or versatile PET/CT systems with improved spatial resolution and sensitivity. Temperature coefficient of $(Lu_{0.7}-Y_{0.3})AlO_3$:Ce light yield is about +1%/°C. Studies of LuYAP temperature dependences indicate benefits of application of these crystals in high temperature environments, for example in well logging. Some basic properties of LuYAP crystals are listed in Table 2.

Table 2. Properties of LuYAP scintillation crystals

Material	YAlO₃:Ce	(Lu₀.₂-Y₀.₈) AP	(Lu₀.₇-Y₀.₃) AP	LuAP
Density, (g/cm³)	5.35	5.9	7.2	8.34
λ_{max}, (nm)	347	360	380	385
Decay time τ, (ns)	30	26(95%), 400(5%)	20(40%), 70(35%), 400(25%)	16(60%), 400(40%)
Light yield (ph/MeV)	16,000	14,000-16,000	12,000-14,000	10,000-12,000

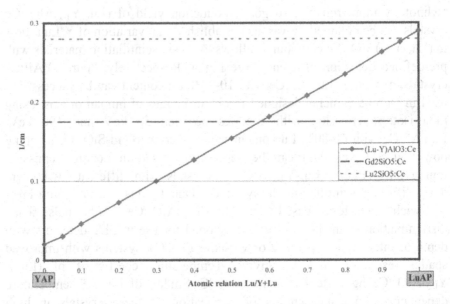

Figure 4. Photo-electric linear attenuation coefficient of $(Lu_x-Y_{1-x})AlO_3$:Ce at 511 keV compared with that of Lu_2SiO_5:Ce (LSO:Ce) and Gd_2SiO_5:Ce (GSO:Ce).

4.   DETECTOR MONITORING

Operation of a detector strongly depends on the stability of its components. Scintillation detector, in general, consists of a scintillator, optical coupling, photodetector, read-out electronics and construction elements. Behavior of every element under irradiation and temperature is specific. To provide energy resolution of a detector as required in modern experiments, one must use external very stable reference light source for monitoring (or stabilization) of the energy scale. Basically, monitoring system should provide test signals very similar to real working signals producing by a radiation. Though in practice it is hardly to meet this requirement, we have developed two types of pulsed light sources that can be used in many applications of fast scintillation detectors.

First, radio-luminescent light pulse source contains YAP:Ce scintillation crystal with deposited [241]Am low activity radioactive source sealed in aluminum housing [10]. Main source characteristics are listed in Table 3 and its sketch is presented in Fig. 5.

Table 3. Radio-luminescent source characteristics

Parameter	Value
^{241}Am activity (5.5 MeV α-particles), Bq	20÷200
Light pulses count rate, cps	10÷100
Count rate tolerance, %	±15%
Light pulse decay constant, ns	30
Light emission maximum, nm	350
Light intensity, photons per pulse	≥ 20,000
Energy resolution, % FWHM	≤ 6.0
Operation temperature, °K	250÷350

The radio-luminescent sources have many advantages: no need in additional electronics, short duration of light flash (if a fast scintillator used), compactness, relative cheapness, possibility of operation in strong magnetic fields. The dispersion of number of photons in flash is minimal, as is mainly determined by photo-statistics. However, disadvantages of radio-luminescent sources, such as impossibility to change a frequency and amplitude of pulses, and sometime inadequate temperature and long-term stability essentially restrict an area of their application.

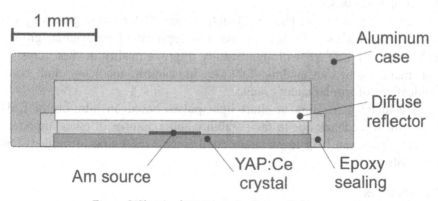

Figure 5. Sketch of YAP-based reference light source.

To overcome mentioned above problems, another pulsed light generator based on a LED was developed [11]. Active quenching of a LED afterglow allows reducing light pulse duration down to 10 ns. Number of photons is achieved as high as ~10^9 per pulse. Stabilization of light pulse intensity is based on optical feedback loop with thermo-stabilized reference photo-detector. Long-term instability of the light pulse intensity about ±0.1% is obtained. Pulser is capable to operate in strong magnetic fields up to 3 T. More characteristics are listed in Table 4.

Table 4. LED-based pulser main characteristics

Parameter	Value
Pulse rate, Hz	$1 \div 10^4$
Number of photons per pulse	$\geq 10^9$
Light pulse duration (adjustable), ns	$10 \div 300$
Light wavelength range, nm	$380 \div 650$
Operation temperature, °K	$270 \div 320$
Temperature stability, %/°C	0.01
Long-term stability, %	≤ 0.1

CONCLUSION

In conclusion we can state, that improved lead tungstate crystal with increased light yield can be grown by mass production technology. Its scintillation with maximum of emission at 420 nm is rather fast, 97% of light is emitted within 100 ns in the 310-250°K temperature range. Light yield of 1 cm³ sample is changed from 41 to 66 phe/MeV in the 296-258°K temperature range. Permanent progress in PWO properties and improvements in APD technology may result in possibility of PWO low energy applications.

Range of perspective $(Lu_x-Y_{1-x})AlO_3$:Ce scintillators were grown by the Czokhralski method. We believe that development of Bridgeman growth technology for LuYAP will significantly improve quality of these crystals and make them outstanding for use in nuclear medicine and other applications of similar energy range.

Developed generators of short light pulses based on ultra-bright LED with optical feedback due to their stability and a number of other important characteristics can help to maintain good energy resolution of modern fast scintillation detectors.

References
1. M. Kobayashi, In: Inorganic scintillators and their applications, Proc. SCINT'2003 Int. Conf., Accepted by Nucl. Instr. Meth. (2004).
2. A. Annenkov, A. Borisevich, Hoffstattler et al. NIM A450 (2000) 71.
3. M. Kobayashi, Y. Usuki, M. Ishi, M. Nikl. NIM A486 (2002) 170.
4. P. Lecoq, I. Dafinei, E. Auffray et al. NIM A365 (1995) 291.
5. V.G. Baryshevsky, M.V. Korzhik, V.I. Moroz et al NIM B58 (1991) 291.
6. M.V. Korzhik, O.V. Missevitch, A.A. Fyodorov. NIM B72 (1992) 499.
7. V.G. Baryshevsky, M.V. Korzhik, B.I. Minkov et al J.Phys. D 5 (1993) 7893.
8. A.A. Fyodorov, A.L. Kholmetskii, M.V. Korzhik et al NIM B88 (1994) 462.
9. W.W. Moses et al. 1994 IEEE Nucl. Sci. Symp., Conf. Record.
10. V.A. Katchanov, V.V. Rykalin, V.I. Solovyanov et al. NIM A314 (1992) 215.
11. A. Fyodorov, M. Korzhik, A. Lopatik, and O. Missevitch NIM A413 (1998) 352.

POROUS DIELECTRIC DETECTORS OF IONIZING RADIATION

M. P. Lorikyan
Yerevan Physics Institute, Alikhanian Brothers Str. 2, Yerevan 375036, Armenia

Abstract: The results of the investigations of the multiwire porous dielectric detector and microstrip porous dielectric detector filled with porous CsI are presented. Detectors were exposed to α-particles with energy of 5.46 MeV and X-rays with energy of 5.9 keV. The detectors perform stably and provide a high spatial resolution

Keywords: Drift and multiplication of electrons, Porous dielectric detector, Microstrip porous dielectric detector

1.   INTRODUCTION

The use of porous dielectric as working media in radiation detectors is a novelty in the radiation detection technique. The operation of these detectors is based on the effect of the (EDM) in porous dielectrics under the action of an external electric field [1-9]. Porous alkali halides, such as CsI and CsBr are used as working materials.

The EDM in porous dielectrics takes place as follows. The ionizing radiation produces electrons and holes in the walls of pores (see Fig. 1). If the energy of these electrons is sufficiently high, they also produce secondary electrons and holes on the surface of the pore walls. The secondary electrons are accelerated in the pores by the applied electric field. The accelerated electrons produce new secondary electrons and holes on the surface of the walls. When the secondary electron emission factor $\delta > 1$ on each wall, this process repeats, producing an avalanche multiplication of electrons (holes) (details see in [7-9]).

H. Wiedemann (ed.), Advanced Radiation Sources and Applications, 329–334.
© 2006 *Springer. Printed in the Netherlands.*

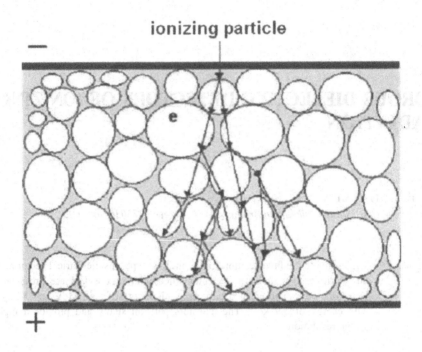

Figure 1. Schematic view of drift and multiplication by ionizing electrons in porous dielectrics under the action of external electric field.

The porous detectors operate in a vacuum of better than $\approx 7 \times 10^{-3}$ Torr. In this report the results of the study of the multiwire porous dielectric detector (MWPDD) and microstrip porous dielectric detector (MSPDD), filled with porous CsI are presented.

2.   DESCRIPTION OF POROUS DIELECTRIC DETECTORS

The schematic view of the multiwire porous dielectric detector is shown In Fig. 2.

The 25 μm diameter anode wires are made of gilded tungsten and are spaced at b=0.25 mm. The cathode is made of Al foil. The gap of the detector is 0.5 mm. It is filled with porous CsI. The density of porous CsI is $\rho/\rho_0 \approx 0.64$ %, where ρ_0 is the density of CsI monocrystal.

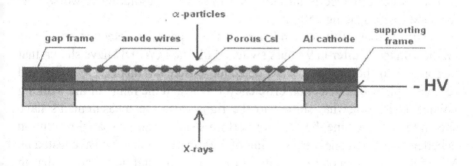

Figure 2. Schematic view of a multiwire porous dielectric detector.

The schematic view of microstrip porous dielectric detector is shown in Figure 3. Golden strips are deposited onto a ceramic plate. The width of strips is 20 μm. The spacing between them is 100 μm. The cathode is made of micromesh. The detector gap is 0.5 mm and is filled with porous CsI.

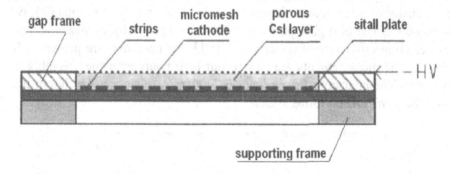

Figure 3. Schematic view of a microstrip porous dielectric detector.

3. DETECTION OF α-PARTICLES

3.1 Low-intensity.

In this experiment the purity of porous CsI is 99.99 %. The intensity of α-particles is 5cm^{-2} s^{-1}.

Investigations of the dependences of the α-particles detection efficiency on the voltage U shown that detection efficiency in one hour after the assembly of MWPDD.is in a weak dependence on U and at this time the detector's performance is unstable and the spatial resolution is worse than the anode-wire spacing 250 μm.

Investigations of the dependences of the α-particles detection efficiency on the voltage U after in 9 hours switched off the MWPDD have shown that in contrast to the previous case, detection efficiency increases rapidly and reaches a plateau at detection efficiency 100%, and the range of the working voltage U is somewhat shifted to the right. Also the measurements have shown that at this time the detector performs stably and its spatial resolution is better than the anode-wire spacing of 250 μm. Note that we have tested an MWPDD with an anode wire pitch of b = 125 μm and have shown that in this case σ ≤ 35μm.

3.2. High intensity

The attempts of using CsI of 99.99 purity for registration of higher-intensity α-particles did not give good results. We assumed that poor results are caused by the bulk charges accumulated at the crystal lattice defects. As the density of these defects depends mainly on the purity of porous CsI, we used porous CsI of higher purity (99.999). The time stability measurements of detection efficiency of α-particles for 43-day operation are presented in Fig. 4. The measurements are carried out in10 hours after the assembly of detector. The intensity of α-particles is $I700cm^{-2}s^{-1}$. One can see that in this case the MWPDD performs stably.

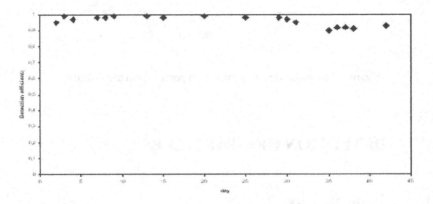

Figure 4. Temporal stability of detection efficiency for α – particles of high intensity.

3.3. The results of detesction of X-rays

The intensity of x - rays is $I = 10 \text{ cm}^2\text{s}^{-1}$. The measurements began on the first day after assembling of detector and lasted for 17 days. Fig. 5 shows the time-stability of the X-quanta detection efficiency.

One can see, that within the experimental errors η_x detection efficiency stays constant after the fourth day. Thus, the multiwire porous detector operates stably also in case of X-rays. Note that X-rays deposit 1000 times less energy in the active material than α-particles do.

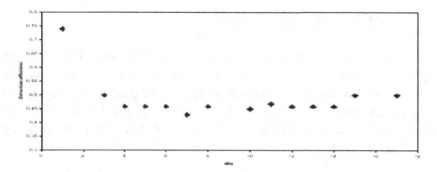

Figure 5. Temporal stability of X - rays detection efficiency for MWPDD.

4. INVESTIGATION OF THE MICROSTRIP POROUS DIELECTRIC DETECTOR

The investigations were curried out for α-particles. The time stability of detection efficiency for 36 days of microstrip detector's operation is presented in Fig. 6.

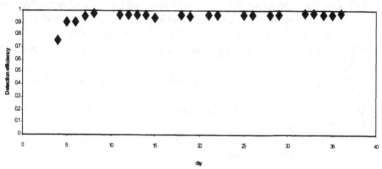

Figure 6. Temporal stability of α - particle detection efficiency for MSPDD.

It is clear from Fig. 6 that during the first 8 days η_α increases and then remains unchanged. The reason of increasing of detection efficiency at the beginning is not clear to us. Note that in the microstrip porous detector, in contrast to a multiwire porous detector, the signal strips are situated on a ceramic plate. We assume that this feature of microstrip detectors causes the difference in the characteristics of the detector.

The investigations have shown that the spatial resolution of this MSPDD is $\sigma_x \approx 30\mu m$.

5.   ACKNOWLEDGEMENTS

The author is grateful to the founders of the International Science and Technology Center and expresses his gratitude to G. Ayvazyan, R. Aivazyan, H. Vardanyan, for their assistance in the process of carrying out these experiments and Professors M. A. Piestrup and Ch. K. Gary for valuable collaboration and Professors F. Sauli and A.K. Odian for rendering of the large support.

References

1. M.P. Lorikyan et al., The investigation of controllable secondary electron emission from single particles,.*Nucl. Instr.and Meth.* **122** (1974) 377.

2. M.P. Lorikyan, N.N. Trofimchuk, Particle detection by mean of controllable secondary electron emission methode, *Nucl. Instr. and Meth.* **140** (1977) 505

3. M.P. Lorikyan and V.G. Gavalyan, Multiwire dielectric X – ray detector., *Nucl. Instr. and Meth* **A 340** (1994) 624.

4. M.P. Lorikyan, V.G. Gavalyan, K.J. Markaryan, Multiwire particle detector based on porous dielectric layers, *Nucl. Instr. and Meth* **A350** 244 (1994)

5. M.P. Lorikyan, Detectors of Nuclear Radiation and Multiply-Charged Particles with Porous Dielectric Working Media, *Uspekhi Fizicheskikh Nauk* **165** (1995) 1323-1333. English version - *Uspekhi Phys.* **38** (11), (1995) 1271-1281.

6. M.P. Lorikyan, The multiwire porous detector, .*Nucl. Instr.and Meth.* **A 454** (2000) 257-259.

7. M.P.Lorikyan. Porous CsI multiwire dielectric detectors. *Nucl. Instr.and Meth* **510** (2003) 150-157.

8. M.P.Lorikyan, New investigations of porous dielectric detectors. *Nucl. Instr.and Meth* **513** (2003) 394-397.

9. M.P.Lorikyan, Study of counting characteristics of porous dielectric detectors. *Nucl. Instr.and Meth* **515** (2003) 701-715.

VIBRATING WIRE SCANNER/MONITOR FOR PHOTON BEAMS WITH WIDE RANGE OF SPECTRUM AND INTENSITY

M.A. AGINIAN[1], S.G. ARUTUNIAN[1], V.A. HOVHANNISYAN[1], M.R. MAILIAN[1], K. WITTENBURG[2]

[1] *Yerevan Physics Institute, 2 Alikhanian Brothers Str., 375036 Yerevan, Armenia*
[2] *DESY, Notkestr. 85, D-22607 Hamburg*

Abstract: Developed vibrating wire scanner showed high sensitivity to the charged particles beam intensity (electron, proton, ions). Since the mechanism of response of frequency shift due to the interaction with deposited particles is thermal one, the vibrating wire scanner after some modification can be successfully used also for profiling and positioning of photon beams with wide range of spectrum and intensity. Some new results in this field are presented.

1. Scanning of Proton Beam

We developed a new method of beam profiling by means of a vibrating wire [1-6]. The operating principle of such vibrating wire scanner (VWS) is based on high sensitivity of natural oscillations frequency of the strained wire from its temperature. At achieved resolution 0.01 Hz (VWS natural frequency is about 5000 Hz) the corresponding temperature resolution is about $2 \cdot 10^{-4}$ K.

In May 2004 in PETRA, DESY, Hamburg experiments for proton beam halo profiling were done (energy of protons 15 GeV, beam current approximately 10 mA, beam horizontal size $\sigma = 6$ mm) [6]. A few pA resolution of deposited proton current on the wire was achieved. This experiment was aimed to profile of proton beam periphery. Results of scan at distances $3 \div 6.5$ σ are presented in Fig. 1. Simultaneously with scanning of the beam by our scanner, signals from two scintillator pickups were measured. The scintillators are used in traditional methods for beams profiling with wire scanners, where beam particles scattered on the wire are measured. The scan was started at distance of 40 mm from the vacuum chamber center. It is seen from Fig. 1 that the signal from our sensor changes

335

H. Wiedemann (ed.), Advanced Radiation Sources and Applications, 335–342.

from the start of movement, while the signals from scintillators begin to increase at distances of 27 mm from the vacuum chamber center.

The scanner was fed from park position toward the vacuum chamber center up to 20 mm. In this experiment the proton beam was shifted towards the scanner park position by distance of 4 mm by means of bump-magnets system.

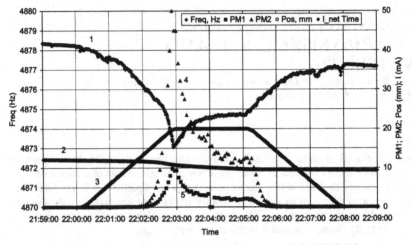

Figure 1. Scan of the proton beam using VWS: 1- frequency of the VWS, 2 beam current, 3 – VWS position relative to the vacuum chamber center, 4 and 5 – scintillator signals.

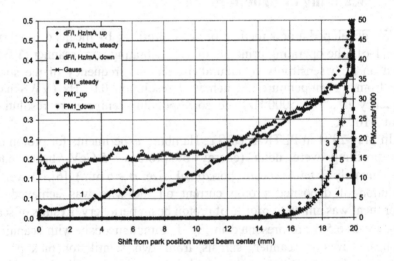

Figure 2. Dependence of frequency shift and scintillator signal on scanner position: 1- frequency at motion toward the beam, 2- frequency at motion from the beam, 3 and 4 – the same for signal from scintillator, 5 – reconstruction of beam profile from Gaussian distribution with $\sigma = 6$ mm

The dependence of the signals on the scanner position is presented in Fig. 2. The signal from the scanner was presented as shift of the frequency from the value in park position, normalized to the mean proton beam current. Some hysteresis effect during the backward movement was observed. As seen from the Fig. 2 the signal from VWS appears at distances 27-40 mm from the vacuum chamber center while there is no signal from the scintillators here. Some contribution in wire heating can be added from the influence of electromagnetic background accompanying the proton beam[1]. The electromagnetic component also can heat the wire. For separation of two mechanisms of heating two wires can be used with materials having strongly different absorption factors for proton and electromagnetic components. Clarification of this problem and corresponding modifications of VWS require additional efforts.

Vibrating wire can serve as an excellent thermometer responding to particles and radiation beams. In this case a detecting unit with a high spatial resolution can be developed on the basis of vibrating wires [1, 2]. High resolution of the method is due to the fact, that the wire oscillations frequency is defined by its tension, which in case of rigid fixation of wire ends has a heightened sensibility to the temperature (up to 10^{-4} K at thermo stabilisation of the base on which the wire is fixed on the level of 10^{-1} K and the resonator's frequency relative sensibility of order of 10^{-6}). Such a wire generator was developed and used for scanning of low-level laser beams (a flux density sensitivity of 5×10^{-4} W/cm^2 has been achieved [2]). In this paper it is suggested to use a fence of vibrating wires. Such a fence will allow simultaneous fast (down to a few milliseconds) non-destructive measurement of the beam profile without complicated mechanical scanning unit. As compared with the traditional way of beam scanning by passive wire and subsequent measurement of secondary radiation/particles the advantage of this method is that the information about the local spatial intensity is concentrated in a specific vibrating wire. The usage of a fence in traditional methods brings forth principal difficulties connected with the spatial resolution of signals from different wires in radiation/particles detectors. High noise-resistance is another advantage of our method, since, unlike analogue measurements, a frequency signal is formed. The device can be used in beam vacuum chambers of accelerators as well as for solving of other problems in accelerator tunnel, e.g. for tracing of alignment laser beams.

---

[1] It seems that this should be also seen by the scintillator but background's influence on heating processes and scintillator detection can differ essentially.

2.   Scanning of Photon Beam

Photon beam passing through the material will also cause heating of the wire material with the advantage of no additional electromagnetic component. Thus the photon beam also can be measured by method of vibrating wire. Some estimates of absorption parameters for different materials and photon energies are presented in Table 1.

Table 1. Absorption parameter μ-1 for photon beams with different energies.

Material	102 eV	103 eV	104 eV
Tungsten	$3.5 \cdot 10^{+5}$	$7.6 \cdot 10^{+4}$	$1.8 \cdot 10^{+3}$
Molybdenum	$5.0 \cdot 10^{+4}$	$5.0 \cdot 10^{+4}$	$8.0 \cdot 10^{+2}$
Silicium	$7.4 \cdot 10^{+4}$	$3.7 \cdot 10^{+3}$	$7.3 \cdot 10^{+1}$

To estimate the wire frequency shift under irradiation of photon beam we numerically solved the model task of thermal conductivity along the wire allowing finding the temperature profile along the wire for different photon beam parameters, wire materials and geometry. For example, we consider the photon beam with parameters of XFEL TESLA [7]:

Wavelength, 1-5 A

Average power, 210 W

Photon beam size (FWHM), 500 μm at distance 250 m from source

Photon beam divergence (FWHM), 0.8 μrad

Bandwidth (FWHM), 0.08 %

Average flux of photons $1.0 \times 10^{+17}$ ph/s

Average brilliance, $4.9 \times 10^{+25}$ ph/(s·mrad^2·mm^2 0.1 % bandwidth).

Taking into account that here the beam diameter is less than 6 mm beam of PETRA it is reasonable to take a wire of less length and diameter. We have an experience with wires of length about 10 mm and diameter 20 μm, these sizes we take for further calculations. Wire initial frequency set (the second harmonics) - 5000 Hz.

Figure 3 represents the temperature profiles of wires from Molybdenum and Silicium at different time moments. It is seen that wire transition time for temperature balance of the beam with the wire is about 1 sec. Also, because the beam size is much less of wire length the temperature profile is almost a triangle.

The lower limit of the sensitivity of the VWS to temperature changes is defined by electronics and the method of signal measurement and is about 0.01 Hz, which corresponds to temperature 2×10^{-4} K. The temperature determines the upper limit, when the wire oscillation generation broke out (approximately 2000 Hz, corresponding temperature shift is about 30 ^{0}C).

Estimates show that sensitivity of our scanner allows scan of beams of energies 102 - 104 eV at distances 1.9 mm - 3.1 mm. The spatial resolution of VWS depends on flux of photons falling on the wire and improves at distances close to the beam center and decreases in beam periphery. Fig. 4 represents the dependence of the spatial resolution of the scanner depending on distance from the beam center. The beam parameters are the same as for the XFEL TESLA [5]. The assumed wire material is Molybdenum.

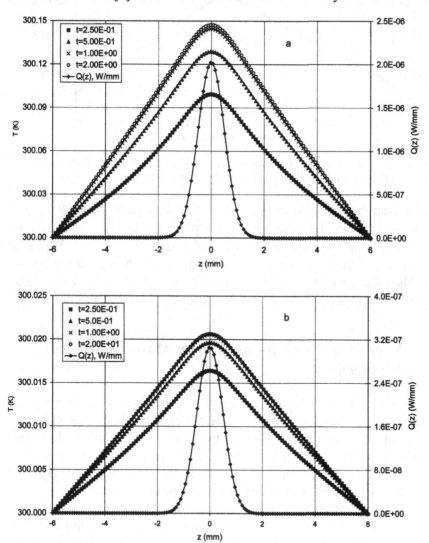

Figure 3. Dynamics and profile of temperature along the wire made of Silicon (a) and Molybdenum (b). Parameters of the photon beam are the same and the VWS in both cases is placed at distance 2.5 mm from the beam center.

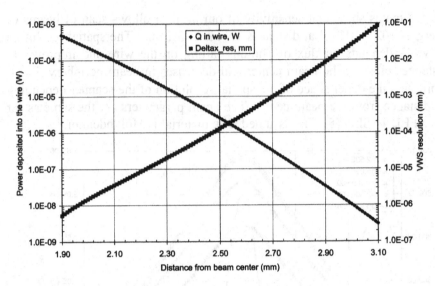

Figure 4. The power from the photon beam deposited into the wire and VWS spatial resolution depending on the VWS position from the beam center.

As seen from the Fig. 4 the spatial resolution of the scanner strongly depends on its position and at scanner's limit positions is less than 0.01 μm (x = 1.9 mm) and 0.1 mm (x = 3.1 mm). The lower limit 0.01 μm of resolution has a formal meaning since the diameter of the wire is much more than this value and such resolution can be reconstructed by VWS signal derivation only if the measured beam has known structure of profile.

3. Experiment with X-ray Source: Material Science Perspectives

To investigate the influence of the penetrative radiation on vibrating wire characteristics we have done an experiment using X-ray apparatus RUP-200-15. For the experiment we chose a two-wire modification with second harmonic frequencies 3680 and 3670 Hz. The second wire was screened by 2-mm lead plate, which allowed neglecting the influence of X-rays and comparing its signal with irradiated one. The base of the pickup was made of low absorbing glass fiber plastic.

The pickup was located at minimal distance from radiation source, where the intensity was about 2000 Roentgen/min. The estimates showed that radiation density at this distance was less than resolution of wire pickup by 2-3 orders. Indeed, in experiment no change of frequency was observed at switching on/off of the X-ray source. However, dose-depending increase of the frequency of first wire was revealed. The incline of the first wire time-frequency graphs was increased twice at the X-ray apparatus voltage switch from 100 kV to 165 kV at 15 mA current. The frequency of the second wire kept the same. At the end of irradiation (total radiation time was 45 min.) the difference between 2 frequencies from 10 Hz increased to 18 Hz. Fluctuation of frequencies during the measurement was less than 0.15 Hz. The X-ray radiation also resulted in change of temperature dependence of irradiated wire from 7.03 Hz/K (before irradiation) to 4.7 Hz/K (after), while for the non-irradiated wire this value changed negligibly (3.26 and 3.1 Hz/K, correspondingly). The results are presented in Fig. 5. The temperature was measured by a semiconductor thermoresistor (type- KTY).

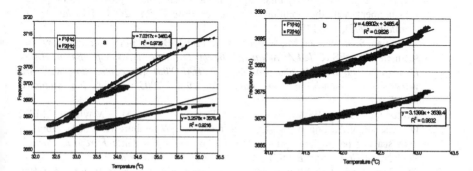

Figure 5. Frequency-temperature dependence of wires before (a) and after (b) irradiation.

Thus, already small doses of radiation can influence the mechanical tensions and/or redistribution of dislocations, and this was fixed by our pickup. Method of irradiation of the wire by X-rays can also be used for studies of radiation quenching and aging of materials. Note, that investigations in this area by traditional methods require long-term experiments, while the high sensitivity of the vibrating wire pickup allows to fix the changes practically on-line.

The presented ideas and results show that measurement of photon fluxes by means of a vibrating wire scanner has a good perspective. Unmovable sensor can also be used for photon beam position monitoring. Preliminary experiments show that vibrating wire electromechanical resonators can be used for express analysis of characteristics of materials, irradiated by photon and particles beams in wide range of parameters.

References

1. Arutunian, S.G., Dobrovolski, N.M., Mailian, M.R., Sinenko, I.G., Vasiniuk, I.E. (1999) Vibrating wire for beam profile scanning, *Phys. Rev. Special Topics - Accelerators and Beams*, 2, 122801.
2. Arutunian, S.G., Avetisyan, A.E., Dobrovolski, N.M., Mailian, M.R., Vasiniuk, I.E, Wittenburg, K., Reetz R. (2002) Problems of Installation of Vibrating Wire Scanners into Accelerator Vacuum Chamber, *Proc. 8-th Europ. Part. Accel. Conf. (3-7 June 2002, Paris, France),* 1837-1839.
3. Arutunian, S.G., Dobrovolski, N.M., Mailian, M.R., Vasiniuk, I.E. (2003) Vibrating wire scanner: first experimental results on the injector beam of Yerevan synchrotron, *Phys. Rev. Special Topics. - Accelerators and Beams*, 6, 042801.
4. Arutunian, S.G., Bakshetyan, K.G., Dobrovolski, N.M., Mailian, M.R., Soghoyan, A.E., Vasiniuk, I.E. (2003) First experimental results and improvements on profile measurements with the vibrating wire scanner, *Proc. DIPAC'2003, (5-7 May, Mainz, Germany 2003),* 141-143.
5. Arutunian, S.G., Werner, M., Wittenburg, K. (2003) Beam tail measurements by wire scanners at DESY, *ICFA Advanced Beam Dynamic Workshop: Beam HALO Dynamics, Diagnostics, and Collimation (HALO'03) (in conjunction with 3rd workshop on Beam-beam Interaction) (May 19-23, 2003 Gurney's Inn, Montauk, N.Y. USA).*
6. Arutunian, S.G., Bakshetyan, K.G., Dobrovolski, N.M., Mayilyan, M.R., Oganessian, V.A., Soghoyan, A.E., Vasiniuk, I.E., Wittenburg K. (2004) Vibrating wire scanner parameters optimization. – *Proceedings of Ninth European Particle Accelerator Conference EPAC'04 (July 5–9, Lucerne, Switzerland).*
7. The Superconducting Electron-positron Linear Collider with an Intagrated X-Ray Laser Laboratory.- *TESLA Technical Design Report, Part V The X-Ray Free Electron Laser.- DESY 2001-011, ECFA 2001-209, TESLA Report 2001-23, TESLA-FEL 2001-05.*

NUCLEAR ASTROPHYSICS MEASUREMENTS ON ELECTRON LINEAR ACCELERATOR

Nina A. Demekhina

Yerevan Physics Institute, .Alikhanyan Br. St.2,Yerevan 375036, Armenia

Abstract A number of photon capture reactions at astrophysical relevant energies is accessible
to measurements at high current electron accelerator in Yerevan Physics Institute.
High intensity bremsstrahlung beam with end energies varied in the range of 3-20
MeV should be used as photon source. Superposition of several bremsstrahlung
spectra can give valuable information about reaction yields with quasi-thermal
photon energies. A number of photonuclear reactions at astrophysical temperatures
$T_9 \sim$ 2-3 related to p – processes can be investigated. Calculation of astrophysical
process rates is possible with reciprocal correlation on the base of measurements of
reverse γ- reaction rates.

1. Introduction

Investigations of nuclear reactions under laboratory conditions that
simulate the stellar environment form the empirical foundation for realistic
and sophisticated theoretical models of complex processes during the
evolution of a wide variety of astrophysical systems and of nucleosynthesis,
responsible for the abundance of element in the universe. Accelerators allow
experimentalists to simulate nuclear interactions at stellar temperatures that
range from 15 million degrees at the core of the Sun to 15 billion degrees in
an exploding supernova. These temperatures actually correspond to energies
much lower than those available in a typical accelerator. Kinetic energy
distributions for nuclei inside a star are described by the Maxwell-Bolzmann
equations. The stellar core temperature is such that most of the nuclei
participating in reactions do not have enough energy to overcome the
Coulomb repulsion. Only the nuclei in the high energy tail of the distribution
can tunnel through Coulomb barriers. The range of effective energies
allowing interactions is called the Gamow window. Reactions at such
energies fall into subthreshold ranges, where the interaction probability is
very low. Therefore, these kinds of measurements are difficult for practical
realization , and experimental data are very scanty.

Currently, experimental nuclear astrophysics develops in two main

H. Wiedemann (ed.), Advanced Radiation Sources and Applications, 343–350.
© *2006 Springer. Printed in the Netherlands.*

directions. First, measurements made at low energy accelerators are being closely linked to the complex physics of stellar evolution and the formation of planetary nebula–shells of glowing gas surrounding stars. Since in this energy range reaction cross sections are low, high beam intensity accelerators are required to simulate stellar nuclear processes in a laboratory. Recent developments in accelerator and detector technology led to the possibility to perform experiments that simulate stellar conditions [1-3].

The second initiative focuses on reactions involving "exotic" nuclei which control stellar explosions such as novae, X-ray bursts and supernovae. Under these conditions nuclei typically capture protons and neutrons much faster than they decay, thereby driving the reaction path far away from the "line of nuclear stability". These series of reactions resulting in unstable isotope creation are usually sources of "exotic nuclei". Most of the experiments related to astrophysical objects are carried out at radioactive ion-beam facilities. Low energies as well as high beam intensities are necessary for such measurements.

Experimental data of photonuclear reactions, playing an important role in several nucleosynthesis processes are concerned energies mostly essentially lower than those available at traditional photon sources, as electron accelerators. The bremsstrahlung photons from accelerated electrons are used usually as high intensity photon source with continuous energy distribution. The energy range relevant to astrophysical reaction rates induced by photons is defined by the product of the photon flux density exponentially decreasing with energy and cross section increasing with energy above reaction thresholds (Gamow-like window).

The present paper describes the possibilities to use an electron accelerator for nuclear reaction measurements at astrophysical relevant energies.

2.   Photonuclear Reactions in Element Synthesis Processes

2.1   P-Processes

Relatively rare proton-rich nuclei, the so–called p-nuclei that lie in mass range between ^{74}Sc and ^{196}Hg on the proton–rich side of the valley of β-stability, are impossible to produce by neutron capture in s- or r- processes alone, in contradistinction to other nuclei heavier than iron. In development of the theory of nucleosynthesis, the p-nuclei production requires a special mechanism termed p-process [3]. This nucleosynthetic scenario involves more or less complicated production sequences of neutron, proton, and α-

particle emission in photon induced reactions, as well as in (p, γ) and (α, γ) reactions. Several possible mechanisms and basic features of p-process including some framework for the conditions required to produce p-nuclei were considered in a number of papers (see ref. in [3]).

According to well known models, p-nuclei can be produced from the "burning" of preexisting more neutron rich nuclei in stellar environment at high enough temperature in the following reactions: proton radioactive captures in hot ($T_9 \sim 2$-3) proton rich environment or photon induced n, p, and α–particle removal, in hot environment also. It was found that lighter p-nuclei could be produced by proton capture reactions while the photodissociations are required for synthesis of heavier nuclei [4]. The p-process in nucleosynthesis develops in the neutron deficient region of a chart of nuclei, where most masses and β-decay rates are known experimentally. The reliable modeling of the p-process flows necessitates the consideration of an extended network of some 20000 reactions linking about 2000 nuclei in the A<210 mass range The situation is by far less satisfactory concerning the laboratory knowledge of the rates of the nuclear reactions involved in the p-process networks. Some predictions of the heavy nucleus production via photodissociation are based on cross sections, most of that are unknown at present. Most of the rate measurements concern radiative captures by stable targets, which cover a small fraction of the needs for p-process calculations. The scarcity of the relevant information makes it mandatory to rely heavily on rate predictions. Such predictions are exclusively based on the Hauser-Feshbach statistical model and various nuclear ingredients required in such a framework [5]. One of the most severe problems in this respect is the lack of experimental data.

The γ-process appearing in production of nearly all heavy (A $\geq$ 100) p-nuclei, operates essentially in the following way: photons from high temperature bath ($T_9 \sim 2$-3) containing previously synthesized heavy nuclei initiate successive (γ, n), and (γ, α) reactions on those nuclei to populate heavy p-nuclei, (γ, p) –reactions participate in p-process nucleosynthesis by indirect way, on the whole. The astrophysical reaction rate in photonuclear interaction $\lambda(T)$ is given by:

$$\lambda(T) \sim \int n_\gamma(E, T) \, \sigma_{(\gamma,x)}(E) \, dE \, ,$$

where $\sigma_{(\gamma,x)}(E)$ is photonuclear reaction cross section., $n_\gamma(E,T)$ is photon density, that can be presented by the Plank distribution in the case of stellar photon flux or in form of bremsstrahlung spectra in the case of electron accelerator. These kinds of reactions on stable targets, accompanied by protons and α- particles emission and creation of proton rich isotopes can be considered for nuclei in range of medium and heavy elements. In the same mass range neutron production rates in photonuclear processes on stable target with residual p-nuclei or neighboring nuclei can be measured. These reaction thresholds are several MeV equal to neutron separation energy. The typical width of the Gamow window is about 1 MeV. Therefore the excitation energy for these reactions is fell into an energy interval S_n +1MeV. Experiments for photon induced reactions in the astrophysically relevant energy range can be carried out in practice using two different technique: monochromatic photons from Compton backscattering of a Laser beam [2] and quasi- thermal photon spectrum, obtained by a superposition of bremsstrahlung beams with different end energies [6-8]. The photon beams at variable energies in the range 3-20 MeV, produced on electron accelerator at 20 MeV energy are suitable for realization such program.

As a first test for the (γ, n) reaction rates on the platinum isotopes ^{190}Pt, ^{192}Pt and ^{198}Pt were derived by using bremsstrahlung beams [6].The reaction products are analyzed using photoactivation technique. The high sensitivity of this registration method allows the measurements of reaction yield for isotopes with very low natural abundance's. These experimental results were almost first data for the reactions relevant for the γ – process at astrophysical energy.

2.2   Photonuclear reactions as reverse processes in nucleosynthesis

Investigations of nuclear reactions at bombarding energies relevant to stellar nucleosynthesis are impossible often with the use of modern technology. This situation occurs, for example, in studies related to hydrogen and helium burning. Here cross sections are extremely small because of Coulomb barrier penetrability. Similarly, the experiments at radioactive ion beam facilities require beam intensities that are beyond the capabilities of present generation radioactive ion accelerators. As a result, frequently indirect methods are applied to deduce cross sections of astrophysical interest. In certain cases the reverse channel yield measurements allow to receive necessary information about direct reaction rate.

2.3   Helium capture process via photon induced reactions

One of the most important reactions in synthesis of light elements is hydrogen burning in a star and formation of a hot dense core of helium, that fuels the nucleosynthesis of heavier elements. Helium burning occurs in the so-called "triple–α" capture process to produce ^{12}C. The subsequent capture of α-particles results in production ^{16}O.

$$^{12}C + \alpha \rightarrow ^{16}O + \gamma$$

After that, the following processes of α-particles capture occur:

$$^{16}O + \alpha \rightarrow ^{20}Ne + \gamma$$

$$^{20}Ne + \alpha \rightarrow ^{24}Mg + \gamma$$

$$^{24}Mg + \alpha \rightarrow ^{28}Si + \gamma$$

$$^{28}Si + \alpha \rightarrow ^{32}S + \gamma$$

$$^{32}S + \alpha \rightarrow ^{36}Ar + \gamma$$

A part of listed reactions can be measured as reverse processes induced by photons at energies exceeding threshold values. The development of the high sensitively registration methods of α- particles with extremely low energies is the significant problem in such experiments.

2.4   Carbon and oxygen "burning" processes

The carbon and oxygen "burning" processes responsible for light element production can be studied also in reverse processes induced by photon beam:

$$^{24}Mg + \gamma \rightarrow ^{12}C + ^{12}C$$

$$^{32}S + \gamma \rightarrow ^{16}O + ^{16}O$$

As a whole, these "burning" reactions are characterized by different number of emitted particles with creation of different residuals, participating in other nuclear reactions.

2.5   Silicon" burning"process

Main products, created in carbon and oxygen burning processes, are ^{32}S, ^{28}Si and ^{24}Mg. The first element undergoes to photo spallation should be ^{32}S because of lower coupling energies for n, p and α-particle in comparison with ^{28}Si. These processes lead to increasing Si content:

$$^{32}S + \gamma \rightarrow ^{31}P + p$$

$$^{30}Si + \gamma \rightarrow ^{29}Si + n$$

$$^{29}Si + \gamma \rightarrow ^{28}Si + n$$

The Si photo spallation begins at higher temperatures above $3 \cdot 10^9 K$:

$$^{28}Si + \gamma \rightarrow {}^{27}Al + p$$

$$^{28}Si + \gamma \rightarrow {}^{24}Mg + \alpha$$

The subsequent photo spallation of these nuclei results in production of lighter elements in reactions:

$$^{24}Mg + \gamma \rightarrow {}^{23}Na + p$$

$$^{24}Mg + \gamma \rightarrow {}^{20}Ne + \alpha$$

$$^{20}Ne + \gamma \rightarrow {}^{16}O + \alpha$$

$$^{16}O + \gamma \rightarrow {}^{12}C + \alpha$$

In some models, the carbon oxygen burning processes are considered as a basic way for the realization models of element abundance in mass $20 \leq A \leq 64$ in solar system [3].

3.   Bremsstrahlung Photon Flux

The availability of intensive bremsstrahlung photon sources on the base of high current electron accelerators presents a powerful tool for nuclear astrophysics measurements. The electron linear accelerator, operating at the Yerevan Physics Institute, produces electron beams with variable energies in the range from 3 MeV up to 20 MeV with an average current of 200-500 µA. The photon flux in an energy range relevant to astrophysical tasks can be on average 10^{13}-10^{14} /s. Measurements of astrophysical reaction rates in photon induced interactions are possible because of the special properties of bremsstrahlung radiation. By the appropriate superposition of several bremsstrahlung spectra with different end point energies, a very good agreement can be obtained with thermal photon spectra, related to the mean energy value, above the (γ, n), (γ, p) and (γ, α) thresholds. In these measurements, the neutron background should be strongly suppressed by replacing the heavy elements materials that form the collimator and bremsstrahlung target by cooper which has a higher neutron emission threshold of 9.9 MeV [9].

Superior sensitivity detection systems should be developed on the base of modern technologies for measurements of a reaction product yields at subthreshold energy range. At the same time, the routine photo activation method using HPGe detectors have been used to provide direct measurements of reaction products [10]. The techniques in beam on and off beam measurements can be employed in these kinds of experiments. Time-of-flight (TOF) techniques should be applicable to the identification reaction products also. A solid state detectors, covered with thinnest plates of target materials seem to be good for detecting particles and light nuclei emitted in reactions. Development of a multiwire proportional chamber by using filling gas as an active target is planned for measurements of α-particles or protons yields in some reactions. Finally, measurements of the reaction yields in silicon targets can be made in the volume of silicone detector. All these briefly mentioned methods request careful consideration.

4.   Conclusion

Information on astrophysical nuclear reactions can be obtained from experiments on high intensity electron accelerator. Energies of photon beam induced nuclear reactions are selected by application of a superposition method at different electron energies. Reaction rate measurements in the range of several *p*-processes can be realized at energies of astrophysical interest, i.e. close to the reaction threshold. The possibility to calculate several reaction rates on the basis of reverse γ-process study will shed light upon some questions related to the nucleosynthesis.

References

1. S.Galanopoulos, P.Demetriou, M.Kokkoris et al., The ^{88}Sr(p,γ)^{89}Y reaction at astrophysically relevant energies, *Phys.Rev. C* **67**, 015801 (2003)

2. H.Utsunomiya, H. Akimune, S.Goko et al., Cross section measurements of the ^{181}Ta (γ,n) ^{180}Ta reaction near neutron threshold and the p-process nucleosynthesis, *Phys.Rev. C* **67**, 015807 (2003)

3. G.Wallerstein, I.Iben, P.Parker et al., Synthesis of elements in stars: forty years of progress, *Rev.Mod.Phys.* **69**, 995 (1997)

4. J.W.Truran, A.G.W.Cameron, The p process in explosive nucleosynthesis, *Astrophys. J.* **171**, 89 (1972)

5. W.Hauser and H.Feshbach, The Inelastic Scattering of Neutrons, *Phys. Rev.* **87**, 366 (1952)

6. P.Mohr, K.Vogt, M.Babilon et al., Experimental simulation of a stellar photon bath by bremsstrahlung: the astrophysical γ process, *Phys.Lett.B* **488**,127 (2000)

7. K. Vogt,P. Mohr,M. Babilon et.al., Measurement of the (γ, n) reaction rates of the nuclides ^{190}Pt, ^{192}Pt , and ^{198}Pt in the astrophysical process, *Phys.Rev.C* **63**, 055802 (2001)

8. K.Vogt P. Mohr, M. Babilon et al., Measurement of the (γ,n) cross section of the nucleus
^{197}Au close above the reaction threshold, *Nucl.Phys. A* **707**, 241(2002)
9. P.Mohr, J.Enders, T.Hartmann et. al., Future Perspectives of a 10 MeV Bremsstrahlung
Facility, 10th Int. Symp., Capture gamma-rays spectroscopy and related topics Santa Fe New
Mexico, 743 (1999)

A COMBINED HIGH ENERGY MŒLLER AND COMPTON POLARIMETER

R.H. Avakian, AE.Avetisyan, P. Bosted[1], K.R. Dallakian,
S.M. Darbinian, K.A. Ispirian and I.A.Kerobyan*
Yerevan Physics Institute, Brot. Alikhanyan Str. 2, 375036, Yerevan, Armenia
[1]*Thomas Jefferson Laboratory, Newport News, Virginia23606, USA*

Abstract: The properties of a polarimeter designed for the measurement of the degree of
               polarization of longitudinally polarized electrons and of circularly polarized
               photons at energies 10-40 GeV, which can be used in various fixed target
               experiments at SLAC and Jefferson Laboratory, have been investigated.

Keywords: Electron beam, photon beam. polarization, polarimetry.

1.   INTRODUCTION

At present polarized high-energy electron and photon beams have found wide applications in collider and fixed target experiments. The planned and beginning experiments at Jefferson Laboratory [1] and SLAC [2] require the development of methods to measure the degree of polarization in electron and photon beams. Longitudinally polarized electron beams allow the generation of circularly polarized high energy photon beams, which requires the development of new methods to measure the polarization of both primary electron and secondary photon beams. The polarization of the photon beam will be determined by measuring the asymmetry of the cross-section for high-energy photons generated by Compton scattering on polarized electrons in a ferromagnetic target with a known degree of polarization of 8%, and using the theoretically calculated value of asymmetry in the cross-section for completely polarized photon beam (P_γ= 100%).

*E-mail:keropyan@mail.yerphi.am

H. Wiedemann (ed.), Advanced Radiation Sources and Applications, 351–356.

For the experiments E-159, E-161 with circularly polarized high-energy photon beam it has been proposed [2] to construct a Compton polarimeter. The Compton polarimeters are connected with the difficulty that the cross-section decreases with the increase of photon energy.

Since the theory of Mœller polarimeter is well known in this work we shall describe only the calculation of the parameters of the proposed Compton polarimeter considering its difficulties including the influence of the Levchuk effect.

2. COMPTON POLARIMETRY

Polarization of circularly polarized photons with energy $x=E/E_e$ produced by longitudinally polarized electrons on amorphous or crystalline target is determined by the formula

$$P_\gamma \approx P_e \frac{x(4-x)}{(4-4x+3x^2)}.$$

A circularly polarized photon spectrum has $1/E_\gamma$ behavior, which means that in spectra there is huge number of small energy photons. In case of coherent photons spectra for special crystal orientation the number of photons near end of spectrum increases 5-6 time, however the number of soft photons still is huge.

To obtain the Compton scattering of photon of coherent bremsstrahlung spectra on polarized atomic electrons of ferromagnetic materials for measuring the circular polarization of high energy part of photons we are going to measure the parameters of the scattered photons and recoil electrons simultaneously. It was easy to understand that unfortunately the small energy photons with huge number of them and high Compton cross section fill up the photon detectors directed at the angles corresponding to the scattering of high energy photons and which makes impossible high energy photon detection. We find out another possibility to measure the Compton scattering of high-energy photon by measuring only angle and momentum of scattered electron in magnetic spectrometer. In this case we avoid from small energy photon of bremsstrahlung spectra. Now the problem is to calculate the accuracy of measurement of electron momentum and scattering angle with respect to angle of initial photon direction. By this measurement we can reconstruct the photon spectra near the peak energy of the coherent bremsstrahlung spectra and compare that with electron spectrum of tagged photon to be sure that we measured the Compton scattering of photons on the peak of spectrum.

To accept the concept of measuring only angle and momentum of scattered electron we can use some methodic from the Mœller polarimeter. We will use the septum pipe inside of magnet and special type of collimator after ferromagnetic target. This collimator allows the passage of the scattered electrons in wide range of θ (polar angle) and limits the passage of electrons in small range of φ (azimuthal angle).

3. THE PROPOSED EXPERIMENTAL ARRANGEMENT

The combined Mœller and Compton polarimeter setup is illustrated in Fig.1. It consists of crystalline target T1, deflecting magnet M1 and tagging hodoscope TH, ferromagnetic target T2, dumping system D, collimator-mask CM, second deflecting magnet M2, top and bottom scintillator counters S, lead preshowers PS, vertical and horizontal hodoscopes HV and HH and lead glass total absorption calorimeters.

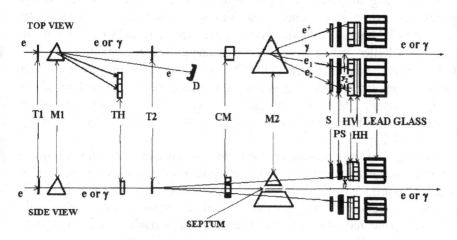

Figure 1. Experimental setup (not in scale) of the Mœller, Compton polarimeter.

The degree of the incoming electron beam longitudinal polarization will be measured in the Mœller regime of the polarimeter when the crystalline target T_1 of the tagging system is removed and the tagging magnet M_1 is switched off. The primary electrons not interacted in T_2, pass the central hole of the collimator mask CM, the septum between the upper and bottom poles of the magnet M_2, vertical and horizontal slits in the scintillation counters S_{1-4}, preshowers PS_{1-4}, 4 hodoscopes HH_{1-4}, HV_{1-4} and 4 lead glass counter arrays LG_{1-4}. The secondary electrons e_1 and e_2 after Mœller scattering in the ferromagnetic target T_2 pass the top and bottom slits of the collimator-mask

CM and are deflected to one side by the magnet M_2. After passing the trigger scintillation counters S they can produce a few additional electrons in the 2-3 rad length preshower $PS_{1,2}$ and their horizontal and vertical coordinates are detected by the hodoscopes $HH_{1,2}$ and $HV_{1,2}$ with an accuracy equal to about the width of hodoscopes' sticks. Their energy is measuring by the magnetic deflection curvature and by the total energy deposition in lead glass counters $LG_{1,2}$ with typical accuracy $\Delta p'/p' \sim 2\%$ and $\Delta E'/E' \sim 4\text{-}8\ \%$, respectively when $E' \approx 3\text{-}10$ GeV. The hodoscopes allow separating the true Mœller electrons hitting a certain widened parabola from the background mainly from Mott scattering events (see below).

In case of CW electron beam (e.g. at Jefferson Laboratory) in additional a photon tagging system is present with magnet M_1 and tagging scintillation detectors TS. The tagging system provides the measurements of the high-energy photons with an energy accuracy of $\Delta E/E = (1\text{-}2)\ \%$. The main part of the primary electron beam is deflected by the magnet M_1 and buried in the dump D.

The circularly polarized photon beam is produced by coherent bremsstrahlung in diamond target T_1 in order to increase the intensity of photons in certain photon energy intervals a few times. If the electron beam is polarized longitudinally the photons are circularly polarized. As in the case of Mœller polarimetry the high-energy tagged photons hit the ferromagnetic polarized target T_2. The secondary particles from the two particle Compton reaction $\gamma\, e \rightarrow e\,\gamma$ pass through the top or bottom slits of collimator-mask CM. The recoil electrons from the same energy of primary photons hit the hodoscope by the parabolic shape detected by $S_{1,2}$, $HH_{1,2}$, $HV_{1,2}$ and $LG_{1,2}$.

The parts of the combined polarimeter, for instance the collimator-mask CM, ferromagnetic target etc., which are quite similar to those of the existing Mœller polarimeters [3,4], will not be discussed here. The target T1 is a crystalline radiator in the precise goniometer system and produces the circularly polarized photon beam. The target T2 is made of a ferromagnetic alloy of Fe ($\sim$ 49 %), Co ($\sim$ 49 %) and Va ($\sim$ 2 %) providing a target of electrons with 8 % polarization and is described in the works [3,4].

The collimator-mask CM made of $\sim$ 20 rad. length tungsten, is similar to that of the work [3,4] and has two vertical, top and bottom slits which define the Mœller and Compton scattering angle, around $\theta^* \approx 90°$ in CMS within a narrow azimuthal angle interval $\Delta\varphi = 0.22$ rad.

4. INFLUENCE OF LEVCHUK EFFECT IN COMPTON POLARIMETER AND RESULTS

As it is well known if the atomic electrons are in rest there is a simple relation between the angle of the Mœller scattering and the scattered electron energy [4]. However, taking into account the motion of the target atomic electrons and their momentum distribution (the Levchuk effect) the relation between the energy of primary photon, recoil electron and its scattered angle will be violated.

The influence of Levchuk effect on the double-task Compton- Mœller polarimeter operation has been investigated. It has been shown that in case of double arm Mœller polarimeter the influence of Levchuk effect is negligible. That's because the Levchuk effect affects on the scattered electron angle, while only the scattered electron energy value is decisive for a double arm polarimeter. Levchuk effect does not make serious influence to the value of asymmetry too due to large angular acceptance and satisfactory energy resolution. To reconstruct the primary gamma quanta energy the accurate measurement of the scattering angle and of the energy of electron is necessary, and in this case the Levchuk effect could play a serious role. In particular, if the error of the scattering angle is 2% (with assumption that the error of energy determination is small) then the error of the primary gamma quanta energy determination is 3%. A formula has been derived to calculate the influence of Levchuk effect correctly, which was used for the final calculation of that influence.

The influence of Levchuk effect in case of Compton polarimetry has been calculated by Monte-Carlo method. While the photon energy is 10 GeV and recoil electron energy is 5 GeV the influence of Levchuk effect to the value of electron angle creates a systematic error of 10% for K-shell, 5% for L-shell, 2% for M-shell and 0.5% for N shell, total error estimated as 2.5%. This estimation has been done without influence of systematic error related to the geometry and sizes of scintillator counters. In case of linear scintillator counters the systematic error based on geometry and sizes was estimated as 5-6%, so that influence of Levchuk effect is negligible in case of Compton polarimetry too.

Some results of Monte-Carlo simulations are given in Table 1.

Table 1. The parameters of the beams, the accuracy of polarization and the times of their measurement.

Type of measurement	Energy, (GeV)	Expected beam intensity	Degree of polarization, P (%)	$\Delta P / P$, (%)	Required time, (min)
Electron longitudinal polarization	48.3 12	5×10^9e/sp 6×10^{11}e/s	80 85	~2 1.5	15 10
Photon circular polarization	66.4 8.4	2×10^6e/sp 2.4×10^8e/s	70 60	3 2	30 35

References

[1]Proceedings from JLAB/USC/NCCU/GWU Workshop on "Polarized Photon Polarimetry", (G. Washington University, June 2-3, 1998).

[2] V. Ghazikhanian et al, SLAC proposal E159, 2000.

[3] H.R.Band et al. A Mœller polarimeter for high-energy electron beams, Nucl. Instrum. Methods Phys. Res. A **400**, 24 (1997).

[4] M.Swartz et al. Observation of target electron momentum effects in single-arm Mœller polarimetry, Nucl. Instrum. Methods Phys. Res. A **363**, 526 (1995).

EXPERIMENTAL STUDIES OF POSITRON SOURCES USING MULTI-GEV CHANNELLED ELECTRONS AT CERN[a] AND KEK[b]

Robert Chehab

LAL-Orsay and IPN-Lyon (IN2P3/CNRS); on behalf of WA103 (CERN) and KEK groups
LAL - Bât. 200, Université Paris-Sud – 91898 Orsay Cedex (France)

Abstract: Enhancements of photon production by multi-GeV electrons moving along the axes of aligned crystals are used to generate high yields of positrons in the same crystals or in amorphous disks put after the crystals. The experiment WA103 used tungsten crystals oriented along their <111> axis, alone or followed by tungsten amorphous disks. The incident electron energy was chosen between 5 and 40 GeV and mainly used at 6 and 10 GeV. Photon as positron production enhanced by channelling and coherent bremsstrahlung have been measured with photon and positron detectors. The positron detector consisted in a drift chamber immersed in a magnetic field and presenting rather large momentum and horizontal angle acceptances of 150 MeV/c and 30°, respectively. The KEK experiment on the KEKB injector used 8 GeV incident electron beam and different kinds of crystals. Tungsten crystals and more recently Silicon and Diamond crystals have been used. The measurements concerned the positrons. The results gathered on both experiments are presented and commented.

Keywords: Channelling, Coherent Bremsstrahlung, crystals, Photons, Positrons, Drift Chamber

1.   INTRODUCTION

Achievement of high luminosities at the interaction point of an e^+e^- linear collider requires high intensity and low emittance beams. Concerning the positrons, conventional sources use high intensity electron beams impinging on thick amorphous targets with high atomic number. As a consequence, a large energy deposition induces important heating problems leading to mechanical stresses and target failure. Moreover, multiple Coulomb cattering occurring in the thick target leads to large emittance positron beams.

357

H. Wiedemann (ed.), Advanced Radiation Sources and Applications, 357–371.

The reduction of the overall target thickness is obtained through the generation of a high number of photons- much higher than with ordinary bremsstrahlung- which are, then, materialized in thin targets. These photon generators –hereafter called radiators- could be magnetic undulators, as first proposed by V.Balakin and A.Mikhailichenko [1] or "atomic" undulators like oriented crystals [2]. The latter has been studied theoretically and experimentally since many years [3, 4, 5, 6]. Proof of principle and experiments were run out in Europe and Japan [7, 8, 9, 10, 11, 12] with different crystals and various incident electron energies.

We report here, on the experiment WA103 carried out at CERN on the X-5 transfer line of the SPS. The crystal chosen was tungsten and the incident energies were between 5 and 40 GeV with the main measurements at 6 and 10 GeV, which correspond to the chosen values for the incident energies of NLC (Next Linear Collider) and JLC (Japanese Linear Collider), respectively. The results gathered at KEK on the KEKB linac are also reported. They concern Tungsten, Silicon and Diamond crystals. Some emphasis will be put on the latter.

2.   WA103 EXPERIMENT

2.1   Experimental set-up

The experiment WA103 is using tertiary electron beams of the SPS having energies between 5 and 40 GeV. The electrons, after passing through profile monitors (delay chambers) and trigger counters, impinge on targets installed on a goniometer. Photons as well as e+e- pairs are produced in the targets, come mainly in the forward direction and travel across the magnetic spectrometer made of a drift chamber and positron counters (see Fig. 1).The photons and high energy electrons and positrons come out in the forward direction. The charged particles are swept by a second magnet (MBPL) while the photons reach the photon detector made of a preshower and a calorimeter.

The beam: the electron pulses of 3.2/5.2 seconds duration and with periods of 14.4/16.8 seconds have intensities of $\sim 10^4$ electrons/pulse. Two energies have been particularly used: 6 and 10 GeV. The channelling condition requires that the incident electron angle be smaller than $\Psi = \sqrt{(2U/E)}$, where U represents the depth of the potential well of an atomic row and E, the electron energy. At 10 GeV and for the <111> orientation of the tungsten crystal, this angle is 0.45 mrad. As crystal effects are lasting slightly above this critical angle, we gave to the trigger system, made of scintillation counters (see Fig. 1), an angular aperture of 0.75 mrad.

The trigger selection is improved by the information provided by the proportional delay chamber (off line selection).

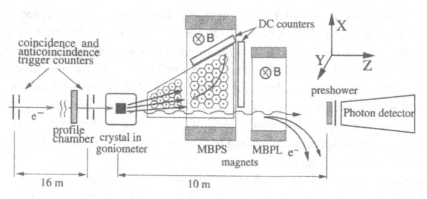

Figure 1 : WA 103 set-up.

The targets: four kinds of targets have been used:
- a 4 mm thick crystal,
- a 8 mm thick crystal,
- a compound target made of 4 mm crystal followed by 4 mm thick amorphous disk,
- an amorphous disk 20 mm thick

The mosaic spread of the crystals was measured by γ-diffractometry and was less than 500 μrad.

Positron detector: it consists in a drift chamber with hexagonal cells, filled with a gas mixture {He(90%);CH4(10%)} and positron scintillation counters. It presents two parts:

- the first part (DC₁) with a cell radius of 0.9 cm, located mainly outside the magnetic field. It allows the measurement of the exit angle.
- The second part (DC₂) with a cell radius of 1.6 cm is immersed in the magnetic field. It allows the measurement of the positron (electron) momentum. Two values of the magnetic field are used: 1 and 4 kGauss in order to investigate the whole momentum region of interest, up to 150 MeV/c.

Photon detector: the photon multiplicity is rather high ~ 200 γ/event at 10 GeV for a 8 mm crystal oriented on its <111> axis. The detector is made of:

- a preshower (0.2 Xo thick Copper disk and a scintillator) gives information on photon multiplicity. It is used for crystal alignment.

- a spaghetti calorimeter [13] with thin scintillation fibbers gives the amount of radiated energy. Preshower and calorimeter are in close contact.

The wires are parallel to the magnetic field. The available space in the bending magnet (MBPS) is restricted vertically to 6 cm. In order to avoid border effects of the metallic walls of the chamber, the useful part of the wires is limited to a central part of 3 cm defined by the dimensions of positron counters put on lateral and back sides of the chamber (see Fig. 1). That sets the angular acceptances, vertically to ± 1.5° and horizontally to 30°. The overall acceptance represents 6 % of 2π solid angle. Achieved coordinate resolution is about 500 μm leading to momentum and angle resolutions of 0.6 MeV/c and 0.25°, respectively.

Reconstruction; positron (electron) tracks are reconstructed in the drift chamber. Signals are coming from particle activated wires ; each hit is represented by a circle centred on the respective wire and the radius is proportional to the drift time. The reconstruction procedure assigns 3 parameters to each track: the e+ or e- trajectory, projected on the horizontal plane is parameterized as a circle with the 3 parameters: r, the circle radius; (xc, zc) the coordinates of the circle centre. A track is determined by a minimum of 3 wires struck by particles. To find a good compromise between the number of fake tracks (reconstructed but not real) and the lost tracks (real tracks not reconstructed) a minimum of 10 as number of wires hit has been defined for the track selection. In these conditions a reconstruction efficiency of 80% has been verified for the worst case (amorphous target 20 mm thick and 10 GeV).

2.2   Results

Reconstructed trajectories: an example of reconstructed trajectory is given on Fig. 2. It concerns a crystal 8 mm thick and incident energy of 6 GeV.

Photon measurement: the preshower gives the average photon multiplicity in a forward cone. On Fig. 3, we compare the measurements with the simulations for a 4 mm crystal and incident energy of 6 GeV.

Positron measurement: energy spectrum and angular distributions have been measured for all the targets and comparisons carried out between crystal and amorphous targets of the same thickness. In the latter case, the crystal was in full random orientation.

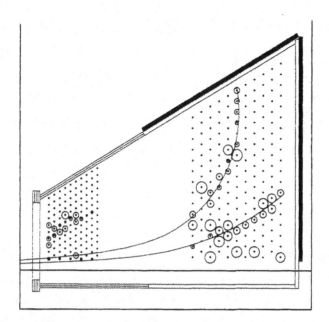

Figure 2: Reconstructed trajectories for 8 mm thick crystal and 6 GeV.

(a) amorphous target 20 mm thick; this measurement was important for two reasons:

- to test our reconstruction programme in the "worst case", i.e. with maximum occupancy in the chamber (at 10 GeV incident energy),

- to make valuable comparisons between simulations and experiment, particularly with GEANT and also to compare the simulations from GEANT to those of an original programme [14].

The results are conclusive as we can see on Fig. 4 where both kinds of simulations are compared to the experiment for the angular distribution in the "worst case". The agreement is quite good.

(b) Crystal target 4 mm thick: simulation and measurement are compared for the crystal on axis and in random orientation. A significant enhancement is observed when the crystal is aligned on its <111> axis; this enhancement is slightly larger than four for an incident beam of 10 GeV, as can be seen on Fig. 5. For an incident energy of 6 GeV, the enhancement is slightly higher than three.

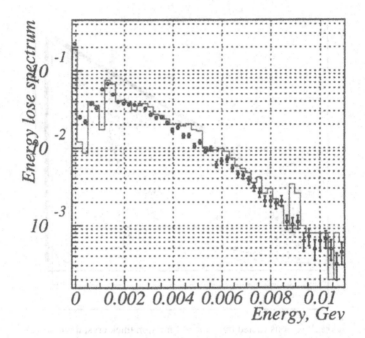

Figure 3. Signal from the preshower for 4 mm thick crystal and 6 GeV.

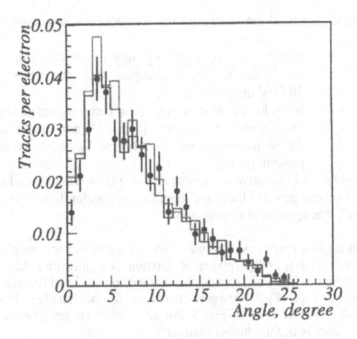

Figure 4. Angular distributions (simulations and experiment) for 20 mm crystal and 10 GeV Points: experiment. Histograms: simulations; blue: SGC, red: GEANT.

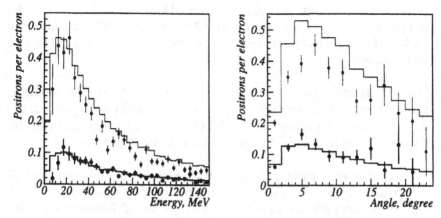

Figure 5. Positron energy and angular distributions for a 4 mm thick crystal and 10 GeV. Blue: crystal. Black: amorphous. Points: experiment; histogram: simulation.

Crystal 8 mm thick and compound target; At 10 GeV incident energy, the observed enhancement in positron yield is slightly larger than 2 for an incident energy of 10 GeV. Comparison of the 8 mm crystal target with a "compound" target made of a 4 mm crystal followed by a 4 mm amorphous disk shows very close results at 6 GeV (Fig. 6). This is showing that after 4 mm in the crystal, the processes are the same as in an amorphous medium. The same observations can be made at 10 GeV incident electron energy.

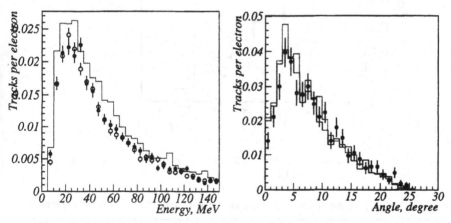

Figure 6. Positron energy distribution for 8 mm thick crystal and compound target at 6 GeV. Empty circles: compound target; filled circles: 8 mm crystal Histogram: simulations.

(c) Distribution in transverse momentum: the transverse acceptance of the positron matching systems being defined by the maximum transverse momentum, it is interesting to consider the transverse momentum distribution of the measured positrons. This is illustrated on Fig. 7

corresponding to 8 mm crystal target with 10 GeV as incident energy. We can see that we have a peak around 3 MeV/c and a FWHM width of 6 MeV/c. Existing matching systems (adiabatic devices) are presenting such acceptances.

(d) Positron yields: given the maximum total and transverse momenta accepted by a matching system put after the positron target, we can determine the accepted positron yield. On Table 1, we give the yields for the case of 8 mm crystal target and 10 GeV, electron energy.

(e) Enhancement of positron production with energy using oriented crystals. The possibility to raise the energy, for the incident electrons, up to 40 GeV (and more) allowed us to make a comparison of the positron yields - measured on the side positron counters – for the three incident energies: 10, 20 and 40 GeV. This can be seen on the rocking curves of Fig. 8.

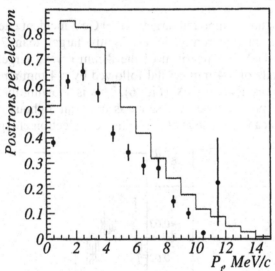

Figure 7. Transverse momentum distribution for 8 mm thick crystal and 10 GeV. Points: experiment; histogram: simulations.

Table 1. Measured positron yields in (pl, pt) acceptance domains for the WA 103 experiment

	$5 < p_l < 25$	$5 < p_l < 30$	$5 < p_l < 40$
$p_t < 4$	1.16 ± 0.04	1.28 ± 0.04	1.43 ± 0.04
$p_t < 6$	1.66 ± 0.05	1.85 ± 0.05	2.13 ± 0.05
$p_t < 8$	2.11 ± 0.07	2.46 ± 0.08	2.90 ± 0.08
$p_t < 10$	2.31 ± 0.08	2.75 ± 0.08	3.32 ± 0.08
$p_t < 12$	2.40 ± 0.08	2.94 ± 0.09	3.67 ± 0.10

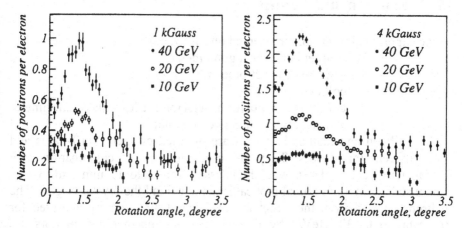

Figure 8: Rocking curves measured on the positron counters for 8 mm crystal and two values Of the magnetic field: 1 and 4 kGauss.

3. KEK EXPERIMENT

3.1 Motivations

Tungsten crystal based positron sources have also been investigated at KEK [11]. The experimental set-up, with a different conception from that of CERN, allowed the measurement of positron yields for different kinds of crystals using an electron *beam* rather intense (~10^9 e-/bunch).

Intense photon generation by channelled electrons can be operated also in light crystals as shown in the first proposition [2]. Instead of using tungsten, lighter crystals as silicon or diamond may be used. Theoretical considerations by Baier, Katkov and Strakhovenko [3] showed the interest to use Diamond, for instance, to get a significant enhancement in the energy radiated and in the photon yield at optimum thickness. If, after these authors, we represent the *efficiency of radiation* as the ratio Lrad/Lch where Lrad and Lch are the radiation lengths of the amorphous material and of the oriented crystal, respectively, this ratio is of 8.4 for Silicon and 5.4 for tungsten at an energy of 10 GeV, close to the energies used in CERN and KEK. From the point of view of radiation resistance, Diamond would be more interesting than tungsten [15] On the other hand, promising results were obtained at Yerevan with diamond crystals using a 4.5 Gev electron beam [16] Using the same experimental apparatus for all the targets, results have been obtained on tungsten, Silicon and Diamond crystals. Some emphasis will be put on the latter.

3.2 Experimental Set-up

The experimental set-up, represented on Fig. 9, comprises:
- a target holder in a goniometer,
- a magnetic spectrometer,
- collimators,
- two kinds of positron detectors: a lead-glass calorimeter and an acrylic Cherenkov counter.

The silicon and diamond crystals are oriented on their <110> axis and have different thicknesses. They are used alone or in combination with amorphous tungsten plates, with thicknesses from 3 to 15 mm with 3 mm steps installed on a horizontal moveable stage behind the crystal targets. The positrons emitted from the target in the forward direction are analysed for momenta up to 30 MeV/c by the 60° bending magnet. Collimators are installed before the detectors. The acceptance of the measuring system (from 1 to 9.10^{-4} MeV/c.sr, for momenta from 5 to 30 MeV/c) is rather small. No measurements on photons are foreseen.

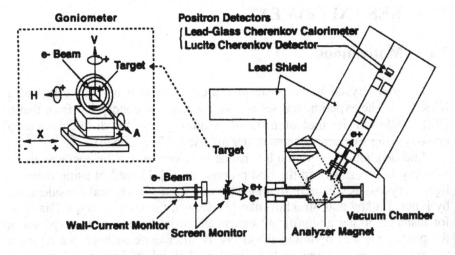

Figure 9: Channelling at KEK: the set-up.

3.3 Results

3.3.1 Tungsten Crystals

Experiments with 2.2, 5.3 and 9 mm thick tungsten crystals oriented on their <111> axis have been worked out using 4 and 8 GeV incident beams. The results show:
- rocking curves widths (FWHM), much larger than the channelling critical angle : crystal effects are present beyond the critical angle,

- positron yield enhancements (crystal/amorphous) for the same thickness going from 5 to 3 and 1.8 when the crystal thickness is increasing from 2.2 to 5.3 and 9 mm, at an incident energy of 8 GeV and for 20 MeV/c positrons. The enhancements are decreasing when the incident energy does. They decrease about 25 % when the energy decreases from 8 to 4 GeV.
- the enhancements in positron yield observed on tungsten crystals at KEK and CERN experiments are coherent between them, taking into account the scaling laws with incident energy and crystal thickness. Even if the angular acceptances of the two positron detectors are very different, the enhancements found remain coherent.
- the experimental results for tungsten crystals at CERN and KEK have been compared to simulations worked out by V.Strakhovenko [17]. Good agreement was found.

3.3.2   Silicon and Diamond Crystals

Experiments with Silicon crystals 9.9, 29.9, and 48.15 mm thick and a 4.57 mm thick Diamond crystal were carried out at KEK. Angular scanning has been performed for all the crystals. As an example, we show on Fig. 10, the scanning operated for 20 MeV/c positrons for the Diamond crystal and the 30 mm thick Silicon crystal. Peaks indicate the <110> axis for both crystals [12]. Mini peaks correspond to the passage by planar orientation [12].

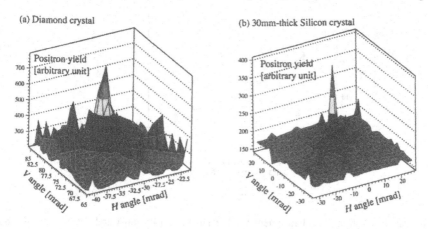

Figure 10. 2-D angular scanning for a 4.5 mm Diamond crystal (a) and a 30 mm Si crystal (b).

Rocking curves: examples of rocking curves obtained for 20 MeV/c positrons at 8 GeV are shown on Fig. 11. It is noticeable that the width FWHM is much larger than the channelling critical angle: 1.4 and 2.4 mrad for diamond and silicon, respectively, compared to Ψ_c = 0.13 and 0.17 mrad, also respectively.

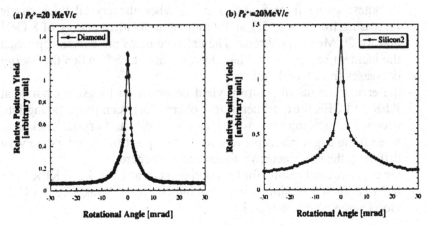

Figure 11. Rocking curves for the 4.5 mm Diamond and the 30 mm Si crystals.

Enhancement in positron yield: the enhancement is measured on the rocking curve between the peak yield and the yield 50 mrad apart from the crystal axis. These enhancements are shown on Fig. 12 for both crystals which thickness is expressed in terms of Xo (123 mm for diamond, 93.6 mm for silicon).

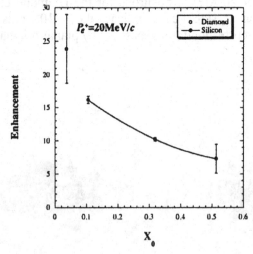

Figure 12. Positron yield enhancement for the Diamond (4.5 mm) and the Si (10, 30 and 50 mm) crystals.

For *combined* targets associating crystals as radiators and amorphous tungsten as *converters*, we can observe a comparison for the enhancement between different crystals on Fig. 13; the total thickness (crystal +amorphous) is put on abscissa.. we can observe that for the light crystals the enhancement (with respect to Bethe-Heitler value) is larger than 1 but smaller than for the "heavy" crystal as tungsten.. The probable reason is that the positrons created by photons generated in light crystals have lower energy due to softer photons than in tungsten crystals (due to smaller potential well) and are more easily scattered and henceforth, less captured in the small acceptance detection system.

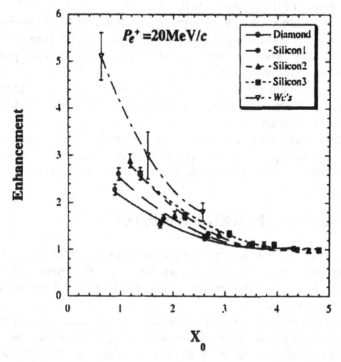

Figure 13. Positron yield enhancements for the combined targets of KEK.

4. SUMMARY AND CONCLUSIONS

4.1 Concerning the CERN Experiment

- A clear enhancement in *photon* and *positron* production is observed for all the crystals aligned on their <111> axis, when they are compared to amorphous targets of the same thickness. The enhancement factors are slightly larger than four and two, respectively, for 4 and 8 mm thick W crystal submitted to 10 GeV electron beam. At 6 GeV, the enhancements are slightly lower.

- Good agreement is found between the results of a 8 mm crystal and a "compound" target made of 4 mm crystal followed by 4 mm amorphous target. This result confirms that crystal effects are essentially present in the first millimetres of the crystal leading to a separation between the *radiator* and the *converter*. Such assembling is interesting to limit the energy dissipation in the radiator, preserving high values of the potential well.
- A large number of *soft* photons is created due to channelling and coherent bremsstrahlung; these photons generate soft positrons, easily accepted by known matching systems which collect preferably positrons up to 30 MeV/c.
- A good agreement is found between simulations and experiment validating these simulations based on crystal processes and allowing further predictions.
- The data collected in this experiment allows 2-D reconstruction in the phase space (p_L, p_T) and positron yield determination in given acceptance domain in longitudinal and transverse momenta.

4.2   Concerning the KEK Experiment

- The results on tungsten crystals are in coherence with those of CERN taking into account scaling laws for target thickness and incident energy. This is true though the acceptance of the two detection systems are quite different.
- The results on light crystals (Silicon and Diamond) show promising results. Real phase space of the produced positrons may be reached through modification of the positron detector towards a larger acceptance.
- Experimental results compared to simulations [17] are in good agreement within the experimental accuracy

4.3   Concerning both Experiments

- Some complementaries are found between different thicknesses and different incident energies, concerning tungsten crystals.
- The KEK set-up is relatively simple and provides rapid indications on yields, enhancements. The measurements being direct, the analysis is straightforward.
- However, the KEK apparatus, due to the very small acceptance does not give indications on the positron phase space.
- The CERN apparatus is conceived to proceed to track reconstruction which is a more complicated task and time demanding

- The large horizontal acceptance of the CERN detection system allows full reconstruction of the positron phase space and precise estimations of the yield in given acceptance conditions.

(a)   CERN co-authors: X. Artru, V. Baier, K. Beloborodov, A.Bogdanov, A.Bukin, S.Burdin, R. Chehab, M. Chevallier, R. Cizeron, D. Dauvergne, T. Dimova, V. Druzhinin, M. Dubrovin, L. Gatignon, V.Golubev, A. Jejcic, P. Keppler, R. Kirsch, V. Kulibaba, Ph. Lautesse, J. Major, J-C. Poizat, A. Potylitsin, J. Remillieux, S. Serednyakov, V. Shary, V. Strakhovenko, C. Sylvia

(b)   KEK co-authors: M. Satoh, T.Suwada, K. Furukawa, T. Kamitani, T. Sugimura, K. Yoshida, H. Okuno, K. Umemori, R. Hamatsu, J. Hara, A. Potylitsin, I. Vnukov, R. Chehab

References

[1] V. Balakin, A. Mikhailichenko Preprint INP 79-85, Novosibirsk, 1979

[2] R. Chehab et al., Proceedings of IEEE PAC, March 1989, Chicago, IL

[3] V.N. Baier, V.M. Katkov, V.M. Strakhovenko Phys.Stat.Solid (b) **133** (1986) 583

[4] V.N. Baier, V.M. Katkov, V.M. Strakhovenko "Electromagnetic proceses at high energies in oriented single crystals" World Scientific Ed. Singapore, 1998

[5] J.F. Bak et al., Nucl.Phys.B **302** (1988) 525

[6] R. Medenwalt et al., Phys.Letters B **242** (1990) 517

[7] X. Artru et al., Nucl.Instr.and Meth. B **119** (1996) 246

[8] R. Chehab et al., Phys.Letters B **525** (2002) 41

[9] X. Artru et al., Nucl.Instr.and Meth.B **201** (2003) 243

[10] K. Yoshida et al., Phys.Rev.Letters **80** (1998) 1437

[11] T. Suwada et al., Phys.Rev. E **67** (2003) 016502

[12] M. Satoh et al., "Experimental study of positron production from silicon and diamond crystals by 8-GeV channeling electrons" To appear in Nucl.Instr.and Meth. B

[13] V. Bellini et al., Nucl.Instr.and Meth. A **386** (1997) 254

[14] V.M. Strakhovenko SGC (Shower Generation in Crystal).
    See V.N. Baier, V.M. Katkov, V.M. Strakhovenko Nucl.Instr.meth. **B103** (1995) 147

[15] X. Artru, V.Baier, R. Chehab, A. Jejcic Nucl.Instr.and Meth. A **344** (1994) 443

[16] R. Avakian et al., Sov.Tech.Phys.Letters **14** (5) 2003

[17] V. Baier, V. Strakhovenko Phys.Rev Special Topics –AB **5** (2002) 121001

SOME RESULTS ON MICROANALYSIS OF GOLD USING MICRO-XRF AT THE ANKA - KARLSRUHE SYNCHROTRON RADIATION FACILITY

Bogdan Constantinescu[1], Roxana Bugoi[1], Rolf Simon[2] and Susanne Staub[2]

[1]National Institute of Nuclear Physics and Engineering "Horia Hulubei", Bucharest, Romania; [2]Institute for Synchrotron Radiation, Forschungszentrum Karlsruhe GmbH, , Germany

Abstract: In the present paper, the possibility to use Synchrotron Radiation – based micro X-Ray Fluorescence (XRF) method to study the micro-inclusions of Platinum Group Elements (PGE) and other high temperature melting point metals in gold archaeological objects is demonstrated. The analyzed samples belonged to different pieces of the Pietroasa hoard. The presence of Ta inclusions was determined indicating the Ural Mountains as a source for the gold ore used to manufacture the objects. Conclusions on how to proceed for a complete examination of different groups of elements (Ir, Os, Ta – Nb, Rh, Ru, Pd – Sn, Sb, Te) are also presented.

Keywords: Synchrotron Radiation; micro X-Ray Fluorescence; Platinum Group Elements; tantalum; micro-inclusions; archaeological gold; Pietroasa hoard; Ural Mountains.

1. THE STORAGE RING ANKA

On the campus of the Forschungszentrum Karlsruhe (Research Center Karlsruhe) the synchrotron ANKA [1] is located in a hall of 3600 m² surface. Its diameter is 35 m with a circumference 110 m. The electrons are produced in the injector on a cathode similar to a television tube and pre-accelerated by a microtron to 56 keV. After the transfer into the booster synchrotron they reach an energy of 500 MeV with which they are injected into the actual

H. Wiedemann (ed.), Advanced Radiation Sources and Applications, 373–380.

storage ring. There, the acceleration to the final energy of 2.5 GeV is made attaining a maximum current of 200 mA. The average lifetime of the electron ray is above 20 hours. Within this time the experiments can be carried out. The ring contains 16 bending magnets and four straight sections scheduled for the installation of insertion devices. To make use of the produced radiation so-called beamlines are added to the storage ring. They lead through an optical hutch where the characteristics of the radiation can be influenced by X-ray mirrors, monochromators and other similar devices. Prepared in such a manner, the radiation is finally used for the actual investigation in the experimental hutch.

Currently ten of these beamlines are available for the following types of examination:
- Absorption
- Diffraction
- Topography
- Fluorescence
- Protein crystallography
- MPI-MF surface diffraction
- Infrared
- Lithography (3 beam lines)
- The ANKA Storage Ring characteristics are the following:

      Energy: 2.5GeV   Max. Current: 0.4A
      Circumference: 110.4m   Structure: 8 DBA
      Emittance: 80 nm-rad
- The ANKA Booster Synchrotron characteristics are:

      Energy: 0.5GeV   Current: 10mA
      Circumference: 26.4m   Structure: 4FODO
- The ANKA Race Track Microtron characteristics are:

      Energy:0.05GeV   Current:10mA

2.   X-RAY FLUORESCENCE ANALYSIS USING SYNCHROTRON RADIATION

In X-ray fluorescence analysis (XRF) the atoms of the sample are excited by means of X-radiation and emit their characteristic X-radiation. The latter is analyzed with regard to its energy or wavelength. Thus the elements contained in the sample can be determined. By means of mathematic modeling, the concentrations of the analyte can be determined from the measured intensities. This method is non-destructive and does not take much effort for the preparation of solid, liquid or pasty samples.

Advantages of Synchrotron Radiation based XRF are the following [2]:

- high intensity
- absolute detection limits in the range of some fg for micro-XRF (elements Na-U) with TXRF down to 10^8 atoms/cm^2
- specific application of monochromatic and polychromatic radiation
- investigation of smallest samples or measurements with the highest lateral resolution through good radiation focusing (5 µm)
- measurements with micro-XRF and total reflection geometry (TXRF) without sample modification
- improved signal-to-noise ratio by radiation polarization
  Typical synchrotron applications of micro-XRF at ANKA are:
- Determination of element distribution in single hairs
- Impurities on semiconductor surfaces
- Element distribution patterns of fluid inclusions in the dimension of 1 mm
- Examinations of aging effects on marked vellum
- Spatially resolved measurements of element concentration in single cells

3. THE ARCHAEOLOGICAL PROBLEM

The study of trace-elements in archaeological metallic objects can provide important clues about the metal provenance and the involved manufacturing procedures, leading to important conclusions regarding the commercial, cultural and religious exchanges between the antique populations [3].

Ancient metallic materials are usually inhomogeneous on a scale of 20 microns or less: they contain remains of imperfect smelting (segregated phases in alloys) and inclusions. Due to their exceptional chemical stability, gold artifacts remain almost unchanged during weathering and aging processes. Gold is usually alloyed with silver as electrum. Gold owes its significance to two important properties: its resistance to corrosion and its extraordinary malleability.

Inclusions of platinum and platinum-group elements (PGE) – Ru, Rh, Pd, Os, Ir and Pt - in gold were released into rivers by the decomposition of rocks and occur in placer deposits in the form of grains and nuggets of complex alloys. Due to their high density compared with gold, they would first occur as hard white spots in gold, and would be a problem to the early smith, since the refining techniques available by that time had little effect upon them. The presence of PGE as inclusions in gold objects can constitute a fingerprint for the ore that the object was manufactured from. The melting point of PGE is very high (even compared with gold melting point); thus, PGE grains remain unchanged during the metallurgical processing of the gold ore. Apart from PGE inclusions, gold alloys can contain low amounts of trace elements, such as Cu, Te, Cr, Nb, Ta, etc., which can be potential fingerprints for base metal deposits.

The hoard discovered in 1837 at Pietroasa, Buzău county, Romania, by two peasants entered the historical literature and the history of arts as "Cloşca cu Puii de Aur" ("The Golden Brood Hen with its Chickens") [4]. This hoard (Fig. 1) is one of the most famous collections of archaeological objects ever found in Romania, due to its fine artistic quality and to the

Figure 1. The Pietroasa hoard.

myths created around it. It is supposed that this treasury belonged to Germanic populations of Visigoths, living in the period of the IVth-V^{th} Century A.D. on the actual Romanian territory. The purpose of this Pietroasa hoard study was to obtain relevant information about the metal provenance. Trace elements and PGE, correlated with known mines fingerprints can help to identify the ore source. The gold provenance of the objects can lead to further historical conclusions. From Pietroasa treasury five small pieces from five different objects are to be analyzed: the large, middle and small fibulae, the dodecagonal basket and a piece of the patera - to be more specific, the central figure representing Cybele.

Due to the exceptional value of the artifacts and to the fact that the micro-XRF study can be carried out on reduced dimension samples, fragments of the original objects were taken. All these samples were small in size (a few square millimeters in area), being obtained by mechanically cutting the artifacts. Cautions were taken in order to obtain the samples from the unimportant, but original zones of the objects, to avoid the deterioration of these precious museum artifacts.

4.   EXPERIMENTAL RESULTS

During the preliminary micro-XRF experiment performed in January 2004, the optimal conditions to detect trace elements which are fingerprints for the natural gold sources: micro-inclusions of Platinum Group Elements (PGE) or high-temperature melting point elements (Ta, Nb, Cr) were established. The archaeological and mineralogical samples were put on two holders (kapton tape as backing on aluminum frame).

At the beginning, it was used for excitation a white beam (broad range of energies), with a beam size of 0.1×0.1 mm. As it can be seen from figure 2 (an XRF spectrum from a sample belonging to the big Germanic style bird type fibula of the hoard), the contribution of Au L rays was huge, covering practically all the trace elements, except for the minor elements Ag (L rays) and Cu (K rays), which are main components of gold alloys.

As a consequence, it was decided to use monochromatic excitation with a value just below the absorption edge of Lα Au=11.7 keV, starting with a value of 11.5 keV. Considering that the Au contribution was still too high, this value was shifted further down to 11.3 keV. The improvement of the detection limits was evident – see figure 3 (a spectrum from the same sample, but using monochromatic excitation), where one can be observe the diminution of Au lines and the presence of various trace-elements (Ta, Ni, Fe, Mn, Cr).

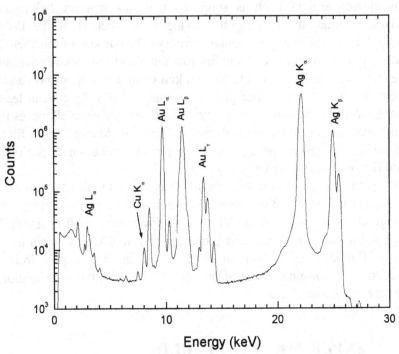

Figure 2. Small fibula fragment XRF spectrum - white beam excitation

The total spectra for three natural gold samples from Transylvania were also acquired; due to the strong contribution of Ag L lines (Transylvanian gold contains up to 25% Ag), the detection of characteristic fingerprints as Sn, Sb, Te was not possible using 11.3 keV as exciting X-ray energy.

Figure 3 demonstrates that 11.3 keV excitation energy is suitable to detect PGE with Z just below the one of gold (Pt, Ir, Os, Ta). However, this energy is too high for the detection of PGE in the fifth period (Ru, Rh, Pd) and Nb (this element always accompanies Ta in minerals). Because it is too far from their absorption edges (L lines) and the gold M lines (2.2 keV) and silver L lines (3.2 keV) dominates this region of the spectrum covering the other elements.

Analyzing the elemental maps obtained using a 20 μm capillary and a sample-detector distance of 15 mm, for the large and small fibulae, inclusions of Ta, Nb and Cr were clearly found. Both Ta and Cr have a high melting point, and they resist the gold processing techniques. The combination Ta – Nb (tantalite – columbite) and Cr in the presence of gold deposits can be considered as fingerprints for gold ores from Ural Mountains (Southern region from Perm to Tchelyabinsk). It follows, that the Germanic 'owners' of the treasuries were coming from the region between Caucasus and Ural Mountains in the second half of the IIIrd Century A.D., bringing along their precious jewelry.

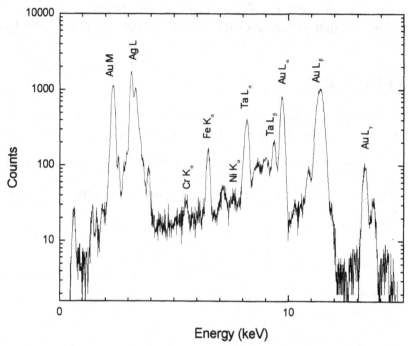

Figure 3. Small fibula XRF spectrum - monochromatic (11.3 keV) excitation.

5. CONCLUSIONS

Micro-XRF method using Synchrotron Radiation proved to be a suitable analytical method for the study of micro-inclusions in gold. The present experiment proved that:

- to detect the trace elements in the region of elements with Z just below the one of gold (Pt, Ir, Os, Ta), 11.3 keV as a value for the excitation energy is a good choice;
- for the fifth period trace-elements (PGE and Nb), an excitation energy value around 2-3 keV is a better choice (e.g., a value of 2.5 keV);
- for the fingerprints of Transylvanian gold (Sn, Sb, Te), an excitation energy of 3.5 keV must be chosen in order to obtain good detection limits;
- for Ag region (from Nb to Te) one can also try to detect the K lines using a higher energy value, around 24 keV.

ACKNOWLEDGEMENTS

The financial support of the Institute for Synchrotron Radiation, Forschungszentrum Karlsruhe GmbH, Germany is gratefully acknowledged.

REFERENCES

1. ANKA Beam line book; http://hikwww1.fzk.de/anka/
2. R. Simon, G. Buth, M. Hagelstein Nucl. Instr. and Meth. in Phys. Res. B **199** 554-558 (2003).
3. R. F. Tylecote, *The early history of metallurgy in Europe* (Longman, London and New York, 1987).
4. Al. Odobescu, *Le tresor de Petrossa: Etudes sur l'orfevrerie antique* (Paris-Leipzig, 1889 – 1890).

INELASTIC PHOTON-ELECTRON COLLISIONS WITH POLARIZED BEAMS

S.R. Gevorkyan, [1] E.A.Kuraev [2]

[1] *Joint Institute for Nuclear Research,141980, Dubna,Russia*

[2] *Yerevan Physics Institute, 375036, Yerevan, Armenia*

Abstract   We discussed the photoproduction of charged particles in pairs $a\bar{a}$ ($a = e, \mu, \pi$) as well as double photon emission processes off an electron accounting for the polarization of colliding particles. In the kinematics when all the particles can be considered as massless, we obtain compact analytical expressions for the differential cross sections of these processes. These results are applied to special cases of pair production by circular and linear polarized photons.

We consider the lowest order inelastic QED processes in photon-electron interaction at high energies.We take into account the polarization of colliding photons and electron and consider the experimental setup,when the polarization of final particles is not measured.The interest to such kind processes is twofold. Linear e^+e^- high energy colliders(planned to be arranged [1, 2]) provide the possibility (using the backward laser Compton scattering) to obtain the high energy photon-electron colliding beams. The problem of calibration as well as the problem of important QED background are to be taken into account for this kind of colliders.The second important reason to investigate such reactions is the well known possibility to use the photoproduction of leptons pairs as a polarimeter process(see,for instance, [3] and references therein)

Despite the fact that at high energy the bulk of particles are produced at very small angles, experimentally it is much easy to detect the particles produced at large angles.Thus we investigate the kinematic in which all invariants determining the matrix elements of the processes under consideration will be much bigger than masses of particles involved in the reaction.

H. Wiedemann(ed.), Advanced Radiation Sources and Applications, 381–386.

We consider the following set of inelastic reaction

$$\gamma(k, \lambda_\gamma) + e(p, \lambda_e) \;\rightarrow\; e(p', \lambda'_e) + a(q_-, \lambda_-) + \bar{a}(q_+, \lambda_+); \quad a = e, \mu, \pi$$

$$\gamma(k, \lambda_\gamma) + e(p, \lambda_e) \;\rightarrow\; e(p', \lambda_e\prime) + \gamma(k_1, \lambda_1) + \gamma(k_2, \lambda_2) \tag{1}$$

Here λ_i are the particle helicities. We will work in the kinematic when all the 4-vector scalar products defined by:

$$
\begin{aligned}
s &= 2pp', & s_1 &= 2q_-q_+, & t &= -2pq_-, & t_1 &= -2p'q_+, \\
u &= -2pq_+, & u_1 &= -2p'q_-, & \chi &= 2kp, & \chi' &= 2kp', \\
\chi_j &= 2k_j p, & \chi'_j &= 2k_j p', & j &= 1, 2
\end{aligned}
\tag{2}
$$

are large compared with all masses:

$$s \;\sim\; s_1 \sim -t \sim -t_1 \sim -u \sim -u_1 \sim \chi_j \sim \chi'_j \gg m^2;$$

$$p^2 \;=\; p'^2 = q_\pm^2 = k^2 = k_j^2 - 0. \tag{3}$$

To obtain the cross sections of above reactions it is convenient to work with helicity amplitudes of corresponding processes [4]. The helicity amplitudes $M^{\lambda_- \lambda_+ \lambda'}_{\lambda_\gamma \lambda}$ for lepton pair photoproduction, $M^{\lambda'}_{\lambda_\gamma \lambda}$ for a pair of charged pion production and $M^{\lambda_1 \lambda_2 \lambda'}_{\lambda_\gamma \lambda}$ for a two photon final state are defined as a usual matrix elements calculated with chiral states of photons and leptons.

The square of matrix element summed over the spin states of the final particles have the form of the conversion of the chiral matrix with the photon density matrix

$$\sum |M|^2 = \tfrac{1}{2} Tr$$

with the photon polarization vector $\vec{\xi} = (\xi_1, \xi_2, \xi_3)$ parameterized by Stokes parameters fulfilling the condition $\xi_1^2 + \xi_2^2 + \xi_3^2 \leq 1$.

The matrix elements of the chiral matrix m_{ij} are constructed from the chiral amplitudes of the process $M^{\lambda_- \lambda_+ \lambda'}_{\lambda_\gamma \lambda_e}$ as

$$
m_{11} = \sum_{\lambda_- \lambda_+ \lambda'} \left| M^{\lambda_- \lambda_+ \lambda'}_{++} \right|^2, \quad
m_{22} = \sum_{\lambda_- \lambda_+ \lambda'} \left| M^{\lambda_- \lambda_+ \lambda'}_{-+} \right|^2,
$$

$$
m_{12} = \sum_{\lambda_- \lambda_+ \lambda'} M^{\lambda_- \lambda_+ \lambda'}_{++} \left(M^{\lambda_- \lambda_+ \lambda'}_{-+} \right)^*, \quad
m_{21} = m_{12}^*.
\tag{4}
$$

We put here only half of all chiral amplitudes which correspond to $\lambda_e = \frac{1}{2}$. The other half can be obtained from these ones by a space parity operation.

The helicity amplitudes can be obtain using the relevant representations for photon polarization vector and Dirac spinors [5]. As a result one obtains for the elements of chiral matrix for the processes under consideration the following expressions:

1. Photoproduction of muon pair

$$m_{11} = \frac{2w}{ss_1}(u^2 + t^2), \quad m_{22} = \frac{2w}{ss_1}(u_1^2 + t_1^2),$$

$$m_{12} = -\frac{4(w_1 - iAw_2)}{(ss_1)^2}\left[(ss_1)^2 + (tt_1)^2 + (uu_1)^2 - 2tt_1uu_1 \right.$$
$$\left. - ss_1(tt_1 + uu_1) + 4i(tt_1 - uu_1)A\right] \quad (5)$$

where

$$A = \epsilon_{\mu\nu\rho\sigma}q_+^\mu q_-^\nu p^\rho p'^\sigma,$$

$$w = -\left(\frac{q_+}{kq_+} - \frac{q_-}{kq_-} + \frac{p}{kp} - \frac{p'}{k.p'}\right)^2,$$

$$w_1 = \frac{w}{2} - \frac{4s_1}{\chi_+\chi_-} + \frac{\chi\chi'}{4s}\left(\frac{w}{2} - \frac{2s}{\chi\chi'} - \frac{2s_1}{\chi_+\chi_-}\right)^2, \quad (6)$$

$$w_2 = \frac{2(\chi_+ + \chi_-)}{s\chi_+\chi_-}\left[\frac{w}{2} + \frac{2s}{\chi\chi'} - \frac{2s_1}{\chi_+\chi_-}\right]$$

2. Electron pair photoproduction

$$m_{11} = \frac{2w}{ss_1tt_1}\left(t^3t_1 + u^3u_1 + s^3s_1\right),$$

$$m_{22} = \frac{2w}{ss_1tt_1}\left(t_1^3t + u_1^3u + s_1^3s\right), \quad (7)$$

$$m_{12} = \frac{4(w_1 - iAw_2)}{(ss_1tt_1)^2}\left[(uu_1)^2 - (ss_1)^2 - (tt_1)^2\right]$$
$$\times \left\{\tfrac{1}{2}\left[(uu_1)^2 + (ss_1)^2 + (tt_1)^2 - 2uu_1(ss_1 + tt_1)\right]\right.$$
$$\left. + 2iA(uu_1 - ss_1 - tt_1)\right\}$$

3. Photoproduction of pions

$$m_{11} = \frac{w}{2}tu, \quad m_{22} = \frac{w}{2}t_1u_1, \quad (8)$$

$$m_{12} = \frac{w_1 - iAw_2}{ss_1}\left[\tfrac{1}{2}(uu_1 - tt_1)^2 - ss_1(uu_1 + tt_1) + 2i(tt_1 - uu_1)A\right]$$

The differential cross section of any pair production has the form

$$\frac{d\sigma}{d\Gamma} = \frac{\alpha^3}{2\pi^2\chi}[m_{11} + m_{22} + \xi_2\lambda_e(m_{11} - m_{22})$$
$$- 2\xi_3\text{Re}(m_{12}) + 2\xi_1\text{Im}(m_{12})],\qquad(9)$$

$$d\Gamma = \frac{d^3p'}{\epsilon'}\frac{d^3q_-}{\epsilon_-}\frac{d^3q_+}{\epsilon_+}\delta^4(p + k - p' - q_+ - q_-)\qquad(10)$$

where ξ_i and λ_e are Stocks parameters and target electron helicity.

The expressions (6)-(10) allows one to calculates the differential cross section of the processes of pair photoproduction off an electron with any polarization of colliding particles. Now it is a simple task to obtain from this expressions the particular cases.

Let us consider the charged particles pair production with circular and linear polarization of photon from unpolarized target. The differential cross section for pair production by circularly polarized photons up to the factor ξ_2-the degree of circular (left or right) photon polarization coincide with the cross section in unpolarized case:

$$\frac{d\sigma_{L(R)}}{d\Gamma} = \frac{\alpha^3}{\pi^2\chi}\xi_{2L(2R)}wZ_{a\bar{a}},\qquad(11)$$

$$Z_{e\bar{e}} = \frac{ss_1(s^2 + s_1^2) + tt_1(t^2 + t_1^2) + uu_1(u^2 + u_1^2)}{ss_1tt_1},$$

$$Z_{\mu\bar{\mu}} = \frac{t^2 + t_1^2 + u^2 + u_1^2}{ss_1}, \quad Z_{\pi\bar{\pi}} = \frac{tu + t_1u_1}{ss_1}$$

Two of these quantities $Z_{e\bar{e}}, Z_{\mu\bar{\mu}}$ can be obtained from the relevant quantities obtained in papers [4] applying the crossing symmetry transformation.

Much more interesting is the case of pair photoproduction by linear polarized photons for which the Stokes parameters are $\xi_2 = 0, \xi_1^2 + \xi_3^2 = 1$. In this specific case the square of full matrix element can be represented as the sum of diagonal term and non diagonal one, where the polarization parameters enter only the non diagonal term in the following form

$$\text{Re}\left[(\xi_3 + i\xi_1)(w_1 - iAw_2)(T_1 + iAT_2)\right] =$$
$$\xi_3\left(w_1T_1 + w_2T_2A^2\right) + \xi_1\left(w_2T_1 - w_1T_2\right)A$$

where the structures $T_{1(2)}$ depend on the type of created pair:

1. For the case of $e^- e^+$ production

$$T_1 = 4 + 8\frac{(uu_1 - ss_1)(ss_1 + tt_1)}{ss_1 tt_1} - 4\left(\frac{tt_1 - uu_1}{ss_1}\right)^2,$$

$$T_2 = 8\left(\frac{tt_1 - uu_1}{s^2 s_1^2} + \frac{1}{ss_1} - \frac{2}{tt_1}\right) \tag{12}$$

2. In the case of $\mu_+ \mu_-$ pair production

$$T_1 = 8\frac{tt_1 + uu_1 - ss_1}{ss_1} - 8\left(\frac{uu_1 - tt_1}{ss_1}\right)^2,$$

$$T_2 = 16\frac{tt_1 - uu_1}{(ss_1)^2} \tag{13}$$

3. For the case of pion pair production:

$$T_1 = 4\left(\frac{uu_1 - tt_1}{ss_1}\right)^2 - 4\frac{uu_1 + tt_1}{ss_1},$$

$$T_2 = \frac{uu_1 - tt_1}{(ss_1)^2}. \tag{14}$$

The same consideration can be done for double Compton effect. The differential cross section for this reaction in the general case, when the both primary particles are polarized can be cast in the following form:

$$\frac{d\sigma}{d\Gamma_\gamma} = \frac{\alpha^3 s}{(2\pi)^2 D}\left\{\chi\chi'(\chi^2 + \chi'^2) + \chi_1\chi_1'(\chi_1^2 + \chi_1'^2) + \chi_2\chi_2'(\chi_2^2 + \chi_2'^2)\right.$$
$$+ 4\xi_1 A(\chi_1\chi_2' - \chi_2\chi_1') - \xi_3(\chi_1\chi_2' + \chi_2\chi_1')(\chi\chi' - \chi_1\chi_1' - \chi_2\chi_2')$$
$$\left.+ \lambda_e\xi_2[\chi\chi'(\chi'^2 - \chi^2) + \chi_1\chi_1'(\chi_1^2 - \chi_1'^2) + \chi_2\chi_2'(\chi_2^2 - \chi_2'^2)]\right\} \tag{15}$$

with $D = \chi^2\chi'\chi_1\chi_1'\chi_2\chi_2'$ and $A = \epsilon_{\mu\nu\rho\sigma}k_2^\mu k_1^\nu p^\rho p'^\sigma$.

For unpolarized target the cross section of double Compton effect in the case of circularly polarized photon is:

$$d\sigma_{L(R)} = \xi_{L(R)}\frac{2s\alpha^3}{\pi^2 D}Z_\gamma d\Gamma_\gamma, \tag{16}$$

with

$$Z_\gamma = \chi\chi'(\chi^2 + \chi'^2) + \chi_1\chi_1'(\chi_1^2 + \chi_1'^2) + \chi_2\chi_2'(\chi_2^2 + \chi_2'^2),$$

$$d\Gamma_\gamma = \frac{d^3 p_1'}{2\epsilon_1'}\frac{d^3 k_1}{2\omega_1}\frac{d^3 k_2}{2\omega_2}\delta^4(k + p_1 - p_1' - k_1 - k_2).$$

The summed on spin states square of the matrix element of the double Compton scattering process can as well be obtained from the ones for three photon annihilation of electron-positron pair,derived in [4].

For the case of double Compton effect under the linearly polarized photons one can obtain for non diagonal part of cross section which as in above case depend on Stoke's parameters the following expressions:

$$\text{Re}\left[(\xi_3 + i\xi_1)(T_1 + iAT_2)\right] = \xi_3 T_1 - \xi_1 AT_2, \qquad (17)$$

with

$$T_1 = \frac{16s}{\chi\chi_1\chi_2\chi'\chi'_1\chi'_2}(\chi_1\chi'_2 + \chi_2\chi'_1)[\chi_1\chi'_2 + \chi_2\chi'_1 + s(s - \chi + \chi')],$$

$$T_2 = \frac{8s}{\chi\chi_1\chi_2\chi'\chi'_1\chi'_2}(\chi_1\chi'_2 - \chi_2\chi'_1)$$

References

[1] Badelek,B. et al.(2001), TESLA Technical Design Report, Part VI, Chapter.1 ed. Telnov.V, DESY 2001-011, 2001-209, TESLA Report 2001-23

[2] Ginzburg,I.F.,Kotkin,G.L.,Panfil,S.L.,Serbo,V.G., Telnov,V.I.(1984) Colliding Gamma Electron And Gamma Gamma Beams Based On The Single Pass, Nucl. Instrum. Meth. **A219, 5**

[3] Akushevic h,I. V., Anlauf,H., Kuraev,E.A., Ratcli⊖e,P .G., Shaikhatdeno v, B.G.(2000) Triplet pro duction by linearly polarized photons, Phys. Rev.A 61,032703.

[4] de Causmaec ker,P. et al.(1982) Multiple bremsstrahlung in gauge theories at high energies.(1)General formalism for quantum electro dynamics Nucl. Phys. B206,53;
Berends,F. et al.(1982) (2) Single bremsstrahlung ibid.,61

[5] Bartos,E.,Galynskii.M.V.,Gev orky an,S.R.,Kuraev,E.A.(2004) The lowest order inelastic QED pro cesses at polarized photon-electron high energy collisions,Nucl. Phys. B676,390

IMPACT PARAMETER PROFILE
OF SYNCHROTRON RADIATION

Xavier Artru

Institut de Physique Nucléaire de Lyon, Université Claude-Bernard & IN2P3-CNRS, 69622 Villeurbanne, France

Abstract   The horizontal impact parameter profile of synchrotron radiation, for fixed vertical angle of the photon, is calculated. This profile is observed through an astigmatic optical system, horizontally focused on the electron trajectory and vertically focused at infinity. It is the product of the usual angular distribution of synchrotron radiation, which depends on the vertical angle ψ, and the profile function of a caustic staying at distance $b_{\mathrm{cl}} = (\gamma^{-2} + \psi^2)R/2$ from the orbit circle, R being the bending radius and γ the Lorentz factor. The *classical impact parameter* b_{cl} is connected to the Schott term of radiation damping theory. The caustic profile function is an Airy function squared. Its fast oscillations allow a precise determination of the horizontal beam width.

Keywords: synchrotron radiation, beam diagnostics, radiation damping

1.   Theoretical starting point

Classically [1, 2], a photon from synchrotron radiation is emitted not exactly from the electron orbit but at some distance or *impact parameter*

$$b_{\mathrm{cl}} \equiv R_{\mathrm{phot}} - R = \frac{R}{v \cos \psi} - R \simeq \frac{\gamma^{-2} + \psi^2}{2} R, \qquad (1)$$

from the cylinder which contains the orbit. R is the orbit radius, ψ is the angle of the photon with the orbit plane and $\gamma = \epsilon/m_e = (1-v^2)^{-1/2} \ll 1$ is the electron Lorentz factor. In this paper we will assume that the orbit plane is horizontal. Equation (1) is obtained considering the photon as a classical pointlike particle moving along a definite light ray and comparing two expressions of the photon vertical angular momentum,

$$J_z = R_{\mathrm{phot}} \ \hbar \omega \ \cos \psi, \qquad (2)$$

$$J_z = \hbar \omega \ R/v. \qquad (3)$$

H. Wiedemann (ed.), Advanced Radiation Sources and Applications, 387–398.

The first expression is the classical one for a particle of horizontal momentum $k_h = \omega \cos \psi$ and impact parameter R_{phot} with respect to the orbit axis. The second expression comes from the invariance of the system {electron + field} under the product of a time translation of Δt times an azimuthal rotation of $\Delta \varphi = v \, \Delta t / R$.

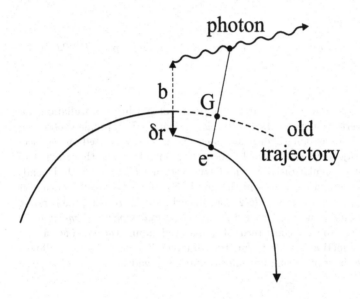

Figure 1.   Impact parameter **b** of the photon and side-slipping $\delta \mathbf{r}$ of the electron.

The impact parameter of the photon is connected, via angular momentum conservation, to a lateral displacement, or *side-slipping*, of the electron toward the center of curvature [1, 2]. The amplitude of this displacement is

$$|\delta \mathbf{r}| = \frac{\hbar \omega}{\epsilon - \hbar \omega} \, b_{\text{cl}} , \qquad (4)$$

so that the center-of-mass G of the {photon + final electron} system continues the initial electron trajectory for some time, as pictured in Fig.1. Whereas b_{cl} is a classical quantity and can be large enough to be observed, $\delta \mathbf{r}$ contains a factor $\hbar$ and is very small: $|\delta \mathbf{r}| \sim (\omega / \omega_c) \, \lambda_C$, where λ_C is the Compton wavelength and $\omega_c = \gamma^3 / R$ the cutoff frequency. Therefore the side-slipping of the electron is practically impossible to detect directly (in channeling radiation at high energy, it contributes to the fast decrease of the transverse energy [3]). However, in the classical limit $\hbar \to 0$, the number of emitted photons grows up like $\hbar^{-1}$ and many small side-slips sum up to a continuous lateral *drift velocity* $\delta \mathbf{v}$ of the electron relative to the direction of the momentum:

$$\delta\mathbf{v} - \frac{\mathbf{p}}{\epsilon} \equiv \mathbf{v} = \frac{2r_e}{3}\,\gamma\,\frac{d^2\mathbf{X}}{dt^2}\,, \tag{5}$$

$r_e = e^2/(4\pi m_e)$ being the classical electron radius[1]. The distinction between $\mathbf{v}$ and $\mathbf{p}/\epsilon$ is illustrated in Fig.2. Equation (5) also results from a suitable definition of the electromagnetic part of the particle momentum [4, 2]. The problematic *Schott term* $(2/3)r_e(d^3X^\mu/d\tau^3)$ of radiation damping can be interpreted [4] as the derivative of the drift velocity with respect to the proper time τ. Thus the measurement of the photon impact parameter would constitute an indirect test of the side-slipping phenomenon and support a physical interpretation of the Schott term.

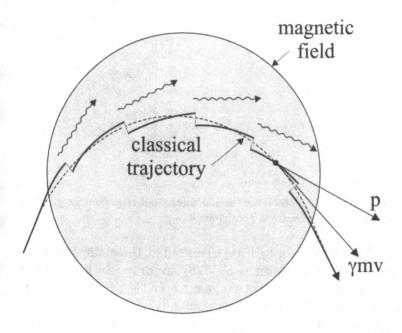

Figure 2. Semi-classical picture of multiple photon emissions. The dashed line represents the classical trajectory. The momentum $\mathbf{p}$ is tangent to the semi-quantal trajectory and not equal to $\gamma m\mathbf{v}$. The classical velocity $\mathbf{v}$ is tangent to the dashed line.

2. Impact parameter profile in wave optics

Using a sufficiently narrow electron beam, the photon impact parameter and its dependence on the vertical angle ψ may be observed through an optical system such as in Fig.3. This system should be *astigmatic*, i.e.

- horizontally focused on the transverse plane P, to see at which horizontal distance from the beam the radiation seems to originate.

- vertically focused at infinity, to select a precise value of ψ.

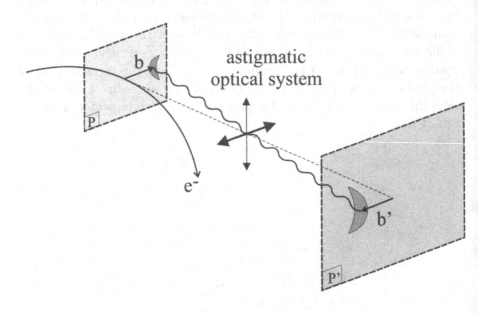

Figure 3. Optical sytem for observing the horizontal impact parameter profile of synchrotron radiation at fixed vertical photon angle.

The horizontal projections of light rays emitted at three different times corresponding to electron positions S_1, S, S_2 are drawn in Fig.4. From what was said before, they are not tangent to the electron beam but to the dashed circle of radius $R + b_{cl}$ at, points T_1, T, T_2. Primed points are the images of unprimed ones by the lens. S_1 and S_2 were taken symmetrical about the object plane P, such that the corresponding light rays come to the same point M' of the image plane P'. When S_1 and S_2 are running on the orbit, M' draws a classical *image spot* on plane P' in the region $x' \geq b'_{cl}$, where $b'_{cl} = G b_{cl}$ and G is the magnification factor of the optics, which from now on we will be taken equal to unity. One can also say that the light rays do not converge to a point but form the *caustic* passing across T'. Classically, the intensity profile in the plane P' behaves like $(x' - b_{cl})^{-1/2}$. Note that if synchrotron radiation was isotropic, emitted at zero impact parameter, and if the optical system had a narrow diaphragm (for the purpose of increasing the depth-of-field), the spot would be located at negative instead of positive x'.

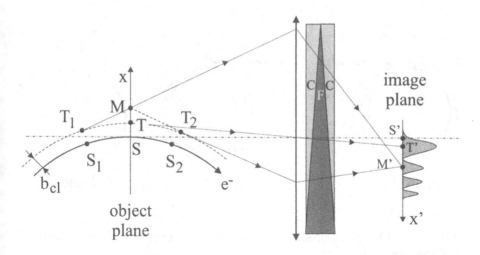

Figure 4. Light rays of synchrotron radiation forming the image of the horizontal impact parameter profile. The profile intensity is schematically represented on the right side. "CFC" is a zero-angle dispersor.

Up to now we treated synchrotron radiation using geometrical optics. However this radiation is strongly self-collimated and the resulting self-diffraction effects must be treated in wave optics. For theoretical calculations, instead of the image spot in plane P', it is simpler to consider its reciprocal image in plane P, which we call the *object spot*. One must be aware that the latter is *virtual*, i.e. it does not represent the actual field intensity in the neighborhood of S. Its intensity is $|\mathbf{E}_{\mathrm{rad}}|^2$ where

$$\mathbf{E}_{\mathrm{rad}} = \mathbf{E}_{\mathrm{ret}} - \mathbf{E}_{\mathrm{adv}} \qquad (6)$$

is the so-called *radiation field* and obeys the source-free Maxwell equations. The distinction between the actual (retarded) field and $\mathbf{E}_{\mathrm{rad}}$ is illustrated in Fig.5.

Using this point of view, one can say that the vertical cylinder of radius $R_{\mathrm{phot}} = R + b_{\mathrm{cl}}$ is the *caustic cylinder* of $\mathbf{E}_{\mathrm{rad}}$ (there is one such cylinder for each ψ). Thus the object spot amplitude can be taken from the known formula [6] of the transverse profile of a wave near a caustic. At fixed frequency ω and vertical angle ψ it gives

$$\hat{\mathbf{E}}_{\mathrm{rad}}(\omega, x, \psi) \propto \mathrm{Ai}\left(2^{1/3}\frac{b_{\mathrm{cl}}(\psi) - x}{b_0}\right), \qquad (7)$$

where $\mathrm{Ai}(\xi)$ is the Airy function [5], $b_0 = R^{1/3}\lambdabar^{2/3}$ characterizes the width of the brilliant region of the caustic and $\lambdabar \equiv \lambda/(2\pi) = \omega^{-1}$. The

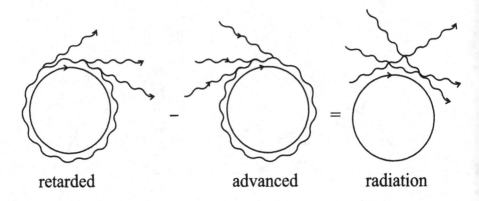

retarded   advanced   radiation

Figure 5.   Schematic representation of the relation beween the retarded, advanced
and radiation fields.

Airy function has an oscillating tail at positive $x - b_{cl}$ and is exponentially damped at negative $x - b_{cl}$. Fig.6 displays the intensity profile

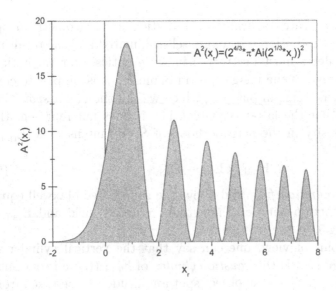

Figure 6.   Intensity of the horizontal impact parameter profile at fixed vertical angle
ψ. The abscissa is $x_r = (x - b_{cl})/b_0$, the ordinate is the value of $\mathcal{A}^2$ in Eq.(17).

$|\hat{\mathbf{E}}_{rad}|^2$, in relative units, as a function of the dimensionless variable
$x_r = (x - b_{cl})/b_0$. The oscillations can be understood semi-classically as
an interference between light rays coming from symmetrical points like

T_1 and T_2 in Fig.4. As said earlier, these light rays come to the same point M' of the image plane, therefore should interfere at this point. The phase difference between the two waves is

$$2\delta = \omega\,(t_2 - t_1) - k_h\,(T_1 M + M\,T_2)\,, \tag{8}$$

where $t_{1,2}$ is the time when the electron is at $S_{1,2}$. Denoting the azimuths of S_1 and S_2 by $-\varphi$ and $+\varphi$, we obtain

$$\delta = (\tan\varphi - \varphi)\,\frac{\omega R}{v} \simeq \tfrac{1}{3}\omega R\,\varphi^3 \simeq \frac{1}{3}\left(2\frac{x - b_{\rm cl}}{b_0}\right)^{3/2}. \tag{9}$$

This is in agreement, up to the residual phase $\pi/4$, with the the large x behavior in $x^{-1/4}\sin(\delta + \pi/4)$ of Eq.(7).

The result (7) can also be obtained by recalling that the photon has a definite angular momentum $J_z = \hbar\omega\,R/v$ about the orbit axis. For fixed vertical momentum $\hbar k_z = \hbar\omega\psi$, we have the following radial wave equation, using cylindrical coordinates (ρ, φ, z):

$$\left[\frac{\partial^2}{\partial\rho^2} + \frac{1}{\rho}\frac{\partial}{\partial\rho}\omega^2 - k_z^2 - \frac{J_z^2}{\rho^2}\right]\mathbf{E} = 0\,, \tag{10}$$

We have neglected the photon spin and approximated the centrifugal term L_z^2/ρ^2 by J_z^2/ρ^2. In the vicinity of the classical turning point $\rho = R + b_{\rm cl}$, we can use a linear approximation of the centrifugal potential and neglect the term $\rho^{-1}\partial/\partial\rho$. This lead us to the Airy differential equation with the argument of (7).

The profile function can also be calculated in a standard way. The electric radiation field $\mathbf{E}_{\rm rad}$ can be expanded in plane waves as

$$\mathbf{E}_{\rm rad}(t, \mathbf{r}) = \int \frac{d^3\mathbf{k}}{(2\pi)^3}\,\Re\left\{\tilde{\mathbf{E}}(\mathbf{k})\,e^{i\mathbf{k}\cdot\mathbf{r} - i\omega t}\right\}, \tag{11}$$

with $\omega = |\mathbf{k}|$. The momentum-space amplitude[2] is given by

$$\tilde{\mathbf{E}}(\mathbf{k}) = e\int_{-\infty}^{\infty} dt_e\,\mathbf{v}_\perp(t_e)\,\exp\left[i\omega t_e - i\mathbf{k}\cdot\mathbf{r}_e(t_e)\right], \tag{12}$$

where $\mathbf{v}_\perp$ is the velocity component orthogonal to $\mathbf{k}$. The electron trajectory is parametrized as

$$\mathbf{r}_e(t_e) = (R\cos\varphi - R\,,\, R\sin\varphi\,,\, 0)\,, \qquad \varphi = v\,t_e/R \tag{13}$$

The energy carried by $\mathbf{E}_{\rm rad}$ through a strip $[x\,,x+dx]$ of plane P, in the frequency range $[\omega, \omega+d\omega]$ and in the vertical momentum range $[k_z\,,k_z+dk_z]$ is

$$dW = 2dx\,\frac{d\omega\,dk_z}{(2\pi)^2}\,|\hat{\mathbf{E}}(\omega, x, \psi)|^2\,, \tag{14}$$

where $\hat{\mathbf{E}}(\omega, x, \psi)$ is the partial Fourier transform

$$\hat{\mathbf{E}}(\omega, x, \psi) = \int \frac{dk_x}{2\pi} \, e^{ik_x \, x} \, \tilde{\mathbf{E}}(\mathbf{k}) \tag{15}$$

evaluated at $k_y \simeq \omega$ and $k_z \simeq \omega \, \psi$. The result of Eqs.(12 - 15) is

$$\frac{dW}{d(\hbar\omega) \, dx \, dk_z} = \frac{\alpha}{8\pi^3} \, \mathcal{A}^2 \left(\frac{b_{cl} - x}{b_0} \right) \left[\mathcal{A}'^2(u) + \left(\frac{\psi}{\theta_0} \right)^2 \mathcal{A}^2(u) \right], \tag{16}$$

$$= \frac{1}{2\pi} \, \mathcal{A}^2 \left(\frac{b_{cl} - x}{b_0} \right) \frac{dW}{d(\hbar\omega) \, (d\Omega/\theta_0^2)}, \tag{17}$$

where $\alpha = e^2/(4\pi\hbar) = 1/137$,

$$u \equiv s^2 \, (1 + \gamma^2\psi^2)/2 \; = \; b_{cl}/b_0 \,, \quad s \equiv (\omega/\omega_c)^{1/3}, \quad \omega_c = \gamma^3/R, \tag{18}$$

$$b_0 \equiv R^{1/3} \lambda^{2/3} \; = s\gamma\lambda \,, \qquad \theta_0 \equiv (\lambda /R)^{1/3} = (s\gamma)^{-1} \tag{19}$$

and $\mathcal{A}(u)$ is a re-scaled Airy function:

$$\mathcal{A}(u) = 2^{4/3}\pi \, \mathrm{Ai} \left(2^{1/3}u \right). \tag{20}$$

Expression (17) relates the (x, ψ) profile to the standard the angular distribution. The two terms of the square bracket of (16) correspond to the horizontal and vertical polarizations respectively.

An example of the (x, ψ) profile is shown in Fig.7. We recall that it is obtained using an astigmatic lens. Therefore it differs from the standard (x, z) profile [7–9] formed by a stigmatic lens. The number of photons per electron and per unit of $\ln \lambda$ in the first bright fringe is, in the $s \ll 1$ limit,

$$\frac{dN_{\mathrm{phot}}}{d\lambda/\lambda} \simeq \frac{1}{70}. \tag{21}$$

The second fringe contains three times less photons.

3.   Applications

Although the impact parameter cannot be sharply defined in wave optics, Eq.(16) keeps a trace of the classical prediction (1): as γ or ψ is varied, the x-profile translates itself as a whole, comoving with the classical point $x = b_{cl}$. The observation of this feature would constitute an indirect test of the phenomenon of electron side-slipping. The ψ^2 dependence of this translation is responsible for the curvature of the fringes. The γ^{-2} dependence may be more difficult to observe: b_{cl} should

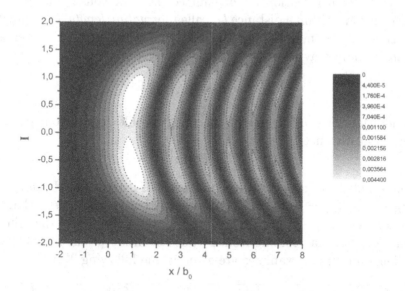

Figure 7. (x, ψ) profile of synchrotron radiation (Eqs.16-17), in the limit $s^3 = \omega/\omega_c \ll 1$. The ith level curve corresponds to the fraction $(i/10)^2$ of the maximal intensity.

be as large as possible compared to the horizontal width of the electron beam, one one hand, and to the FWHM width of the first fringe, $\simeq 1.3\, b_0$, on the other hand. Since the profile intensity (16) decreases very fast at large $u = b_{cl}/b_0$, a sensitive detector is needed.

The following table summarizes the various length scales which appear in synchrotron radiation.

bending radius	R			
long. distances	$l_{-1} = R/\gamma$	$l_0 = R^{2/3}\lambdabar^{1/3}$		$l_2 = \gamma^2\lambdabar$
transverse distances	$b_{-2} = R/\gamma^2$		$b_0 = R^{1/3}\lambdabar^{2/3}$	$b_1 = \gamma\lambdabar$
wave lengths	$\lambdabar_c = R/\gamma^3$			$\lambdabar$

The four quantities of a given row or line are in geometric progression of ratio γ^{-1} or s^{-1} respectively. Going along (or parallel to) the diagonal, the ratio is $\theta_0 = (s\gamma)^{-1}$. The subscripts of the different l's and b's are the powers of γ.

Synchrotron radiation is not emitted instantaneously, but while the electron runs within a distance l_f, called *formation length*, from point S_i. Thus $2l_f$ is a minimum length of the bending magnet. A conservative estimate of l_f may be

$$l_f = \mathrm{Max}\{2l_{-1}, 3l_0\}. \tag{22}$$

In addition, the mth fringe of the (x, ψ) profile comes from points S_1 and S_2 at distance

$$l^{(m)} = (3m\pi)^{1/3} l_0 \tag{23}$$

from point S. To observe it, the magnet half-length should therefore be larger than $l_{\min} = l_f + l^{(m)}$. The distance between the object plane and the lens should be larger than this $l_{\min}$, plus a few l_{-1} so that the lens can accept the ray coming from T_2 but not intercept the near field.

Some numerical examples are given in the following Table.

γ	200	200	6000	1000
R	0.8 m	80 m	4000 m	3 m
$l_{-1} = R/\gamma$	4 mm	0.4 m	0.67 m	3 mm
$b_{-2} = R/\gamma^2$	20 μm	2 mm	0.11 mm	3 μm
$\lambda_c = R/\gamma^3$	0.1 μm	10 μm	19 nm	3 nm
λ	0.1 μm	0.1 μm	0.1 μm	3 μm
		(visible domain)		(infrared)
$s = (\lambda/\lambda_c)^{-1/3}$	1	4.6	0.57	0.1
$\theta_0 = s^{-1}\gamma^{-1}$	5 mrad	1.1 mrad	0.29 mrad	10 mrad
$l_0 = s^{-1}l_{-1}$	4 mm	86 mm	1.2 m	30 mm
$b_0 = s^{-2}b_{-2}$	20 μm	93 μm	0.34 mm	0.3 mm

A practical application of the fringe pattern of Fig.7 is the measurement of an horizontal beam width. Since the successive fringes are more and more dense, they can probe more and more smaller widths. For a Gaussian beam of r.m.s. width σ_x, the contrast between the mth minimum and the $m + 1^{th}$ maximum is

$$a_m = \exp\left[-2(3\pi m)^{2/3} (\sigma_x/b_0)^2\right]. \tag{24}$$

The vertical beam size and the horizontal angular divergence have practically no blurring effect on the observed profile. This is not the case for the vertical beam divergence. However, as long as this divergence is small in units of θ_0, the blurring is small.

The x scale parameter b_0 grows with λ. Therefore a too large passing band of the detector will blur the fringes. For instance the contrast of the mth fringe is attenuated by a factor $2/\pi$ when the relative passing band is $\Delta\lambda/\lambda = 1/(2m)$. It is possible to restore a good contrast for a few successive fringes by inserting a dispersive prism ("CFC" in Fig.4) at some distance before the image plane, such that the λ-dependent deviation by the prism compensates the drift of the mth fringe. This allows to increase the passing band, hence the collected light, without loosing resolution.

4. Conclusion

The analysis of synchrotron radiation simultaneously in horizontal impact parameter x and vertical angle ψ, which, to our knowledge, has not yet been done, can open the way to a new method of beam diagnostics. Only simple optics elements are needed. There is no degradation of the beam emittance (contrary to Optical Transition Radiation) and no space charge effect at high beam current (such effects may occur with Diffraction Radiation). In addition, the observation of the curved shape of the fringes and the precise measurement of their distances to the beam would give an indirect support to the phenomenon of electron side-slipping and to a physical interpretation of the Schott term of radiation damping.

Notes

1. we use rational electromagnetic equations, e.g. div $\mathbf{E} = \rho$ instead of $4\pi\rho$.
2. From now on we omit the subscript "*rad*" of $\mathbf{E}$.

References

[1] X. Artru, G. Bignon, Electron-Photon Interactions in Dense Media, H. Wiedemann (ed.), NATO Science series II, vol. 49 (2002), pages 85-90.

[2] X. Artru, G. Bignon, T. Qasmi, Problems of Atomic Science and Technology 6 (2001) 98 ; arXiv:physics/0208005.

[3] X. Artru, Phys. Lett. A 128 (1988) 302, Eqs.(15-16).

[4] C. Teitelboim, Phys. Rev. D1 (1969) 1572; D2 (1970) 1763. See also C. López and D. Villarroel, Phys. Rev. D11 (1975) 2724. I thank Prof. V. Bordovitsyn for pointing me these references.

[5] M. Abramowitz and I. Stegun *Handbook of Mathematical Functions*, Dover, 1970.

[6] L. Landau and E. Lifshitz, *The Classical Theory of Fields*, 7.7 (Addison-Wesley 1951).

[7] A. Hofmann, F. Méot, Nucl. Inst. Meth. 203 (1982) 483.

[8] R.A. Bosch, Nucl. Inst. Meth. in Physs. Res. A 431 (1999) 320.

[9] O. Chubar, P. Elleaume, A. Snirigev, Nucl. Inst. Meth. in Physs. Res. A 435
     (1999) 495.

STATUS AND HIGHLIGHTS OF THE CANDLE PROJECT

V. TSAKANOV

Center for the Advancement of Natural Discoveries using Light Emission
CANDLE, Acharian 31, 375040 Yerevan, Armenia

Abstract: CANDLE is a 3 GeV third generation synchrotron light facility project in Republic of Armenia. The report includes the main considerations that underlie the Design Report of the project. An overview of the machine and beam physics of the facility is given.

1. Introduction

The development of the CANDLE synchrotron light facility project had been presented in number of the workshops and conferences [1-3]. In this report the progress on machine and beam physics study are given. The basic approaches that under lied the facility design, the competitive spectral flux and brightness, the stable and reproducible photon beams and the high level of the control with user-friendly environment are discussed.

2. Design Overview

The CANDLE general design is based on a 3 GeV electron energy storage ring, full energy booster synchrotron and 100 MeV S-Band injector linac (Fig. 1). The main storage ring parameters are compiled in Table 1.

The full energy booster synchrotron operates with the repetition rate of 2 Hz and the nominal pulse current of 10 mA. The nominal current in the storage ring is 350 mA. The storage ring of 216m in circumference has 16 DBA type periods. The harmonic number of the ring is h=360 for the accelerating mode frequency 499.654 MHz. The design of the machine is based on conventional technology operating at normal conducting conditions. In total 13 straight sections of the storage ring are available for

H. Wiedemann (ed.), Advanced Radiation Sources and Applications, 399–404.

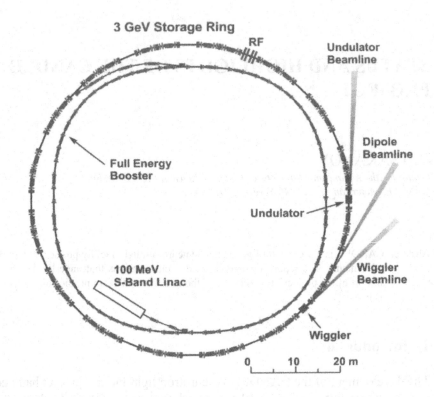

Figure 1. The CANDLE facility layout.

Table 1. Main Parameter of Storage Ring

Energy E (GeV)	3
Circumference (m)	216
Current I (mA)	350
Number of lattice periods	16
Tunes Q_x/Q_y	13.2/4.26
Horiz. emittance (nm-rad)	8.4
Beam lifetime (hours)	18.4

insertion devices (ID). The photon beams from the dipoles and the conventional ID's are covering the energy range of 0.01-50 keV. If the users demand requires the extension of the photons energy to hard X-ray region, the superconducting wigglers may be installed. Figure 2 presents the CANDLE spectral brightness for the dipole, undulator and wiggler sources.

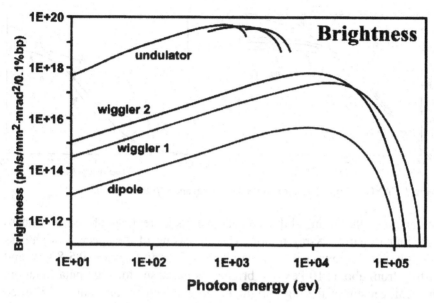

Figure 2. CANDLE spectral brightness.

3. Beam Physics

The CANDLE design has 16 identical Chasman-Green type cells with non-zero horizontal dispersion 0.18m in the middle of the long straight section that provides 8.4 nm-rad horizontal emittance of the beam. The rms energy spread in the beam is at the level of 0.1%.

The optimization of the optical parameters of the lattice to obtain a high spectral brightness of the photon beams from insertion devices keeping large the dynamical aperture of the ring was an important issue of the lattice optics study. The results of the study for CANDLE show [4] that for a real electron beam the spectral brightness in short wavelength range is high for the lattice with large beta value in the middle of straight section. In longer wavelength range, the high brightness implies the small beta lattice.

Figure 3 shows the dependence of the normalized brightness on the emitted photon energy for different horizontal and vertical beta values at the source point. Dashed line corresponds to the theoretical optimal beta values associated with each photon energy [4].

The improvement of the brightness with small horizontal beta is visible only for the photon energies below 0.1 keV. Starting from 0.5 keV the brightness increases with larger betatron fuction. In vertical plane the beam emittance is given by the coupling of the horizontal and vertical oscillations that for CANDLE is expected at the level of 1%. The small vertical

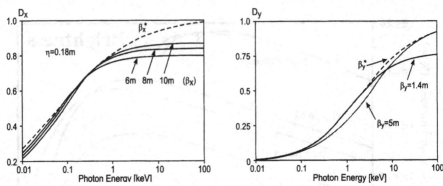

Figure 3. CANDLE brigthness for various horizontal and vertical beta values.

emittance of the beam shifts the characteristic regions of the brightness behaviour to harder X-ray region. The increasing of the spectral brightness with low beta is now visible in the photons energy range of 0.5-8 keV and starting from about 10 keV the brightness increases for high beta function. The achievement of the high brightness in whole spectrum range of emitted photon implyes the low vertical beta design in comparison with horizontal beta. Taking into account the requirement to have sufficient dynamical aperture of the CANDLE storage ring, the horizontal and vertical betatron functions in the middle of the straigth sections are optimized to 8.1m and 4.85m respectively.

The compratevely high beta values in middle of straight section are significantly improving the dynamical aperture of the ring. (Fig.4) prooviding facility stable operation [5].

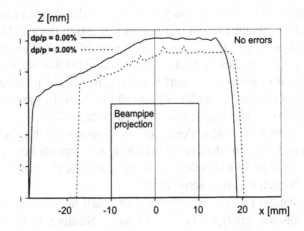

Figure 4. The storage ring dynamic aperture.

4. Beamlines

The increasing demand of synchrotron radiation usage worldwide drives the scenario for the first stage beamlines that are an integrated part of the facility construction [2]: LIGA (dipole), General Diffraction and Scattering Beamline (dipole), X-ray Absorption Spectroscopy Beamline (dipole), Soft X-ray Spectroscopy Beamline (Undulator), Imaging Beamline, Small Angle X-ray Scattering (Wiggler).

LIGA beamlines. The designed LIGA beamlines will cover the experiments with three photon energy regions: 1-4 keV for fabrication of X-ray masks and thin microstructures up to 100 µm height (LIGA I), 3-8 keV for standard LIGA microstructures fabrication with resists height up to 500 µm (LIGA II) and 4-35 keV for deep lithography (LIGA III).

General Diffraction and Scattering Beamline. The general diffraction & scattering beamline is based on the dipole source and will produce a moderate flux of hard (5-30keV) focused or unfocused tunable monochromatic X-rays sequentially serving two experimental stations: EH1 for roentgenography routine or time resolved, low or high temperature structural studies of polycrystalline materials, for reflectivity investigations of thin films and multi-layers; EH2 for single crystal structure determination in chemistry and material science, for charge density studies and anomalous dispersion experiments.

XAS Beamline. The XAS capabilities using the CANDLE bending magnet radiation will cover a photon energy range up to about 35 keV. This region covers the K edges of elements such as Si, S, P and Cl, which are of high technical interest. Using double crystal monochromator and gold coated total reflection mirror, with low expansion water-cooled substrate, the beamline will be able to operate in hard X-ray region, which will allow users to measure EXAFS of all elements either at K or L_3 edges. XAS data could be recorded in transmission, fluorescence and electron yield mode at normal and grazing incident angels.

Imaging Beamline. The Imaging beamline using 3 T permanent wiggler radiation will provide a high flux "white" or tunable monochromatic coherent radiation in 6-120 keV photon energy range at about 150m from the source. The experimental program will include: phase contrast imaging; diffraction-enhanced imaging; hard x-ray microscopy; holographic imaging and tomography; micro-focusing techniques; high resolution X-ray topography and diffractometry; X-ray micro-fluorescence; high resolution X-ray inelastic scattering; properties under the high pressure and high temperature conditions.

SAXS Beamline. The small-angle scattering instrument at CANDLE is advanced Bonse-Hart camera. The scientific applications include small and ultra small angle X-ray scattering and nuclear resonance. The primary optical elements of the beamline are vertically and horizontally adjustable slits, high heat load monochromator with liquid nitrogen cooled crystal (Si) cases, high resolution monochromator which limits the bandpass of the monochromatic photons from the first monochromator to meV range.

Soft X-ray Spectroscopy Beamline. By combining spectroscopic chemical information with high spatial resolution, X-ray micro-spectroscopy provides new research opportunities. These capabilities are achieved the best possible way by using high brightness photon beams provided by advanced undulators in third generation light sources. Two types of microscopes are proposed to be build on undulator source at CANDLE – a zone plate based scanning transmission X-ray microscope and a photoelectron emission microscope.

5. Summary

The next stage of the project development implies an extensive prototyping program, the RF, magnet and vacuum test stands establishment, number of machine and user international workshops. The international collaboration is highly appreciated and we express our gratitude to all colleagues for their interest, support and cooperation.

References

1. V Tsakanov et al, Proc. of EPAC2002, Paris 2002, pp. 763-765.
2. M. Aghasyan et al, SRI2003, San-Francisco, USA, 2003.
3. V. Tsakanov et al, Proc. Of EPAC2004, Lucerne 2004.
4. M. Ivanyan, Y. Martirosyan, V. Tsakanov, Nucl. Inst. Meth. A **531/3**, 651-656, 2004.
5. Yu. Martirosyan, Nucl. Inst. Meth. A **521**, 556-564, 2004.

NON-LINEAR BEAM DYNAMICS STUDY IN LOW EMITTANCE STORAGE RINGS

Y.L. MARTIROSYAN

Center for the Advancement of Natural Discoveries using Light Emission,
CANDLE, Acharian 31, 375040 Yerevan,, Armenia

Abstract.   We report on some results of the beam dynamical parameters dependence on insertion devices gap height and magnetic field in storage rings. The numerical examples are presented for the CANDLE light source design.

1.   Introduction

The stored beam of electrons or positrons in storage ring consists of hundreds of bunches which are subject to external electromagnetic fields, radiation fields and other various influences. As a source of radiation, the user demand imposed very strong requirements on beam quality in storage ring – that are high brightness (which means very small electron beam emittances) and long beam lifetime. The provision of such characteristics requires a detailed study of higher order effects related to magnetic fields of main magnets as well as insertion devices (ID) and collective effects associated with beam density.

2.   Simulation Method

Despite the efforts in development of the symplectic mapping tools in beam dynamics study in storage rings and damping rings that incorporated undulators and wigglers [2], the particle numerical tracking simulation due to modern high accuracy post-processing analysis tools (early indicators for stability and dynamic aperture calculations, interpolated FFT and frequency map analysis, etc) remains one of the basic tools to study the single particle non-linear dynamics in accelerators. This approach is applicable effectively for the beam tracking studies in light sources, where due to small damping times one can limit tracking turns number up to few hundred [3]. In this

H. Wiedemann (ed.), Advanced Radiation Sources and Applications, 405–409.

paper, we will focus our study mainly on two important aspects of beam dynamics in storage ring: ID's caused tune shift and dynamic aperture reduction. Fringe field effects investigations for CANDLE light source using the same approach were reported in [4].

As a basic to start we take the particle trajectory equations in natural coordinate system (alternatively one can use the Hamiltonian formalism):

$$x'' = h_0^2 x + h_0 + \frac{h_0 x x' + 2h_0 x'^2}{1 + h_0 x} + \frac{e}{pc}\sqrt{1 + \frac{x'^2 + y'^2}{(1 + h_0 x)^2}} *$$

$$\left\{ x' y' B_x - [(1 + h_0 x)^2 + x'^2] B_y + (1 + h_0 x) y' B_s \right\};$$

$$\tag{1}$$

$$y'' = \frac{h_0 x y' + 2h_0 x' y'}{1 + h_0 x} + \frac{e}{pc}\sqrt{1 + \frac{x'^2 + y'^2}{(1 + h_0 x)^2}} *$$

$$\left\{ [(1 + h_0 x)^2 + y'^2] B_x - x' y' B_y - (1 + h_0 x) x' B_s \right\};$$

where $h_0 = \dfrac{e}{pc} B_y (x = 0, y = 0)$ is the orbit curvature in the dipole magnet for the reference particle and $B_x(x = 0, y = 0) = 0$ for the plane design orbit, (*)-denotes multiplication. The integration is performed for natural equation of particle trajectory (1) without any simplification. High order terms that describe the curvature effects (third term in r.h.s for x-motion and first term in r.h.s of y-motion in (1)) and kinematic correction terms (square roots in (1)) that entail nonlinear effects purely due to dynamics and independent of the fields (so-called kinematic correction) are also included in the trajectory equation in natural way.

One of the sources of strong nonlinearities in low emittance synchrotron light sources are insertion devices (IDs) such as small gap undulators and high field superconducting wigglers (wavelength shifters)[5]. The effects of ID's on the beam dynamics are both linear and nonlinear. The linear effects are the linear tune shift and optic functions distortion (beta-beating). The major nonlinear effect of insertion devices is the reduction of dynamic aperture of the electron beam and distortions of phase portraits. In our study we included into the trajectory equations (1) the ID's magnetic field components of the form:

$$B_x(x,y,s) = \frac{k_x}{k} B_0 \sinh(k_x x) \sinh(k_y y) \cos(ks)$$

$$B_y(x,y,s) = B_0 \cosh(k_x x) \cosh(k_y y) \cos(ks) \tag{2}$$

$$B_s(x,y,s) = -\frac{k}{k_y} B_0 \cosh(k_x x) \sinh(k_y y) \sin(ks)$$

with

$$k_x^2 + k_y^2 = k^2 = \left(\frac{2\pi}{\lambda}\right)^2 \tag{3}$$

where λ is the period length of the ID and B_0 is its on-axis magnetic field. These expressions do not include high order field harmonics that described effects from magnet pole geometry.

The comparatively slow computation speed is the only disadvantage of the proposed direct numerical integration method, which can be overcome by using parallel computing technique or other possibilities [6].

3. Numerical Simulation Results for CANDLE

The CANDLE (see [1] for details) storage ring lattice consists of 16 identical double bend achromatic cells, of which 13 straight sections are available for installation of ID's with the length up to 4m. As typical examples we consider undulator magnet with the period length 22 mm and number of periods N=72 and superconducting strong wiggler with the period length 17cm and number of periods two.

3.1. Dynamic Aperture Reduction

In Fig. 1, the dynamic aperture simulations obtained by above described method is depicted. As shown Fig. 1, there is small dynamic aperture reduction in the peripheral part of the stable region for the undulator with a gap of 8mm (1T magnetic field) and the strong dynamic aperture decreasing in vertical direction caused by the wiggler which has a maximum magnetic field of 6T. So, in case of operation with strong wigglers, local correction of tunes and betatron function is mandatory.

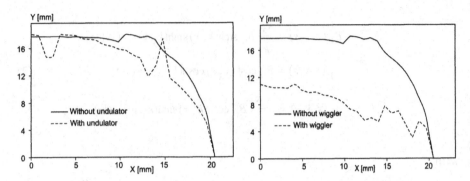

Figure 1. Dynamic aperture reduction from the ID's: left - for the small gap undulators; right - for the strong field wiggler.

3.2.   Tune Shift

We find the tune shift from the undulator magnet tolerable down to a gap height as small as 8mm. In case of wiggler magnet, the same is true for a magnetic field value up to 3T. Beyond these limits one needs additional quadrupole magnets for optical corrections. All this assumes all straight sections are occupied with the same type of insertion device.

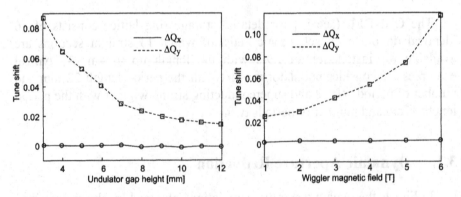

Figure 2. CANDLE tunes shift dependence: left- undulators gap height; right – magnetic field value of strong wiggler.

4. Summary

The CANDLE optics distortion, caused by strong wigglers should be locally corrected by installation of quadrupoles. The dynamic aperture in presence of IDs is sufficient for obtaining a reasonable beam lifetime. The investigation of IDs field errors and misalignments on dynamic aperture and beam lifetime is in progress.

References

1. Tsakanov, V. et al, these proceedings.
2. Wu Y, et.al, Proc. of PAC 2001, Chicago, p.459.
3. Irwin, J. and Yan, Y.T., Handbook of Accelerator Physics and Engineering, A. Chao and M. Tigner, editors, p. 87, (1999).
4. Martirosyan, Y., NIM A, 521 pp.556-564, (2004).
5. Walker, R and Diviacco, B., (2000), Synchrotron Radiation News, vol.13, #1
6. Martirosyan, Y., Ivanyan, M., Kalantaryan, D., Proc. of EPAC'04, Lucerne (2004).

STUDY OF THE MULTI-LEVEL EPICS BASED CONTROL SYSTEM FOR CANDLE LIGHT SOURCE

G. AMATUNI, A. VARDANYAN

Center for the Advancement of Natural Discoveries using Light Emission, CANDLE, Acharian 31, 375040 Yerevan, Armenia

Abstract. The basic considerations that underlie the control system design for CANDLE light source are presented. The design is based on the multi-level EPICS system that is successfully applied for the number of operating facilities. The control system architecture and software are discussed.

1. Control System Architecture

The main requirements to the CANDLE facility [1] operation are driven by the necessity to provide the long-term stability of the electron beam and the facility operation in multimode regime. The three modes of operation foreseen for the project: the single bunch, multi-bunch and top-up injection modes. The central part of the accelerator is a control system, which interacts with hardware components and provides exchange, storage, manipulation and display of data. The main subsystems of the accelerator to be controlled and monitored are the magnet system, RF system, vacuum system, injection system, diagnostic equipment, timing system, insertion devices, interlock system, water cooling, electrical system, personal safety, front ends. Control system of CANDLE accelerator will be a distributed system, with multi-level functional architecture (Fig. 1). Basic element is the communication interface - use of the networks/buses for communication between levels. At CANDLE it will be used Ethernet network with TCP/IP protocol and mostly CAN buses [2].

The designed control system is structured into hardware and software layers that manipulate data on different levels of abstraction. The bottom layer interacts with the electrical signals where the processors have to implement real-time control. The top layer provides the human interface

411

H. Wiedemann (ed.), Advanced Radiation Sources and Applications, 411–414.

where operators can control the accelerator. The layers in between maintain the parameter database and provide data collection and distribution, networking and monitoring.

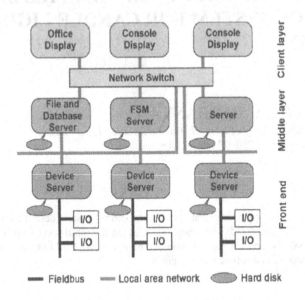

Figure 1. CANDLE Control System Architecture.

2. CANDLE Linac Simulations

In purpose to define the bottom layer, i.e. the design and constructional features of the 500 MHz and 3 GHz RF system of the CANDLE linac control system, simulations were done using Matlab/Simulink program [3]. The CANDLE linac consists of an electron gun modulated by 500 MHz to produce 1 ns electron bunches, 500 MHz and 3GHz bunchers, pre-accelerating cavity and the main accelerating section at 3 GHz. An important feature of the control system is a high level synchronization of amplitude-phase characteristics of the accelerated beam. This requires strict stabilization of the RF frequency, amplitude and phase. A digital feedback system had been adapted to simulations for more flexibility of the control algorithms. Simulations defined the 9 MHz sampling rate and digital I/Q detection. Figure 2 shows the main blocks and interconnection structure of the linac RF system.

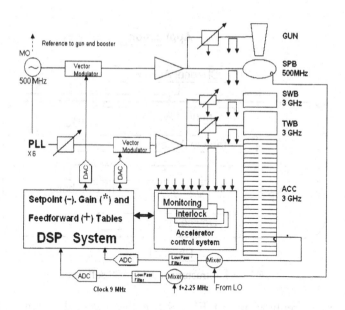

Figure 2. Scheme of the RF control system for Linac.

The C6000 DSP with 64-bit address and data buses is chosen for 9 MHz sampling rate and large output capability. It is planned to use ADCs and DACs which have external timing inputs and internal memory and are suitable for communications with Texas Instruments DSPs. The ADC board consists of four independent ADC channels with 14-bit A/D converters operating at 9 MHz sampling frequency. The DAC board has two channnels. Control system for whole Linac requires 2 DSPs for 500 MHz and 3 GHz RF control. All the DSPs, DACs and ADCs will be located in four crates with VME bus interface.

3. Control System Software

There are many implementations of control systems that have been developed at various laboratories: DOOCS at DESY (Hamburg, Germany), TACO/TANGO, at ESRF (Grenoble, France) and EPICS – Experimental Physics Industrial Control System, developed at APS. After detailed study of these three control system software and taking into consideration the experience gained in other light sources the EPICS was chosen as the base for CANDLE control system software. The EPICS system has a multi-layer type (Fig.3).

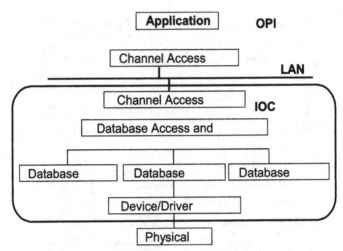

Figure 3. Structure of the EPICS software.

The main realizations of EPICS use the approach of the VME/VXI–based I/O controllers and Motorola 68K CPUs running VxWorks operation system, the real-time Linux versions such as RTEMS and RTOS and UNIX and Windows workstations as operator interfaces. EPICS supports the CAN buses. The advantages of the EPICS are the definition of a standard IOC structure, an extensive library of driver software for a wide range of I/O cards.

4. Summary

The CANDLE control system architecture and software are defined. The simulations for the linac RF system are performed that will used for on-line diagnostics and control of RF during the tuning and operation.

References

1.   V. Tsakanov et. al, CANDLE Design Report, ASLS-CANDLE R-02-001, 2002.
2.   V. Ayvazyan, G. Amatuni, A. Yayloyan, K. Rehlich, S. Simrock, Design of the CANDLE Control System, EPAC 2002, Paris, France, June 02-07, 2002.
3.   A. Vardanyan, G. Amatuni, B.Grigoryan , EPAC 2004, Lucerne, July 04-09, 2004.

MAIN APPROACHES TO THE COOLING SYSTEM FOR THE CANDLE LIGHT SOURCE

S. TUNYAN, G. AMATUNI, V. AVAGYAN

Center for the Advancement of Natural Discoveries using Light Emission
CANDLE, Acharian 31, 375040 Yerevan, Armenia

Abstract. Different systems of CANDLE synchrotron light facility will generate a substantial amount of heat that must be removed insuring the temperature stability of the equipment and the temperature maintain within the specified tolerances. To obtain the temperature stability and maintain the specified temperature, certain cooling equipment must be provided to precisely balance the heat generation rates. Main approaches for achieving the required goals of the cooling system are presented in this paper.

1. Introduction

The different systems and components that will be included in CANDLE will generate a substantial amount of heat that must be removed providing temperature stability of the equipment and its maintaining within the specified tolerances. While a portion of this heat will be radiated into surrounding spaces, the major part of the heat will be removed by means of the complex water cooling system. Numerous magnets, power supplies, accelerating structures and vacuum chambers comprise the bulk of the components requiring cooling in the storage ring (SR). Similarly, the magnets, accelerating structures, power supplies, and other devices will be in the Linear Accelerator (Linac) and Booster Ring (BR). These components require the removal about 5.0 megawatts of heat during the nominal operation of the facility.

H. Wiedemann (ed.), Advanced Radiation Sources and Applications, 415–420.

2.   Description of the cooling system

The CANDLE Water Cooling System will be distributed from 3 unit-type cooling towers and one chiller module where all heat will be rejected and the water will be cooled to a baseline temperature (presumably 26°C) that will be maintained within a stable range of ±0.5 °C. This water will be pumped out to the main heat exchangers at a relatively low pressure (approximately 3 Bar). The water processed up to required parameters from main heat exchangers will be pumped to two local pumping stations where the pressure will be boosted and the temperature will be controlled to its final temperature (presumably 32°C) with a tolerance of ±0.25 °C. This basic configuration is commonly called an integral primary-secondary system (Fig. 1). Two pump stations of the secondary system based on two water manifolds will distribute the cooled water to facility subsystems. One of these manifolds with hydraulic pressure of about 15 Bar will maintain the cooling of the SR components (magnets, vacuum chamber, photon shutters and absorbers). The second collector with pressure of about 6 Bar will cool other subsystems (power supply, booster ring, beam line). The same collector will cool ten heat exchangers, installed in the main building, by separating ten stand-alone cooling loops (SACL) from the secondary cooling system. Eight of the

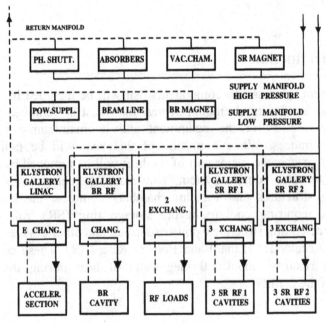

Figure 1. Primary Cooling System.

independent contours, after the additional temperature stabilization and the temperature increase up to the required value, will cool RF cavities and accelerating sections. Other two independent circulars will cool the RF loads. Hydraulic division of these two circulars is required for the change of water quality. One general manifold will provide return of warm water to main heat exchangers. The block diagram of the secondary water system is shown in Fig. 2.

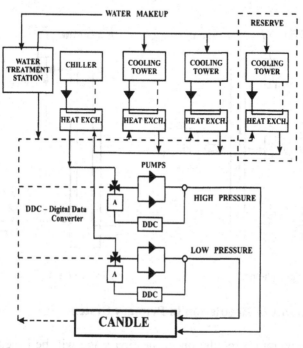

Figure 2. Secondary Cooling System.

3.   Temperature Regulation

Temperature regulation of such water systems as accelerator water cooling system requires consideration of hydraulic parameters and, on the whole, represents an intricate task. Even the use of computer programs does not allow precise solutions as they are based on some modeling and approximation. In addition, there are many nuances caused by the equipment installation, which cannot be expected at the stage of calculation. Any modifications in the scheme require recalculation of the new hydraulic system of pipes, exchangers etc. The rational approach is based on modular construction of water network. The system is divided into modules; the module of the low level enters into the module of the higher level. The number of parallel water loadings, equipped by balancing valves, enters into the module. On the outlet of the module the balancing valve -the partner- is installed. In this approach the single modules can be replaced, removed or added without the destruction of integrity of the overall design. The CANDLE water cooling system temperature regulation will be achieved in three stages. The Block scheme and the Logical scheme of temperature regulation are given in Fig. 3 and Fig. 4 respectively.

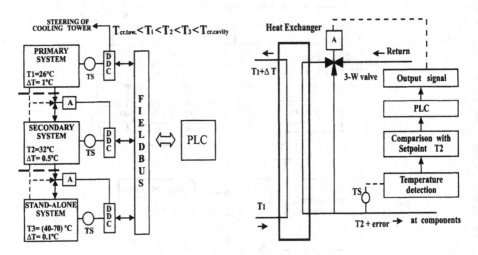

Figure 3. Block Scheme.   *Figure 4.* Logical Scheme.

3.1. Temperature Regulation in Primary System

The temperature regulation in the first stage will be based on the direct temperature adjustment of cooling towers. We anticipate using three identical modules with the forced principle of ventilating cooling, each with capacity of 4000 l/min. The Digital Data Converter (DDC) through inbuilt Temperature Sensor (TS) will control the water temperature of each heat exchanger that working in pair with cooling tower. The DDC must be handled by the main Programmable Logic Controller (PLC) of the synchrotron and according to the developed algorithm should provide temperature regulation by changing the operating mode of the cooling tower. Such method of regulation presumably should provide in the first stage the temperature stability of $\pm$ 0.5°C. The water temperature in this stage will be set by the lowest limit that can be economically achieved via evaporation in this geographic area during the hottest and the most humid summer conditions. In the current stage of the project we foresee to provide 26°C. In addition, to decrease the temperature baseline at hottest and the most humid summer conditions, it is foreseen to use a chiller.

3.2. Temperature Regulation in Secondary System

The secondary cooling system will include two separate subsystems. Two pump stations located near the central heat exchangers, using the general entrance collector will distribute water on subsystems of synchrotron with various pressures. Each separate water manifold will be equipped with independent temperature regulation system. The temperature regulation in secondary circuits will be based on a method of the steering mixing of the cooled and warm (return) water. The simple logical scheme is depicted in Fig. 4. The bridge is connected across the return manifolds and primary mains, and water flow is regulated using a three-way mixing valve. This valve responds to an output from a programmable controller and modulates to maintain supply water temperature set point. In this stage we assume to provide 32°C working temperature with accuracy of ±0.25°C. Such method of adjustment will allow attuning, if necessary, two separate secondary water contours on different temperature points and on different temperature stabilities. The second stage of temperature stabilization will be sufficient for the majority components of the accelerator.

3.3. Temperature Regulation in stand-alone System

There will be six stand-alone water cooling systems with identical hardware components for six SR RF cavities, and two for BR RF cavity and Linac section. The optimum temperature of each Stand-Alone Closed Loop (SACL) system will be defined by a standard procedure. Water temperature will vary, and the set point will be determined at the maximum value of the beam energy. Each SACL will have a unique set point in the range of 40-70°C. Absolute knowledge of the temperature is not essential since the set point is chosen by direct measurement, while the long-term stability is important. The temperature regulation in these high-power RF components should be achieved with accuracy of ±0.05°C. The two other stand-alone systems for RF loads don't need temperature regulation, but the flow must be sufficient for high power. Some additive such as ethylene glycol will be added to the circulating water for better absorption of the RF power.

4. Summary

The CANDLE Water Cooling System design main approaches are formulated. The main concepts are based on the experience gained in other synchrotron light sources.

References

1. E. Swetin, M. Kirshenbaum, C. Putnam, "Cooling Water Systems for Accelerator Components at the Advanced Photon Source", MEDSI, 2002.
2. L. Pelligrino, "The DAΦNE Water Cooling System", LNF-INFN.

PROPOSED FIRST BEAMLINES FOR CANDLE

M. AGHASYAN, A.GRIGORYAN, R. MIKAELYAN
Center for the Advancement of Natural Discoveries using Light Emission
CANDLE, Acharian, 31, 375040 Yerevan, Armenia

1.   Introduction

The research highlights of synchrotron radiation usage in biology, medicine, chemistry, materials and environmental sciences, pharmacy, electronics and nano-technology, promoted the design and construction of a number of third generation light sources at the intermediate energies of 2.5-3.5 GeV [1].

The CANDLE general design is based on a 3 GeV electron energy storage ring, full energy booster synchrotron and 100 MeV S-Band injector linac [2, 3]. With nominal machine operation parameters, the horizontal emittance of the electron beam is 8.4 nm-rad. The resulting electron beam lifetime in the ring is 18.6 hours. The magnets design is based on normal conducting technology, although the superconducting wigglers may be installed in the ring to extend the photon spectral range to hard X-ray region. The optimization of the facility performance is based on the following criteria:

* The high spectral flux and brightness of the photon beams in 0.01-50 keV   range.
* The sufficient dynamical aperture to provide stable operation and long beam lifetime;
* The flexible operation with reproducible electron beam positioning.

For time resolved experiments, the ability to accommodate many different bunch patterns (single and multi-bunch operation) is incorporated into the design.

2.   Beamline Instrumentation

The increasing demand of synchrotron radiation usage worldwide drives the scenario for the first stage beamlines that are an integrated part of the facility construction. Based on the existing synchrotron radiation usage statistics, the following beamlines have been preliminary selected to build at the first phase of CANDLE construction: LIGA (dipole), General Diffraction

H. Wiedemann (ed.), Advanced Radiation Sources and Applications, 421–424.

and Scattering Beamline (dipole), X-ray Absorption Spectroscopy Beamline (dipole), Soft X-ray Spectroscopy Beamline (Undulator), Imaging Beamline, Small Angle X-ray Scattering (Wiggler). The technical design of the proposed beamlines is based on the advanced instrumentation techniques developed for number of modern synchrotron light sources [1] with reasonable freedom in optics to adjust the beamline with the user demanded requirements for a particular experiment.

2.1. LIGA Beamlines

The designed LIGA beamlines will cover the experiments with three photon energy regions: 1-4 keV for fabrication of X-ray masks and thin microstructures up to 100 μm height (LIGA I), 3-8 keV for standard LIGA microstructures fabrication with resists height up to 500 μm (LIGA II) and 4-35 keV for deep lithography (LIGA III). The parameters list for LIGA beamlines is presented in Table 1.

Table 1. LIGA beamline design parameters.

Source		Bending magnet
Accepting angle		8-10 mrad
Energy range	LIGA I	1-4 keV
	LIGA II	3-8 keV
	LIGA III	4-35 keV
Horizontal spot size		120 mm

2.2. General Diffraction and Scattering Beamline

The general diffraction & scattering beamline is based on the dipole source and will produce a moderate flux of hard (5-30keV) focused or unfocused tunable monochromatic X-rays sequentially serving two experimental stations: Experimental Hutch 1 for Roentgenography routine or time resolved, low or high temperature structural studies of polycrystalline materials, for reflectivity investigations of thin films and multi-layers; Experimental Hutch 2 for single crystal structure determination in chemistry and material science, for charge density studies and anomalous dispersion experiments [3]. The design parameters of the beamline are given in Table 2.

Table 2. General Diffraction and Scattering Beamline Parameters.

Source	Bending magnet
Horizontal angular acceptance	6 mrad
Mirror 1 with vertical collimating Rh coated	Radius of curvature 10.5 km
Double crystal monochromator	Si(111) flat crystal, Si(111) sagitally focusing crystal

Radius of sagitally focusing crystal	1.9 m
Mirror 2 Vertical focusing Rh coated	Radius of curvature 4.8 km
Energy range	5-30 keV
Energy resolution	$2*10^{-4}$ at 10 keV
Photon flux (at 10 keV)	$6.1*10^{12}$ ph/s
Spot size	$0.04*0.02$ cm^2

2.3. XAS Beamline

The XAS capabilities using the CANDLE bending magnet radiation will cover a photon energy range up to about 35 keV (Ec =8.1 keV). Using a differential pumping system this beamline can be operated with no window before the monochromator and will produce outstanding intensity in soft X-ray region of the spectrum. This region covers the K edges of elements such as Si, S, P and Cl, which are of high technical interest. Using double crystal monochromator and gold coated total reflection mirror, with low expansion water-cooled substrate, the beamline will be able to operate in hard X-ray region, which will allow users to measure EXAFS of all elements either at K or L3 edges. XAS data could be recorded in transmission, fluorescence and electron yield mode at normal and grazing incident angels. The parameter list for XAS beamline is presented in Table 3.

Table 3. XAS beamline design parameters.

Source	Bending magnet
Horiz. angular acceptance (mrad)	2
Energy range (keV)	1.8-35
Energy resolution ($\Delta E/E$)	$\sim10^{-4}$ with Si (1,1,1)
Photon flux (photons/sec)	10^9 (unfocused) / 10^{11} (focused)
Spot size (mm^2)	10x1 (unfocused) / 0.5x0.5 (focused)

2.4. Imaging Beamline

The Imaging beamline using 3T permanent wiggler radiation will provide a high flux "white" or tunable monochromatic coherent radiation in 6-120 keV photon energy range at about 150m from the source (monochromator designed to be located at 145m from the source) [4]. The experimental program will include: phase contrast imaging; diffraction-enhanced imaging (DEI); hard x-ray microscopy; holographic imaging and tomography; micro-focusing techniques; high resolution X-ray topography and diffractometry; X-ray micro-fluorescence analysis including trace element analysis; high resolution X-ray inelastic scattering; research structure and properties of materials under the high pressure and high temperature conditions by the energy-dispersive X- ray diffractometry and R & D of the new X-ray optics elements.

2.5.   SAXS Beamline

The small-angle scattering instrument at CANDLE is advanced Bonse-Hart camera [5]. Instrument design parameters presented in table 4. The scientific applications include small and ultra small angle X-ray scattering and nuclear resonance. The primary optical elements of the beamline are vertically and horizontally adjustable slits, high heat load monochromator with liquid nitrogen cooled crystal (Si) cases, high resolution monochromator which limits the band pass of the monochromatic photons from the first monochromator to meV range.

Table 4. SAXS beamline design parameters.

Source	3 T wiggler
Horizontal acceptance (mrad)	2
Energy range (keV)	5-35
Energy resolution (ΔE)	~meV

3.   Conclusion

The proposed set of the hard x-ray beamlines for the CANDLE facility is the subject of further discussions after the establishment of the Scientific Advisory Committee. Prior to construction, the number of dedicated workshops is planed to develop and identify the experimental program and corresponding instrumentation according to user demand.

Acknowledgements

We are thankful to J. Hormes, R. Titswort, T. Mappes and A.A. Kirakosyan for very useful discussions and suggestions.

References

1. J. Corbett, T. Rabedeau, Synch. Rad. News 12, 22, 1999.
2. V. Tsakanov et al, Rev. Sci. Instr. 73, 1411-1413, 2002.
3. V. Tsakanov et al, CANDLE Design Report, ASLS-CANDLE R-001-02, July, 2002.
4. A. Grigoryan et al, "Imaging beamline", proceedings of Semiconductor Electronics fourth national conference, Armenia, Tsakhadzor, p. 226, 2003.
5. U. Bonse, M. Harts, Z. Phys. 189, 151-162, 1966.

POWER CONVERTERS FOR CANDLE LIGHT SOURCE

V. JALALYAN, S. MINASY
Center for the Advancement of Natural Discoveries using Light Emission CANDLE, Acharian 31, 375040 Yerevan, Armenia

Abstract.   The report presents the main consideration for the power converters choice for the CANDLE light source. The main block scheme, ramping diagram and design features of 2 Hz repetition rate booster synchrotron power supply are discussed.

1.   Introduction

The power supply for the magnetic system of the CANDLE light source [1] provides three types of the outputs (Fig. 1):

- DC power for the unipolar (dipoles, quadrupoles, sextupoles) and bipolar (corrector magnets) units of the storage ring;
- 2 Hz ramped power for the booster magnets (dipoles, quadrupoles);
- Pulsed power to supply the kickers and septum magnets.

The total average power consumption of accelerator complex is about 5000 kVA. To maintain the powering of the magnetic system it is foreseen to use 160 power converters, including 128 individual, commercially available units for corrector magnets. The power range for converters is 170W –700 kW, the lower case standing for the corrector magnets and upper case standing for the bending magnets. All power supplies are supposed to be high precision units.

H. Wiedemann (ed.), Advanced Radiation Sources and Applications, 425–429.

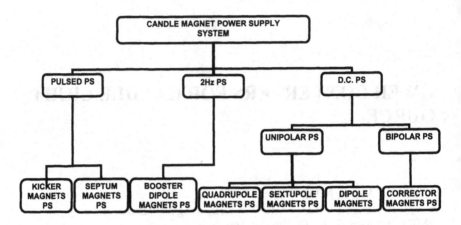

Figure 1. CANDLE magnets power supply (PS) system structure

2.   Booster Bending Magnet Requirements

The booster magnets power supply system has to provide a stable magnet ramping according to the beam energy increase from the injection energy of 100 MeV to the final energy of 3 GeV. The 100 MeV beam is injected into the booster with the dipole magnets field of 24.25 mT. At the maximum energy of 3 GeV, the magnetic field in dipoles reaches 0.7272 T [2].

Usually the booster magnets power supply at the frequency of 10Hz is fulfilled by means of resonance "white circuit". The resonance circuit is created by capacitor bank and inductance of the magnets windings, between which there is an exchange of electromagnetic energy. This circuit is proved at high repetition frequencies and has a good quality Q. In our case the repetition frequency is 2Hz and the resonance "white circuit" is unacceptable as it has a very small value of reactive energy cumulative in the booster magnets (less than 30 kJ), which is evidently insufficient [3]. It should be taken into account also that the energy of the booster electromagnets per unit length should be high than 3kJ/m. The circumference of the booster ring is 192m.

The magnetic structure of the booster lattice is the separate function type, so the dipoles, quadrupole and sextupole families are powered by separate supplies in a strongly synchronized phase to keep the beam parameters in accordance with design specifications. Table 1 presents the main parameters of the booster magnet power supply system. The main stability requirements to the magnet power supply system, including ripples, are:

Dipole magnets - 0.01%
Quadrupole magnets - 0.1 %
Sextupole magnets - 0.5%

TABLE 1. Main parameters of the magnet power supply

Magnets	Quantity	I (A)	V,dc (V)	R (Ω)
Bending	48	1447/50	610	0.203
Quadrupole family QF	32	191	206.3	1.08
Quadrupole family QD	32	191	206.3	1.08
Sextupole family SF	30	25	15	0.6
Sextupole family SD	30	25	15	0.6
Correction dipoles hor/ver	48 / 48	22 / 9.8	14 / 6	0.7 / 1.4

3. PS Block Diagram and Principal Description

The booster ring contains 48 identical dipole magnets. These magnets are electrically connected in series and require a single SCR (Silicon Controlled Rectifier) power supply that is capable for operation in both rectification mode and inversion mode. It is decided to use so-called SLS type scheme [4].

The operating repetition rate of the synchrotron is 2 Hz, and 220ms is allowed for the magnet current to increase from 50 A to 1450 A. As soon as

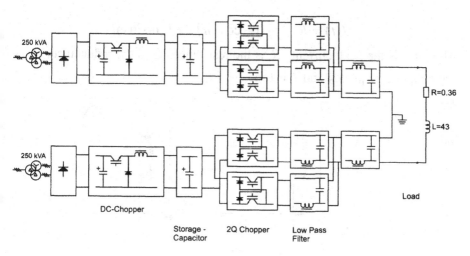

Figure 2. Principal block-scheme of the booster dipole magnet power supply.

beam extraction has been accomplished, the SCR's of the power supply are phased to inversion, and the energy stored in the magnets is returned to the

AC power line as the magnet current decreases. Principal block-scheme of the booster dipole magnet power supply is given in Fig. 2. The scheme consists of two series DC-choppers. Such cascade connection allows the transformation of direct current through the line. The accumulating capacity is feeding two parallel-connected 2Q choppers. In series connected windings of the 48 booster dipole magnets have 0.2Ω and 48 mH loading. The calculated summed current in the circuit equals 1450A, when the maximal frequency is 2Hz. It is necessary to provide 610V peak voltage of the power supply for the given peak current. The Insulated Gate Bipolar Transistors are foreseen as key elements in the chopping circuits. These transistors allow switching frequency 10 kHz, which is necessary for the operation of the device. When the synchrotron booster is operating with the pulsed frequency of 2Hz, the power supply is varying from 700kW to -210kW (minimal value). The average power for the dipole magnets is 170kW.

The voltage swing is limited to less than 1000V. This is achieved by using two power supply units with grounded midpoint. The rectifier is divided into two parts. Each part is presented as a 12-phase rectifier, which provides minimal pulsations of the problem voltage [5].

Figures 3 and 4 show the current, voltage and power waveforms for triangular and sinusoidal ramping of the booster magnets respectively. The current I and voltage U waveforms are given by the following modulations:

$$I(t) = I_m \cdot (1 - \cos \omega t)$$

$$U(t) = R \cdot I_m \cdot (1 - \cos \omega t) + LI_m \omega \sin \omega t,$$

where I_m is the peak current, R is the resistance, L is the inductance, ω is the ramping frequency and t is time.

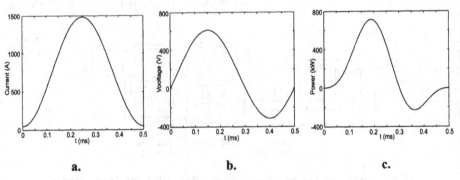

     a.   **b.**   **c.**

Figure 3. Sinusoidal ramping of current (a), voltage (b) and power (c).

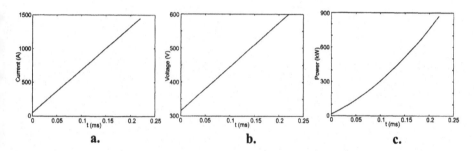

Figure 4. Triangular ramping of current (a), voltage (b) and power (c).

4. Conclusions

The main specifications for the CANDLE booster synchrotron power supply system are given. The SLS type power supply scheme has been adapted for the 2Hz repetition rate booster. The operation modes for triangular and sinusoidal ramping diagrams are given.

The system allows the application of the controllable power supply for the facility operation in top-up injection mode.

References

1. V. Tsakanov *et al*, CANDLE Design Report, ASLS-CANDLE R-001-02, July, 2002.
2. V. Khachatryan, EPAC 2004, 5-9 July, Lucerne, Switzerland, 2004.
3. 1-2 GeV SRS Conceptual Design Report, LBL, Berkley, July 1986
4. SLS Handbook, PSI, Villigen, 1999
5. G.Irminger, M. Horvat, F. Jenni "A 3 Hz, 1MW peak Bending Magnet Power Supply for the SLS", 5 November, 1998.

Figure 7. [illegible] samples of current [illegible] (...power...).

4. CONCLUSIONS

The basic specifications for the CANDLE booster synchrotron power supply system are given. The SLS type power supply scheme has been adopted for the 3 Hz repetition rate Booster. The operation modes, the [illegible] and principal circuit diagrams are given.

The system allows the application of the controlled power supply for the facility operation in top-up injection mode.

REFERENCES

1. V. Tsakanov et al. CANDLE Design Report, ASLS-CANDLE R-001-02, July, 2002.
2. M. [illegible], EPAC 2004, 5-9 July, Lausanne, Switzerland, 2004, [illegible].
3. 1-2 GeV ALS Conceptual Design Report, LBL, Berkeley, July 1986.
4. [illegible] Handbook, [illegible], 1990.
5. G. [illegible], [illegible] 3 Hz, 1MW peak [illegible] Magnet Power Supply for the SLS, [illegible] November 1998.

THE INSERTION DEVICE ADAPTED VACUUM CHAMBER DESIGN STUDY

S. TUNYAN, V. AVAGYAN, M. AGHASYAN, M. IVANYAN
Center for the Advancement of Natural Discoveries using Light Emission, CANDLE, Acharian 31, 375040 Yerevan, Armenia

Abstract. In order to expand the CANDLE radiation capability in hard X-ray region, it is foreseen to install 3 Tesla wiggler for the X-ray imaging beamline at CANDLE. Its design will be mainly based on a 3 Tesla permanent magnet wiggler designed and assembled at ESRF. The paper presents the study of the vacuum chamber thermal loading caused by the installation of similar wiggler with 10 periods. The analysis has enabled to determine the position and construction of the absorbers providing possibility to use similar critical insertion devices in future.

1. Introduction

The CANDLE facility is third generation synchrotron light source project that will provide high brilliance X-ray beams for scientific research. The need for a light source with higher radiation intensity and critical energy requires the utilization of insertion devices with higher field strengths. In order to expand the CANDLE radiation capability in hard X-ray region, it is foreseen to install 3 Tesla wiggler for the X-Ray imaging beamline at CANDLE [1]. Its design will be mainly based on 3 Tesla permanent magnet wiggler designed and assembled at ESRF[2]. Similar wiggler, but with 10 periods, will be a type of device at CANDLE, which achieves wide beam fan and maximum peak flux density that can serve more than one beamline. On the other hand, such increase of the period number will expand the beam horizontal size up to critical concerning the aperture of the storage ring vacuum chamber. In this report we present the study of the vacuum chamber thermal loading caused by the installation of 3T wiggler.

H. Wiedemann (ed.), Advanced Radiation Sources and Applications, 431–436.

2.   3-Tesla Wiggler and Storage Ring Characteristics

The radiation characteristics of synchrotron light from insertion devices depend on the type of device and the machine parameters. The main parameters of the CANDLE facility and 3T wiggler are listed in Table 1. The photon beam power density distribution profile is shown in Fig.1. The spatial distribution of beam from wiggler has a complex form, Gaussian in the vertical plane and parabolic in the horizontal direction. The fan of radiation from 3T wiggler is about 20.8 mrad in the horizontal plane, and of the order of 0.34 mrad in the vertical direction. The maximum total power and the beam horizontal opening are very high, that forces us, already in the current stage of the project, to carry out more detailed analysis of power distribution in the storage ring, as well as to consider possible consequences concerning vacuum chamber walls, which will be made of stainless steel.

Table 1. 3T Wiggler Parameters

Beam Current	350 mA
Beam Energy	3.0 GeV
Magnetic Field	3.0 T
Number of Periods	10
Period Length	21.8 cm
Beam Critical Energy	17.95 KeV
Wiggler Gap	11 mm
Beam Total Power	39.1 kW
Deflection Parameter	61
Peak Power Density	9.2 kW/mr²

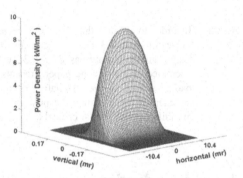

Figure 1. 3T wiggler beam power density.

The CANDLE storage ring lattice consists of 16 identical Double-Bend Achromatic (DBA) cells, 13 of which will be used for insertion devices (ID). One of the photon beam emission tracts is shown schematically in Fig. 2.

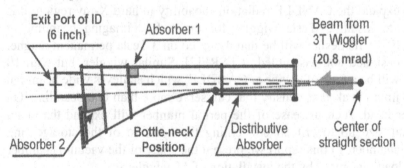

Figure 2. Insertion device synchrotron light emission tract.

The first 6-inch exit port in each cell, mounted on ID radiation emission axis, should provide outlet of radiation power from the storage ring vacuum chamber. The both absorbers located near the exit port will be used to protect vacuum chamber walls from radiation of bending magnets. For the horizontal emission fan of 3T Wiggler, there is an opening of about 67 mm on both sides of the central line at the end of the vacuum chamber. Such an angular opening from the center of the straight section is rather conformed to the exit aperture of the vacuum chamber (70 mm on both sides). Taking into account the use of an interlock system that will abort the electron beam at a significant deviation from a design orbit, we could avoid additional protection of vacuum chamber walls. However, the detailed analysis presented below obviously shows that the synchrotron light of the critical insertion device similar to 3T Wiggler can leave a trace onto the storage ring vacuum chamber walls, even without miss-steering.

3.   Evaluation of Heat Load

While solving the problem of thermal loading on vacuum chamber caused by the 3T wiggler, we were interested basically in two parameters: fan horizontal opening and beam power density. Our evaluation is based on the well-known formula for the distribution of radiated power from an electron in a sinusoidal trajectory, which applies with reasonable approximation to undulators and, to a lesser extent, wigglers [3]:

$$\frac{dP}{d\theta d\psi} = P\frac{21\gamma^2}{16\pi K}G(K)f_k(\gamma\theta,\gamma\psi),$$

where P is the total (angle-integrated) radiated power, γ is Lorenz relativistic factor, K is the deflection parameter of device. The function $G(K)$ tends to unit at large K, as in our case. The function $f_k(\gamma\theta,\gamma\psi)$ describes the angular distribution and can be expressed in integral form for numerical evaluation:

$$f_k(\theta,\psi) = \frac{16K}{7\pi G(K)}\int_{-\pi}^{\pi}\sin^2\alpha\left[\frac{1}{D^3} - \frac{4(\gamma\theta - K\cos\alpha)^2}{D^5}\right]d\alpha,$$

where $D = 1 + (\gamma\psi)^2 + (\gamma\theta - K\sin\alpha)^2$ and $\alpha = 2\pi Z/\lambda$, $0 \le Z \le N\lambda$ ($N\lambda$-wiggler length). Sub-integral expression of angular distribution function strongly depends on magnetic structure, and at the transition to Cartesian coordinates requires the consideration of design distance. The high value of the deflection parameter and short design distance has led us to consider this multipole wiggler as a system of separate sources. With such assumption the superposition method has been applied for the solution illustrated in Fig. 3.

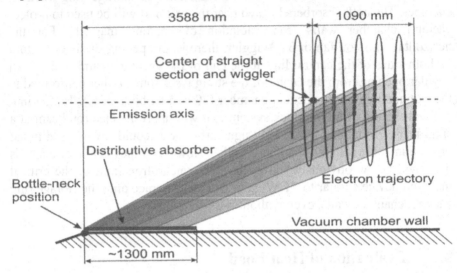

Figure 3. Target setting of heat load evaluation.

Correctness of similar assumption is beyond question, as in case of $K \gg 1$ the radiation emitted in the poles of the wiggler is incoherently superimposed and interference effects are less important [4]. The subsequent analysis shows, that some part of radiation power is dissected on the walls of the vacuum chamber. Special attention should be paid to the photon Absorber 1 (see Fig. 2), located at the exit of the photon beam. It is also worth to note that the area of the vacuum chamber near Position Bottle-neck can run the risk to be attacked by photon beam from the 3T Wiggler. The last statement has compelled to install distributive absorber in this zone. The thermal load in these absorbers depends on the source distance and on the level of miss-steering. In turn, the definition of the real level of miss-steering is an intricate problem. However, in this stage an attempt to estimate the values of thermal load on absorbers at the reasonable levels of electron bunch deviation from the design orbit has been made. The superposition method has been applied for calculation of thermal loading on photon absorbers. The matrix superposition of power density values, reinforced by each separate source, revealed spatial distribution of thermal loading in the impact area. As evident in Figure 4, the radiation of first five sources is allocated on distributive absorber. The same profile of power density is allocated on Absorber 1. The peak power densities on these absorbers at normal angle incidence are shown in Table 2.

Table 2. Peak power density values (W/mm²)

Orbit offset	Absorber 1	Distributive Absorber
none	46	126
0.32mm	50	131
1.0mm	56	150
1.8mm	61	166

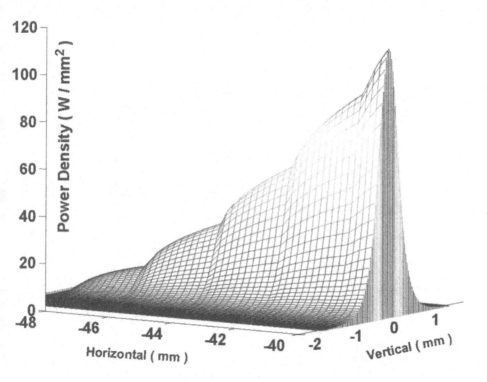

Figure 4. Spatial power density profile.

4.   Conclusion

Detailed analysis of power distribution generated by ID in CANDLE storage ring has been carried out. Thermal load profile and values on two absorbers of the storage ring vacuum chambers are determined and calculated. Subsequent Finite Element Analysis shows that the temperature due to heat load generated by 3 Tesla permanent multipole wiggler is significantly smaller than the maximum permissible limit of absorber's material.

References

1. V. Tsakanov et al, CANDLE Design Report, ASLS-CANDLE R-001-02, July, 2002.
2. J. Chavanne, P.Van Vaerenbergh and P.Elleaume, NIM A 421 (1999) 352-360.
3. S. L. Hulbert and G. P. Williams, BNL, Upton, New York, USA, July 1998.
4. V. Saile, Properties of Synchrotron Radiation, IMT, Karlsruhe, Germany.
5. S. Tunyan et al, "Absorbers Design for the CANDLE Storage Ring", MEDSI 2004.

Index